SOVIET WARSHIPS 1945 TO THE PRESENT

▼*Minsk* on a sortie in the Pacific Fleet area in 1984. The Yak-38 nearest the camera is the Forger B training variant, and the white-painted Ka-25 helicopter is the Hormone C which is employed for SAR duties. Note the port-side SS-N-12 launchers which are in the elevated position. (TASS, 1984)

SOVIET WARSHIPS 1945 TO THE PRESENT

REVISED AND EXPANDED EDITION

JOHN JORDAN

ARMS AND ARMOUR

Contents

Arms and Armour Press
A Cassell Imprint
Villiers House, 41–47 Strand, London
WC2N 5JE.

Distributed in the USA by Sterling Publishing Co. Inc., 387 Park Avenue South, New York, NY 10016–8810.

Distributed in Australia by Capricorn Link (Australia) Pty. Ltd, P.O. Box 665, Lane Cove, New South Wales 2066.

British Library Cataloguing in Publication Data
Jordan, John, *1947–*
Soviet warships: 1946 to the present.
2nd ed
I. Title
623.8250947
ISBN 1-85409-117-4

Line illustrations by the author.

Designed and edited by DAG Publications Ltd. Designed by David Gibbons; edited by Michael Boxall; layout by Anthony A. Evans; typeset by Wyvern Typesetting, Bristol; camerawork by M&E Reproductions, North Fambridge, Essex; printed and bound in Great Britain by The Bath Press, Avon.

Line Drawings

Maps

Preface

THE FIRST EDITION of *Soviet Warships* was published in 1983. Since that time there have been important political developments, and new types of surface warship have entered service. It was felt that the scale of these developments was such as to justify a complete re-write, rather than a simple updating of the original.

Many of the changes in the scope and organization of the book reflect the format of its companion volume, *Soviet Submarines*, published in November 1989. Although the major emphasis is on the surface ships laid down in the Gorshkov era, there is more detailed coverage of the immediate post-war period. In particular, the author attempts to establish the close ties between the size and shape of the Soviet surface fleet and the shipyards allocated to naval construction by Soviet policy-makers.

Whereas the earlier edition treated each major class of surface warship as a separate entity, the new edition groups together those types which began construction at broadly the same time in order to clarify the policy decisions which were responsible for their conception. The result is less a linear progression, more a series of mini-revolutions in Soviet maritime thinking. This provides a marked contrast with developments over the same period in Western navies, which have been essentially evolutionary.

To illustrate the book the author has prepared a complete set of new line-drawings. The original profile views have been extended and updated, and are now complemented by plan views of all the major types. The drawings of weapons systems are also new, being based on the extensive range of close-up photographs now in open publication.

Photographic coverage has been both renewed and extended in order to incorporate the many high-quality photographs which have become available from Western defence agencies during the 1980s. With the progressive withdrawal of the Soviet surface fleet into operational areas close to home, it appears unlikely that this 'golden age' of photography will be repeated, opportunities being increasingly limited to inter-fleet deployments.

John Jordan,
January 1992

Introduction

MUCH HAS CHANGED since the appearance of the first edition of this book in 1983. Since that time the Soviet Navy has undergone a major expansion in size and capabilities. With the passage of time it has become easier to trace the path of Soviet naval development and its long-term goals. This process has been assisted by changes in the political climate in the USSR, which have resulted in a new openness with regard to military capabilities and intentions.

It is perhaps ironic that with the Soviet Navy on the verge of acquiring the final trappings of a maritime superpower in the form of aircraft-carriers capable of operating high-performance aircraft, it now finds itself under greater threat than at any time during the past 30 years. The Soviet Union is on the verge of political and economic disintegration, and it is difficult to envisage what sort of Soviet Navy will finally emerge.

Even before the current crisis, the parlous state of the Soviet civilian economy and the relaxation of political tension between the Soviet Union and the West were combining to force doctrinal changes and major reductions in military spending. Two of the three shipyards building major combatants for the Soviet Navy, the Baltic and 61 Kommuna Shipyards, were turned over to mercantile construction in 1989, the keels of vessels already laid down being broken up *in situ*. And in 1990 it was officially stated that the carrier programme would be terminated after completion of the third ship, *Ulyanovsk*. In March of 1991, Admiral Feodor I. Novoselov, the Deputy Commander-in-Chief of the Navy for Shipbuilding, was quoted in the journal *Red Star* as saying that expenditure for shipbuilding had dropped by 9–10 per cent in 1990 and by 23 per cent in 1991 by comparison with 1989, without taking inflation into account.[1]

In addition to the cuts in the Navy's building programme, funds allocated for ship maintenance are reported to have been reduced by 50 per cent,[2] and research and development appropriations have been cut by 40 per cent.[3] The first measure will affect the Soviet Navy's ability to update older warships. The second is bound to have a serious impact on the development of the Soviet Navy before the year 2000; designs will stagnate, and obsolescent weapons systems will continue to be fitted because they will be the only systems available.

Political disintegration is bound to have far greater consequences. Even if operational control of existing naval forces continues to be exercised by a central Soviet defence ministry, there remain many unanswered questions regarding the future funding of the defence forces by the republics, many of which will experience great difficulty in creating balanced economies outside the tight centralized framework which currently exists. The division of the existing military-industrial complex, including shipyards and naval bases, between the larger republics such as Russia and the Ukraine will have major consequences which are as yet difficult to predict.

Few of these effects are yet apparent. In contrast to the other Soviet armed services, which are already suffering the consequence of the cutbacks in military production, the Soviet Navy has yet to reach the zenith of its power. In part this is because naval programmes are inherently more difficult to stop in their tracks. The carrier programme which is only now yielding fruit is reported to have been approved as early as 1976; the Rocket Cruisers of the *Kirov* and *Slava* classes are taking 6–10 years to build. Even the destroyers of the *Udaloy* and *Sovremenny* classes now entering service were laid down in 1986–7, not long after Gorbachev had come to power, and before the major arms limitation agreements were finalized.

However, until quite recently there had also been strong indications that the Soviet Navy would not suffer the same reductions as the other armed services. Many elderly, obsolescent ships have been retired, but they have been replaced by larger, more capable vessels which have continued to roll off the production lines in a steady stream. The Soviet Navy has become leaner and fitter as a result.

President Gorbachev's concept of defensive 'sufficiency' seemed to imply that while major cuts would be applied to the ground forces and the strategic missile forces as a result of arms limitation agreements, cuts in naval forces could not take place unless there were comparable reductions in the size and capabilities of the US Navy. With the steadfast refusal of the United States to discuss naval force levels, it appeared that a strong Soviet Navy would continue to be necessary in order to protect the Soviet landmass against an undiminished threat of assault from the sea.

Indeed, in retaliation for the American refusal to negotiate on naval arms limitations the Soviets have been steadily transferring air and ground force units previously allocated to the Central Front to Soviet Naval Aviation (AV-MF) or Naval Infantry command. By mid-1991 these transfers included three motorized rifle divisions, several regiments of fighters and large numbers of ground-attack aircraft.

These adjustments to Soviet force structures will no doubt continue, but in the longer term the move to strengthen forces in the maritime theatres of operations at the expense of the ground forces will be undermined by the reduced funding allocation for new surface ship construction.

A Crisis of Doctrine

Of equal importance to the political and economic effects of the current crisis are the changes in maritime strategy and operational doctrine which may result.

The Soviet Navy, which forms the subject of this book, has developed within the relatively narrow framework of Marxist–Leninist thinking. The history of the defence of the Soviet Union from the October Revolution of 1917 up to the political upheavals of August 1991 has been dominated by the concept of a People's Army charged with defending the Motherland against hostile external forces.

The Soviet Navy – even the powerful ocean-going Navy of today – has throughout performed only a supporting role. Missions have focused almost exclusively on territorial, as opposed to maritime, objectives. Early missions of the defence of the Soviet coastline and the direct support of land operations have in recent years expanded to include the defence of Soviet territory against long-range bombers and missile submarines, and the creation and support of a large force of Soviet SSBNs to provide a counter-threat to an enemy separated from the Soviet Union by two oceans.

Still absent from the catalogue of Soviet naval missions are traditional maritime objectives such as the protection of mercantile trade and economic interests overseas, and the ability to deploy and sustain power projection forces in distant waters. The Soviet peacetime naval 'presence' may have grown over the past two decades, but the objectives have been political rather than military, and the ships deployed have been designed with other, quite specific missions in mind.

This is not to say that deep, and frequently acrimonious, doctrinal disagreements have not taken place both within the Soviet Navy and between the Soviet Navy and the Army-dominated High Command. However, these have focused not on aims and objectives, which have been closely defined by Marxism–Leninism, but on the *most effective means* of furthering aims and fulfilling objectives.

Reduced to its simplest terms the debate has been between those advocating a specific (and not exclusively naval) response to each of a number of threats posed by hostile naval forces to Soviet territory, and advocates of a Soviet 'fleet' capable of defeating an enemy fleet on the high seas and thereby neutralizing any threat that it poses to Soviet territory. The former have tended to favour 'dissimilar' forces, often comprising large numbers of low-value naval and aviation units, and operating under a 'combined arms' operational-tactical doctrine with tight central control from the shore. The latter have favoured the construction of coherent Soviet fleets which, while not a precise mirror-image of their Western counterparts, have 'balanced' offensive/defensive capabilities, and can operate – at least at the tactical level – independently of shore-based control.

The concept of combined arms operations under tight central control is a natural extension of Soviet Ground Forces doctrine, and has therefore been more readily acceptable to the Soviet High Command. The 'balanced fleet' concept, which first surfaced in the immediate post-war period as a response to the threat posed by the powerful British and US Navies, and emerged again in the 1970s in the writings of Admiral Sergei Gorshkov and others, has been regarded with suspicion by those in the Soviet establishment who fear a resurgence of the imperialist ambitions of the Tsarist Navy.

Under Admiral Chernavin the Soviet Navy has withdrawn into its shell. Distant deployments have all but ceased and exercises now take place almost exclusively within Soviet waters. These changes have served to reduce the 'political' impact of the Soviet Navy on the world stage, and in consequence have made it more difficult for potentially hostile navies to justify their bids for ever-higher levels of funding. However, these changes have also served to reaffirm traditional Soviet naval doctrine, with its emphasis on Combined Arms and shore-based operational-tactical control.

It now seems probable that the yoke of Marxism–Leninism will be lifted from the entire apparatus of the Soviet state. This is bound to provoke a fundamental reassessment of the role of the Soviet Navy, and will enable discussion of operational-tactical issues to take place without fear of the dead hand of the Communist Party establishment. Even if we assume that the Soviet

▶ Despite the decommissioning of large numbers of elderly, obsolescent vessels, the Soviet Navy is continuing with its programme of 8,000-ton destroyers, which will form the backbone of its surface forces in the year 2000. Depicted here is *Gremyashchy* of the *Sovremenny* class. (MoD, Bonn)

Navy will survive as a single entity (which, in the current crisis, is by no means certain), there can be little doubt that within ten years it will be a very different force, underpinned by a very different philosophy from that which has prevailed since the October Revolution of 1917.

335

62

1. The Post-war Era

◀**The first Soviet post-war destroyer, *Skory*, was derived from the war-built types. The main armament of four 130mm guns, housed in two twin semi-automatic mountings, could be elevated to only 40 degrees and was therefore incapable of engaging aircraft. Note the heavy conventional torpedo armament and the rudimentary radar outfit. (Skyfotos, 1957)**

◀**The larger companion to *Skory* was the *Sverdlov*-class cruiser. Derived from the war-built *Chapaev* class, *Sverdlov* incorporated a substantial amount of German technology but was obsolete before the first ship was completed. Note the triaxially stabilized mountings for the secondary 100mm anti-aircraft guns and the German-style Round Top high-angle fire control director abaft the first funnel. This early photograph of *Zhdanov* reveals the backward state of Soviet radar technology of the day. Note also the quintuple bank of torpedo tubes amidships, which was not removed until the 1960s. (Skyfotos)**

FOR THE TWO DECADES that followed the Revolution, Soviet mainstream thinking about the role of the Navy revolved around the philosophy of the 'Young School', in which traditional high-seas 'battlefleet' strategies were rejected in favour of coastal defence, employing large numbers of relatively cheap units such as torpedo-boats, submarines and land-based bombers operating under tight central control from command posts on land. The two traditional Soviet fleets in the Baltic and Black Seas, together with newly organized squadrons (later 'fleets') in the Arctic and the Pacific, were therefore closely integrated into the land-based command structures of their respective areas.

In the late 1930s Stalin became convinced of the need for conventional battlefleets to supplement the coastal defence forces. The 1937–42 Five Year Plan provided for the construction of battleships, cruisers and fleet destroyers. The Soviet squadrons in the Arctic and the Pacific were accorded fleet status and were to be supported by the newly constructed shipyards at Molotovsk (later renamed Severodvinsk) and Komsomolsk-an-Amur respectively. In spite of considerable assistance from Italian designers, this ambitious programme proved to be beyond the capabilities of Soviet industry, and although the Type VII series of fleet destroyers was duly completed, together with a handful of heavy cruisers of the *Kirov* class, construction of the giant 60,000-ton battleships of the *Sovietsky Soyuz* class proceeded very slowly, and was finally abandoned in October 1940.

The strategic rationale for Stalin's pre-war battlefleet is unclear. The Soviet Union, a continental land power endowed with rich quantities of natural resources and with almost exclusively 'internal' communications, did not need to contest the high seas with the major maritime powers. Rather, the big-ship programme of 1937–42 appears to have been the result of preoccupations regarding political and military status, which could be resolved only by the possession of the traditional trappings of a Great Power. Soviet battlefleets could at least dominate the Baltic and Black Seas, thereby increasing Soviet political influence in the two regions crucial to Soviet security.

Stalin's naval ambitions were cut short by the *Blitzkrieg* of July 1941. The construction of a prestigious new battlefleet was overtaken by far more urgent priorities. Every ton of steel was required for guns, tanks and aircraft. Moreover, work at the traditional shipyards of Leningrad and Nikolayev ground to a halt as first Leningrad was subjected to a siege which was to last for three years, and then Nikolayev was occupied by the Wehrmacht. Construction at the new shipyards in the Arctic and the Far East was hamstrung by a shortage of materials. Both Molotovsk and Komsomolsk were located far from the industrial infrastructure that was to have provided them with turbines, weaponry and other essential components, and they were therefore able to complete only those ships for which the materials had already been assembled.

Undeterred by the almost total destruction of the major Soviet shipyards, Stalin revived his grandiose naval ambitions at the end of the Second World War, and as early as July 1945 declared his intention to build a great sea-going fleet. However, this declaration was now intended for different ears. The potential enemy was no longer Hitler's Germany, and the focal point of Soviet naval operations was no longer the Baltic. The Soviet Union was faced by the world's two foremost maritime powers, Great Britain and the United States, and the Soviet Navy would have to compete in all four fleet areas in order to protect Russian soil against amphibious assault backed up by wide-ranging carrier task forces.

The new fleet planned by Stalin was surprisingly conservative. The lack of naval experience during the Great Patriotic War, in which Soviet naval activity was generally confined to fire support, small-scale landings, mining and largely ineffectual anti-submarine operations close to Soviet bases, undoubtedly stymied Soviet thinking, resulting in a programme which constituted a return to unfulfilled pre-war designs. Although the Soviet Navy's new opponents were the Americans and the British, the new ships were closely patterned on those of the German Navy and incorporated large quantities of captured German technology.

In some respects it might be said that Soviet technological backwardness was so great that the technology involved in building major surface warships assumed a higher profile than the strategy which those ships would be charged with implementing. If a ship is designed around German state-of-the-art artillery, fire control systems and propulsion machinery, the end result is something looking very much like a German Second World War cruiser or destroyer, and capable of executing only a very similar – and limited – maritime strategy.

Thus the Soviet post-war naval programme centred around conventional cruisers and destroyers, with two 40,000-ton 'battlecruisers' armed with a combination of 12in guns and guided missiles being laid down in 1951–2. According to remarks made by Admiral Chernavin during an interview with the Soviet press in late 1989, studies for aircraft carriers were begun in 1952, but these clearly made little progress before the death of Stalin in 1953, after which time the entire naval programme was to be subjected to an extensive review. Had carriers been built, it is unlikely that they would have constituted the focal point of fleet operations as in the British and US Navies. It is more likely that their primary mission would have been the support of the Soviet surface fleet, in the manner of the German and Italian carriers of the Second World War.

Four Fleets of Equal Status

Stalin had long harboured the ambition to raise the status of the Northern and Pacific Fleets to a level comparable with that of the traditional Baltic and Black Sea Fleets. In the aftermath of the Second World War this became a pressing requirement, in that the Arctic and the North-west Pacific constituted the two 'open flanks' of the Soviet Union against which the full power of Western carrier task forces and amphibious fleets could be projected.

Fortunately, the newly constructed shipyards at Molotovsk and Komsomolsk had remained intact throughout the conflict, being spared from damage by their very remoteness from the Eastern Front. As soon as they could be supplied with the necessary plans and components they would be ready to play a full part in the post-war naval construction programme. The impressive facilities at Molotovsk, created by the use of forced labour during the 1930s, included a massive under-cover building dock, 335m long by 137m wide, which had been designed to permit the simultaneous construction of two battleships side by side.

When the first major class of post-war Soviet destroyers, the 72-ship *Skory* class, was laid down beginning in 1948, the hulls were distributed equally between four shipyards: Molotovsk in the Arctic, Komsomolsk in the Far East, Zhdanov in the Baltic and 61 Kommuna in the Black Sea. The Zhdanov Shipyard at Leningrad had played a major part in the pre-war destroyer programmes. The 61 Kommuna Shipyard at Nikolayev also participated in the pre-war destroyer programmes and had laid down two cruisers of the

▼The Guard Ships of the 'Riga' class were built in response to traditional Soviet concerns: the domination of territorial waters and the provision of fire support for the ground forces in coastal regions. Three single 100mm guns in shielded mountings constitute the main armament. Anti-submarine weaponry on this early variant comprises an MBU-900 multiple rocket-launcher (similar to the US Mousetrap) forward of the bridge structure, and four BMB-2 mortars on the quarterdeck. (MoD, 1975)

Chapayev class in 1939–40, one of which had been completed post-war.

The *Skory* was a solid, unremarkable design based on the pre-war *Ognevoi* class. It featured twin 130mm mountings forward and aft, two quintuple banks of 533mm torpedo tubes, and mine rails capable of accommodating 50 mines. Contemporary Western destroyer designs emphasized anti-air capability with dual-purpose main gun batteries and radar-controlled anti-aircraft weapons. However, the main battery of the *Skory* was capable only of direct surface fire, and the rudimentary radars provided only a limited degree of aerial surveillance. Anti-submarine capability was restricted to a small high-frequency 'searchlight' sonar, together with two depth-charge throwers and two depth-charge racks. Low-pressure, low-temperature steam propulsion machinery was a further indication of the essentially conservative nature of the design.

The Soviet designation of these ships was *Eskadrenny Minonosets*, literally 'Fleet Minelayers'. This was the traditional Soviet destroyer classification and one should not read too much into its literal meaning; likewise, the Western contemporaries of the *Skory* had long ceased to be 'Torpedo-Boat Destroyers'. However, the name does carry with it associations which make it clear that the tactical employment envisaged for Soviet destroyers was somewhat different from those we generally associate with their Western counterparts.

Western destroyers had always been intended to accompany capital ships on the high seas, and to protect them first against enemy torpedo-boats, then against submarines and aircraft. The very different operational regime obtaining in the enclosed waters of the Baltic and Black Seas, however, meant that the Russian 'destroyer' developed as a fast, hard-hitting ship capable of carrying out raids on enemy shipping and minelaying operations off enemy ports and harbours. The *Skory* was simply a continuation of traditional Soviet destroyer design, hence the heavy conventional armament and the almost total absence of effective anti-aircraft or anti-submarine weapons.

The other major element in the first Soviet post-war naval programme was the *Sverdlov*-class cruiser. Again, these ships were developments of a pre-war design, the *Chapaev*, although the secondary battery of 100mm guns, together with the fire control arrangements, incorporated much German technology. There were primitive radar sets for air surveillance, but there was only a limited capability for blind fire. Traditional Soviet concerns were reflected in the provision of quintuple banks of torpedo tubes, and an extensive network of mine rails on the quarterdeck for an estimated 140–200 mines.

The *Sverdlovs* were large ships by any standard:

▼The major difference between *Skory* and the 'Kotlin' (seen here) was the recognition of the threat from the air. While offensive anti-surface capabilities remained unchanged, the 'Kotlin' featured a dual-purpose main battery of 130mm guns with radar fire control. The guns were in a triaxally stabilized mounting similar in conception to the 100mm twin AA mounting of the *Sverdlovs*, and could elevate to 85 degrees. The mixed small-calibre anti-aircraft outfit of *Skory* was superseded by a uniform battery of four quadruple 45mm mountings of new design, controlled by two Hawk Screech tellerform radars. Note also the Slim Net E-band air surveillance radar at the top of the mainmast, from which the later Head Net series of radars was to be developed. (Skyfotos)

with an overall length of 210m and a beam of 22m, they displaced fully 16,000 tons (standard). This apparently precluded their construction at the Komsomolsk Shipyard, where insufficient depth of water meant that large surface ships had to be towed 450 kilometres downriver for fitting out. The cruisers intended for the Pacific Fleet therefore had to be built in the traditional shipyards of the Baltic.

The lead building yard was the Ordzhonikidze Shipyard, Leningrad, which had three slipways capable of accommodating ships of this size; nine hulls were laid down from 1949 to 1954. Because of the requirement to complete additional hulls for the Pacific Fleet, four hulls were also laid down at the Marti Shipyard, Leningrad. Three ships were laid down at the Molotovsk Shipyard and a further five at the Marti South Shipyard, Nikolayev.

All these shipyards except Molotovsk had had recent experience of building large surface warships. Ordzhonikidze had laid down the first of the giant battleships of the *Sovietsky Soyuz* class in 1938, and had completed two cruisers of the *Kirov* class and two of the succeeding *Chapayev* class. The Marti Shipyard, Leningrad had laid down the first of the *Kronshtadt*-class battlecruisers in 1939, and completed *Zhelezniakov* of the *Chapayev* class during the immediate post-war period. Marti, Nikolayev was the most important of the big-ship yards, and besides laying down one unit of the *Sovietsky Soyuz* class and one of the *Kronshtadt* class during the pre-war period, had since completed two *Kirov*s and a single *Chapayev*.

Other Ships in the Programme

The third major element in the post-war surface programme was the Guard Ships of the 'Kola' and 'Riga' classes. The concept of the Guard Ship (*Storozhevoi Korabl'* or SKR) dated from the inter-war period, and was closely associated with the coastal defence theories of the Young School. These were essentially second-rate destroyers, slower than their 'fleet' counterparts, and with a heavy gun armament complemented by a single bank of torpedoes and rudimentary anti-submarine weapons.

The six 'Kolas' were built from 1949 to 1951 at the Kaliningrad Shipyard, formerly the German Schichau Shipyard. This yard had been responsible for the design and construction of the German fleet torpedo-boats of the 'T' series prior to and during the Second World War, and the influence of the war-built vessels can be seen clearly in the architecture of the 'Kola'.

The 'Riga' which followed was a smaller, simpler vessel which in its conception derived more closely from Soviet pre-war Guard Ships. A total of 64 units was completed from 1952 to 1958, the majority being

◀Two elements of Stalin's post-war surface fleet, the 'Kotlin'-class destroyer *Blestyashchy* (foreground) and the *Sverdlov*-class cruiser *Admiral Suvorov*, photographed off Singapore in 1971 *en route* to the Indian Ocean. In spite of some upgrading of electronics and weapons systems, the capabilities of these ships in a modern maritime environment were strictly limited. Note the RBU-6000 rocket-launchers on the quarterdeck and the new platform for twin 25mm AA guns amidships on *Blestyashchy*, one of eleven units modernized from 1960 to 1962. (MoD)

built at Kaliningrad, but with contributions to the programme from 61 Kommuna, Nikolayev, and from Komsomolsk in the Far East.

Roughly contemporary with the 'Riga', and strikingly similar in terms of architectural style, were the fleet destroyers of the 'Kotlin' class, which followed the *Skory* on to the slipways from 1953 onwards. Thirty-six were planned, construction to be shared between Zhdanov, 61 Kommuna and Komsomolsk. Interestingly, Molotovsk was omitted from the programme, in part because it was engaged in building cruisers of the *Sverdlov* class, but also presumably because of the yard's increasing importance to the post-war submarine programme. By the early 1950s Molotovsk was already building the large diesel-powered 'Zulu' and would soon become the lead yard for the nuclear submarine designs then at an early stage of development.

The shortfall in building capacity created by the unavailability of Molotovsk was compensated by placing the additional orders with Zhdanov, which unlike the other two yards was not also involved in the Guard Ship programme.

The 'Kotlin' was a major improvement on the *Skory* in a number of respects. It was the Soviet Navy's first truly 'post-war' design and therefore benefited from technological advances which were not available to its predecessor. The main armament of two twin 130mm guns was similarly disposed, but the new triaxially stabilized gunhouses allowed aerial targets to be engaged at up to 85 degrees' elevation with radar-directed fire. In place of the mix of 85mm, 37mm and 25mm anti-aircraft guns of the *Skory* there was a uniform battery of four quadruple 45mm mountings. The 'Kotlin' also introduced high-pressure steam plants to the Soviet Navy.

Considering that some units of the class were completed as late as 1958, the 'Kotlin' remained conservative in conception. Its closest Western counterparts were the British *Daring*, which was essentially a late-war design for operations in the Pacific, and the French *Surcouf*, which although incorporating the latest in radar-controlled artillery was based on a pre-war design. Even the latter ships, however, possessed vastly superior anti-submarine capabilities to the 'Kotlin', which in its early days was fitted only with depth-charge throwers and racks. The offensive armament of the 'Kotlin' still focused on the anti-surface torpedo and the mine.

This deficiency was apparently to have been remedied in the projected 'Kruplin'-class destroyers intended to succeed the 'Kotlin'. Breyer[1] credits the provisional design as being armed with two twin 76.2mm gun mountings (utilized for the 'Kynda', 'Kashin' and 'Petya'), four quadruple 57mm AA mountings (fitted in the 'Kildin' and 'Krupny' intermediate missile destroyers), two triple banks of 533mm torpedo tubes, and no fewer than four of the new RBU-2500 multiple anti-submarine rocket-launchers.

Khrushchev: The New Broom

The death of Stalin in March 1953 brought with it a power struggle accompanied by a series of major policy reviews. Almost immediately the battlecruiser programme was terminated and several *Sverdlov* hulls were dismantled on the slipways. Of the twenty ships launched, only those in the final stages of fitting out were to be completed, while others were towed away and their construction suspended pending the outcome of the major defence review set in motion in 1954 under the leadership of Georgi Malenkov. The large destroyer and submarine programmes, which could be more easily reconciled with traditional Soviet coastal defence strategies, were for the moment spared but would come under increasing scrutiny.

In February 1955 serious policy disagreements within the Politburo resulted in the downfall of Malenkov. He was replaced as First Secretary of the Communist Party by the energetic Nikolai Khrushchev, who proceeded to proclaim a 'Revolution in Military Affairs' under which the massive conventional forces favoured by Stalin to protect the Soviet Union would be replaced by smaller, 'high-technology' forces based on a combination of missiles and nuclear warheads.

This turnabout in military policy was prompted not only by technological developments (the Soviets exploded their first hydrogen bomb in August 1954), but by the huge economic cost of Stalin's conventional military programmes. Khrushchev saw nuclear weapons not only as a more effective way of defending the Soviet Union, but as an opportunity to reduce military expenditure, thereby releasing both resources and manpower to expand the civil economy.

Khrushchev was later to describe the big surface ships of Stalin's naval programme as 'floating coffins'. Of equal significance, however, is his description of the naval shipyards as 'metal eaters'.[2] The termination of the 'big-ship' construction programme would enable the major shipyards of Ordzhonikidze (renamed the Baltic Shipyard); Marti, Leningrad (renamed the Admiralty Shipyard); and Marti South, Nikolayev (renamed the Black Sea Shipyard); to be transferred to mercantile construction. Between them these yards possessed eight of the largest building ways in the Soviet Union, and their conversion from military use was to be a major factor in the build-up of the Soviet merchant fleet which was to establish a pre-eminent

◀ The 'Kildin' was the first of two intermediate 'Rocket Ship' designs. The derivation from the 'Kotlin' is readily apparent. The prominent quadruped lattice foremast carries the Flat Spin missile targeting radar. Note the bulky launcher for SS-N-1 Strela missiles on the quarterdeck, with its prominent magazine. This is *Bedovy*, distinguished from her sister-ships by her retention of the original 'Kotlin' second funnel. (US Navy)

position in world shipping from the 1970s, and which was to become a major foreign currency earner for the Soviet Union.

The cruiser programme was terminated as a matter of urgency for largely economic reasons. The stay of execution granted to the submarine and destroyer programmes, however, would hold only for as long as it would take for plans to be revised in the light of the 'Revolution in Military Affairs', as the new destroyers were every bit as poorly equipped as their cruiser counterparts to play their part in that revolution. In the event only 26 of the projected 36 'Kotlins' would be completed to the original design. Five further hulls already on the building ways would undergo major modifications, and would serve as prototypes for new types of warship.

'Kildin' and 'Krupny'

In early 1956 the Soviet naval Commander-in-Chief, Admiral Kuznetsov, who had been closely identified with the 'big-ship' programme begun under Stalin, retired through ill-health. He was replaced by the relatively youthful Sergei Gorshkov, who was charged by Khrushchev with the task of dragging the Navy into the modern age.

Khrushchev wanted guns and torpedoes replaced by missiles, which could strike the enemy at far greater ranges than had previously been possible and, if armed with nuclear warheads, to far greater effect. In the view of Khrushchev the launch platform did not need to be elaborate because it did not need to 'fight' in the traditional sense. It simply needed to get into a position in which to launch its missiles, under shore-based direction and control, and then to retire out of range before the enemy was able to counter-attack. Attacks would, in accordance with established Soviet tactical doctrine, be co-ordinated with similar strikes by missile-armed submarines and land-based bombers of Soviet Naval Aviation.

The first major Soviet missile ships were the four *Raketny Korabl'* (lit. 'Rocket Ships') of the 'Kildin' class, which were completed in 1959–60. All were conversions of 'Kotlin' hulls, with a single missile-launcher and a magazine measuring 13m by 7m displacing the after gun mountings and the second bank of torpedoes. The after quadruple AA gun mounting was relocated in place of the forward 130mm mounting, so that the missile system represented the sole anti-surface capability of these ships. The forward bank of torpedo tubes was replaced by two twin trainable mountings, and two of the new RBU-2500 anti-submarine rocket-launchers were fitted forward of the bridge. A second air search radar antenna, Flat Spin, supplemented the Hair Net antenna fitted in the 'Kotlin'.

The first unit converted, *Bedovy*, was in such an advanced stage of completion that she retained a number of 'Kotlin' features which were modified or eliminated in the other three ships. In particular she retained the distinctive 'Kotlin' funnels and the Wasp Head fire control system, and the quadruple AA mountings were of 45mm calibre. The other three ships had slimmer funnels, no Wasp Head, and were fitted with the new 57mm gun mountings.

The SS-N-1 Strela missile (NATO designation: Scrubber) had been under development since the mid-1950s, and a prototype launcher is reported to have been installed aboard the *Sverdlov*-class cruiser *Admiral Nakhimov* for trials in the Black Sea in the late

The SS-N-1 Strela Missile

Soviet designation: ?
In service: 1959
Length: 7.6m(24ft 11in)
Diameter: 1m(3ft 3in)
Span: 4.6m(15ft)
Weight: 3,200kg(7,050lb)
Warhead: 500kg(1,100lb) HE/nuclear
Propulsion: turbojet
Speed: Mach 0.9
Range: 185km(100nm)
Guidance: command/active radar homing
Fire control: Top Bow

▶The larger hull of the 'Krupny' enabled SS-N-1 launchers and magazines to be fitted forward and aft. Note the proximity to the bow of the forward launcher, which could presumably be loaded and fired only in relatively stable sea conditions. (Skyfotos)

1950s. Essentially a derivation of the German V-1, it was powered by an air-breathing turbojet. Western Intelligence has generally credited the missile with a maximum range of 275km (150nm), but this has to be regarded as theoretical because there was no possibility of over-the-horizon targeting at the time the missile entered service. The range of the SS-N-1 would therefore have been restricted by the acquisition range of the ship's surveillance radars against surface targets – a maximum of 55km (30nm).

The missile was run out on to the missile-launcher from its armoured hangar and was then warmed up ready for launch. One source states that the warhead, which could be either high-explosive (HE) or nuclear, was designed to detach from the missile as it approached its target, so as to achieve an underwater hit.[3]

Two of the 'Kildin' class were completed at Zhdanov; *Bedovy* at 61 Kommuna, and the fourth ship at Komsomolsk. They were quickly followed on to the slipways by another intermediate design, based on the hull of the 'Kruplin'-class destroyer which had been projected as the successor to the 'Kotlin'. The larger hull allowed a second SS-N-1 launcher to be fitted forward, the magazine being an extension of the bridge structure. Reference sources generally credit the 'Krupny' with increased magazine capacity, each of the two magazines being reported to house 8–10 missiles, as compared with only 4–6 for the single magazine of the 'Kildin'.

The fourth AA mounting was moved back to its original location abaft the second funnel, and triple torpedo tubes replaced the twin mountings of the 'Kildin'. New models of air surveillance radar were fitted and the modified Wasp Head director deleted from the later 'Kildins' was restored, presumably to provide fire control for the missiles.

Eight ships were completed in 1960–1, construction being distributed between the Zhdanov, 61 Kommuna and Komsomolsk Shipyards, as with the 'Kotlin' and 'Kildin'.

The SS-N-1 missile system was conspicuously unsuccessful. Its performance was unreliable and reloading the launcher – particularly the forward launcher of the 'Krupny' which extended almost to the bows – must have been nightmare in any sort of a sea, given the lack of freeboard and the narrowness of the hull. The missile was withdrawn from service in the late 1960s and eleven out of the twelve units originally fitted with it were converted for other roles.

Technical Failure and Tactical Promise

The early Soviet missile ships were clearly a failure in operational terms. The missile, which constituted their main armament, was defective and the traditional narrow destroyer hulls in which the system was fitted could not provide the stable platform demanded by the relatively primitive launch and reloading procedures.

The 'Kildin' and 'Krupny' classes, however, were never intended to be anything more than intermediate types, and many of their inherent defects would be eliminated in the purpose-built 'Rocket Cruisers', the keels of which were already being laid as the first 'Kildins' were being fitted out. The conversions provided the Soviet Navy with early experience in anti-surface missile operations, and thereby ensured that by the time the new purpose-built ships were completed they would represent a relatively mature concept.

2. The Anti-Carrier Strategy

THE INTERMEDIATE ROCKET SHIPS of the 'Kildin' and 'Krupny' classes were intended to operate on the perimeter of the Soviet coastal defence zones. Their missile strikes against hostile naval forces attempting to penetrate Soviet sea space would be backed up by, and if possible co-ordinated with, attacks by land-based bombers armed with the AS-1 Kennel missile which entered operational service at about the same time. Within the coastal defence zone itself, flotillas of fast patrol boats armed with the short-range SS-N-2 Styx missile or with conventional torpedoes would be vectored to intercept enemy forces that succeeded in breaching the perimeter.

It had long been recognized by the Soviets that the principal threat to the integrity of their maritime defence zone, particularly in the Northern and Pacific Fleet Areas, was the Western carrier task force. However, with the advent of the new US Navy 'super-carriers' of the *Forrestal* class, designed to operate the A-3 Skywarrior nuclear strike bomber, this threat took on a new dimension. The Skywarrior, with a mission radius of approximately 1,200nm (2,225km), could be launched against Soviet cities from outside Soviet sea space. The threat posed by the Western carrier task forces could therefore no longer adequately be countered by a coastal defence strategy. Anti-carrier forces would have to intercept the attackers outside the perimeters of the maritime defence zones, before they reached their launch points.

The solution adopted, and confirmed in the 1959–60 policy review, was to create a 'Combined Arms' force comprising purpose-built surface warships, submarines and land-based maritime bombers, all equipped with long-range 'stand-off' anti-ship missiles. Enemy forces detected using an extensive High-Frequency Direction-Finding (HF/DF) network would then be relocated and tracked by land-based reconnaissance aircraft which would provide target data to the Soviet missile units either directly or via a central control post. Data links between the aircraft tracking the enemy forces on radar and the missile units would make it possible to launch missiles at their maximum range and to provide mid-course corrections based on the latest target position.

◀*Admiral Fokin* is seen during exercises in the Pacific. The SA-N-1 launcher arms are raised to the vertical loading position. Note the proximity of both the SA-N-1 launcher and the RBU-6000 anti-submarine rocket-launchers to the bows of the ship. (TASS, 1979)

The 'Rocket Cruisers'

The centre-piece of the new surface ship construction programme was the *Raketny Kreyser* (lit. 'Rocket Cruiser'), the first of which was laid down at the Zhdanov Shipyard in June 1959.

Unlike the intermediate 'Kildin' and 'Krupny' classes, the 'Kynda' was designed from the keel up. This was an important advantage for the Soviet designers who were not constrained by traditional hull-forms and were therefore able to design a ship around the new weapons systems.

There was, however, a counter-side. Having painfully acquired the skills and experience necessary to build successful conventional warships, the designers would now have to take something of a leap into the unknown. The Rocket Cruiser was a completely new type with its own particular demands and without any precedent in East or West.

Western missile ships, which were now beginning to appear in numbers, were designed primarily for air defence. Their missile magazine rings were generally located within the hull, the launchers being loaded on to the launcher arms either vertically or at an angle. These arrangements were in essence little different from those that obtained for their conventional gun-armed predecessors.

The large dimensions of the new Soviet anti-ship missiles, however, meant that missile handling at sea was practical only if the reloads were carried on the same level as the launcher and were transferred horizontally. As each of the missiles weighed between five and nine times as much as the average contemporary surface-to-air missile (the estimated weight of the SS-N-3 Shaddock missile is 5.4 metric tonnes), this was clearly going to present problems of weight distribution within the ship.

If one adds to this the problem of accommodating a multitude of increasingly specialized radar antennae to perform the functions of air and surface surveillance, and fire control for each of the missile and gun systems, it is immediately apparent that topweight would be a major design headache. In the new Rocket Cruisers the problem would be compounded by the need to carry at the highest level the bulky radars required for missile tracking and guidance in order to extend the radar

'Kynda' Class: Profile and Plan

The drawing shows *Grozny* as she appeared in the early 1970s.

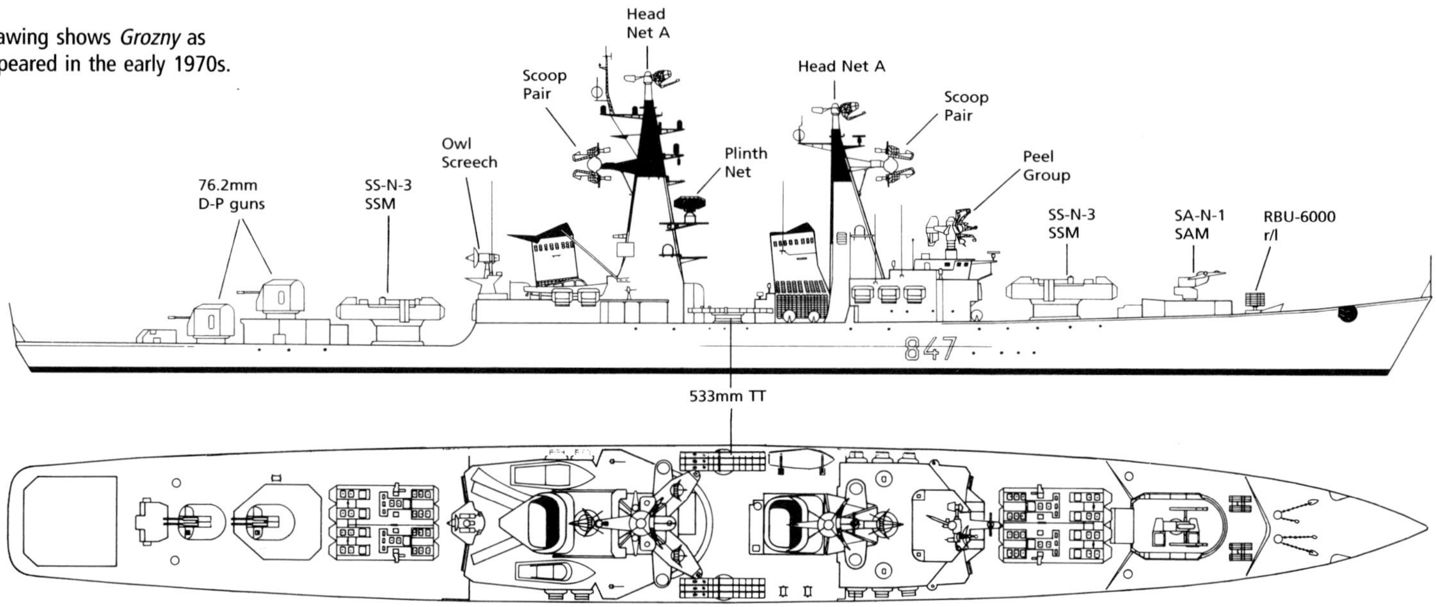

horizon. Moreover, much of this equipment was still in the process of development while the ships were building, compelling the ship designers to work with parameters for equipment weight and configuration that were at best provisional.

The shipyards that had built the cruisers of the *Sverdlov* class were no longer available to the Soviet Navy, so the new ships would have to be laid down in the traditional destroyer yards. This placed firm constraints on maximum hull size and precluded a solution whereby the topweight inherent in the SS-N-3 missile system could be accommodated in a large ship comparable to the *Sverdlov*. The 'Kynda', built exclusively at the Zhdanov Shipyard, had an overall length of only 142m (four metres more than the 'Krupny'), although a relatively broad beam of 15.8m brought a greater degree of stability than the narrow 'destroyer' hull-form of the intermediate types.

The adoption of lightweight high-pressure steam plant using advanced pressure-fired boilers, enabled the ships to attain a maximum speed well in excess of 30 knots, but exacerbated the topweight problem in that the weight of the machinery, located low in the hull, provided inadequate compensation for the concentration of heavy items of equipment above the level of the upper deck.

The SS-N-3 Shaddock

The SS-N-3 missile which constituted the main armament of 'Kynda' had been under development since 1954. Unlike the SS-N-1 Strela which armed the 'Kildin' and 'Krupny' classes, and the SS-N-2 Styx which equipped the fast patrol boats of the 'Komar' and 'Osa' classes, it had been developed primarily as a strategic weapon comparable to the US Navy's Regulus for installation aboard submarines. Armed with a nuclear warhead and with a maximum range estimated at 750km (400nm), it was intended for use against coastal cities and naval installations.

The strategic SS-N-3C variant duly entered service aboard converted diesel submarines of the 'Whiskey' class and the large nuclear-powered boats of the 'Echo I' class in 1960. Indeed, Norman Friedman[1] has suggested that the Rocket Cruiser may originally have been envisaged as a *strategic* weapons carrier, and that the trainable quadruple launcher installed on 'Kynda' was specifically designed to launch the SS-N-3C, which was launched under inertial guidance.

This theory, while eminently plausible, must remain open to question. It seems inconceivable that the Soviets would have regarded the operation of missile-carrying surface ships so close to an enemy's coastline as a practical proposition. The US Navy,

'Kynda' Class
Soviet Designation: Raketny Kreyser (RKR) – Rocket Cruiser

Construction

Name	Builder	Laid down	Launched	In service
Grozny	Zhdanov Shipyard	June 1959	Apr 1961	June 1962
Admiral Fokin	Zhdanov Shipyard	Aug 1960	May 1962	Aug 1963
Admiral Golovko	Zhdanov Shipyard	Dec 1960	Nov 1962	July 1964
Varyag	Zhdanov Shipyard	Sept 1961	June 1963	Feb 1965

Displacement (tons)
Standard: 4,400
Full Load: 5,500

Dimensions
Length: 143m(469ft) overall
Beam: 15.8m(52ft)
Draught: 5.4m(17ft)

Propulsion
2-shaft geared steam turbines; 4 turbo-pressurized boilers; 100,000shp=34 knots max.

Complement
304

Weapons*
ASuW: 2 quadruple SS-N-3 launchers (16 Shaddock missiles); 2 triple banks of 533mm(21in) torpedo tubes
AAW: 1 twin-arm SA-N-3 launcher (16 Goa missiles); 2 twin 76.2mm(3in) D-P guns
ASW: 2 RBU-6000 rocket-launchers

Sensors*
Surveillance radars: 2 Head Net A; 2 Don-2
Fire control: 2 Scoop Pair (SS-N-3 guidance), 2 Plinth Net (SS-N-3 data link – not in *Varyag, Golovko*); 1 Peel Group (SA-N-1); 1 Owl Screech (76.2mm)
Sonars: keel-mounted Herkules (Wolf Paw) HF
Electronic warfare: 2–4 Guard Dog (not in *Fokin, Varyag*); 1 Bell Clout, 1 Bell Slam, 2 Bell Tap; 4 Top Hat

*Weapons and sensors as originally designed. For final fit and later modifications see text.

The SS-N-3B Shaddock Missile

Soviet designation: P-35
In service: 1962
Length: 10.2m(33ft 5in)
Diameter: 0.98m(1ft 3in)
Span: 5m(16ft 5in)
Weight: 5,400kg(11,900lb)
Warhead: 1,000kg(2,200lb) HE/350kT nuclear
Propulsion: liquid-fuel rocket with two solid-fuel boosters
Speed: Mach 1.4
Range: 460km(250nm)
Guidance: command/active radar homing
Fire control: Scoop Pair

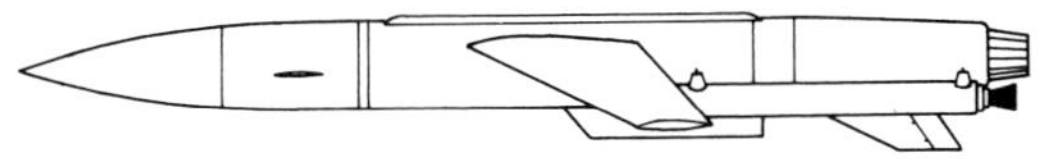

wedded to a strategy of sea control and power projection, could contemplate possible scenarios in which the Regulus missile could be fired against the shore from surface ships, but submarine launch was the only realistic option for the Soviet Navy outside the Baltic and Black Seas.

Moreover, the trainable launcher of 'Kynda' has a precedent in the SS-N-1 launcher installed in 'Kildin' and 'Krupny'. Even though the SS-N-3B anti-ship variant of the Shaddock missile carried by 'Kynda' could be controlled in flight, a trainable launcher had the effect of minimizing post-launch manœuvres to direct the SS-N-3 towards its target, and made it easier to gather the large, ponderous missile into the beam of the tracking radars which would provide post-launch guidance. In contrast, it was the strategic missile-launchers aboard the submarines that were not trainable.

The development of two anti-ship variants of the SS-N-3 missile, one for installation in submarines, the other for installation in the Rocket Cruisers, appears to have been prompted by the defence policy review of 1959–60, which placed the strategic mantle firmly in the hands of the newly formed Strategic Rocket Forces. The Soviet Navy, deprived of a strategic role, was compelled to fall back on its traditional mission of defending Soviet territory against seaborne assault, and the naval construction programme dating from this period took place within the framework of the anti-carrier strategy.

The Tu-16 Badger-C maritime strike bomber, armed with two AS-2 Kipper anti-ship missiles, entered service with Soviet Naval Aviation in 1960–1. The 'Echo II' variant of the 'Echo'-class nuclear submarine, armed with eight SS-N-3A anti-ship missiles, followed in 1962, and was to be complemented by a diesel-powered counterpart, the 'Juliett', armed with four SS-N-3A missiles. *Grozny*, the first of the 'Kynda'-class Rocket Cruisers, entered service in June of the same year. Development of the Bear D maritime reconnaissance bomber, which was to provide the target data for these forces, took a little longer, the first units entering service with the SNA in 1964.

Both the 'Echo II' and the 'Kynda' were armed with eight ready-use anti-ship missiles, which suggests that this number was viewed by the Soviet Navy as a standard 'tactical unit'. Four Badger-C maritime bombers each armed with two AS-2 missiles, or a pair of four-missile 'Julietts' would constitute a tactical unit

▲The 'Kynda' was the Soviet Navy's first purpose-built 'Rocket Cruiser'. The novelty of conception of these ships taxed the designers to the utmost. Virtually every item of equipment, from the bulky quadruple missile-launchers down to the anti-submarine rocket-launchers, was new to this class. (Skyfotos)

◀The massive quadruple launchers for the SS-N-3 Shaddock anti-ship missile are trained to port and fully elevated in this view of *Varyag* on exercises in the Pacific in 1970. The reload hatches can be seen on the forward face of the bridge structure. (US Navy)

▲A key element in the missile targeting equation is the Tu-95 Bear D land-based reconnaissance aircraft, converted from a strategic bomber. The Bear D is fitted with a plethora of electronic surveillance equipment, of which the most prominent is the underbelly Big Bulge surface surveillance radar. A data link enables the Bear D to transmit the picture on its own radar screen to ships like 'Kynda', which can then compare the Bear's radar picture with that received on a downlink from the missile homing radar and transmit the necessary guidance commands to the missile. Here, a Bear D of the Soviet Northern Fleet is being 'escorted' by an F-14 Tomcat from USS *Nimitz* during exercise Teamwork 80. (US Navy)

equivalent to a single 'Kynda' or 'Echo II'. A complement of eight missiles would allow a mix of nuclear and conventional warheads, enabling the ships to respond effectively to any tactical eventuality.

The large size of the Soviet air- and surface-launched anti-ship missiles, which were in effect unmanned aircraft, was necessary to provide the required stand-off range. It also made possible a sufficiently large explosive payload to disable a major surface vessel such as an aircraft carrier.

The alternative nuclear warhead would almost certainly have been fuzed for air burst, because this would result in considerable damage to aircraft parked on the flight deck, and the effect of electro-magnetic disturbance on radar and radio transmission would be to 'blind' the task force, leaving it vulnerable to more conventional forms of attack.

Accommodating the Missiles

The SS-N-3 launchers adopted for the 'Echo' and the 'Kynda' differed from the SS-N-1 launchers of the 'Kildin' and 'Krupny' in that the missiles were already housed in the launch tubes, which elevated for firing. This meant that the missiles were not only fully protected from the elements, but could be launched in rapid succession towards their target in order to saturate enemy defences.

The 'Kynda' was also provided with reloads – one missile per tube – although it was probably envisaged that the ship would turn away following the initial attack to enable reloading to be carried out beyond the range of enemy retaliation.

The need to accommodate two of the bulky quadruple launchers together with reloads largely dictated the layout of the ships. The extreme sheer which had been such a marked feature of the destroyer hulls of the 1950s was abandoned in favour of a flatter, more stable platform better suited to the accommodation of large missiles stowed horizontally.

Both quadruple banks of missile tubes were located well inboard, far from the extremities of the ship. This was in part dictated by the reloading arrangements, the reload tubes being housed within the ship's superstructures. However, it also proved to be a neat solution to the problems experienced in reloading the exposed launchers of 'Kildin' and 'Krupny' in heavy seas, in that the forward bank of tubes was spared the worst of the heaving and pitching accelerations and was relatively 'dry'.

The missile transfer arrangements were also markedly simpler, the reloads being located on the same level as the launch tubes and therefore able to be moved directly into the tubes on a system of rollers, thereby eliminating the need for hoists.

Missile Guidance

The missile tracking radars (NATO designation: Scoop Pair) are carried high on the forward side of the foremast and the after side of the mainmast respectively. The Scoop Pair antenna comprises two identical lattice reflectors mounted above and below a spherical housing, the whole assembly being fully stabilized to compensate the motions of the ship.

The flight profile of the SS-N-3 missile involves a steep climb after launch. It then cruises at high altitude, enabling the Scoop Pair radars to continue tracking well beyond horizon range. The missile is flown in the direction of the last-known position of the enemy ship or task force, and uses its own active radar to relocate and home in on the target in the terminal phase of flight.

A video data link enables the missile to transmit its radar picture back to the launch ship,[2] thereby allowing the latter to select the target on the basis of the relative size of the radar 'blips' and the relative position of the ships that make up the enemy's force. The process of target designation is facilitated by the high flight profile of the missile. If the missile is to be effective at anything approaching its maximum range of 250nm (460km), however, additional target data inputs are required from aircraft such as the Bear D, which is reportedly equipped with a similar video data link which allows its own radar picture to be monitored by the launch ship.

Without these additional data inputs from the Soviet Ocean Surveillance System, evasive manœuvres by the enemy force might well mean that the missile homing radar encountered only empty seas. Lengthy pre-launch missile preparations followed by a flight time of perhaps 15–20 minutes would then combine to create a 'dead time' sufficient for an enemy force to have moved 15nm (28km) from its last-known position. Effective mid-course guidance from the launch ship therefore depended on continuous (or at least regular) real-time target updates from off-board sources.

Air Defence

The requirement to conduct anti-carrier operations outside the Soviet coastal defence zones where surface ships could not be guaranteed protective cover from land-based fighter patrols, meant that a new approach to air defence had to be adopted. While the radar-controlled multi-barrel anti-aircraft mountings of 45mm and 57mm calibre which had entered service during the 1950s were relatively effective for self-defence against aircraft armed with conventional iron bombs, the advent of air-launched stand-off missiles presented a threat of a completely different order.

The only weapon with sufficient range to engage aircraft armed with stand-off missiles was the surface-launched area defence missile, which was then at a relatively early stage of development. The Terrier system had entered operational service with the US Navy as early as 1955, with the long-range Talos following two years later, but both had experienced serious problems of accuracy and reliability. Research and development in the Soviet Union lagged some way behind, and appears to have been directed exclusively towards land-based missiles, the flight profiles and guidance systems of which were not ideally suited to operations in a maritime environment.

The standard US Navy air-launched stand-off weapon of the period was the ASM-N-7A Bullpup (redesignated AGM-7A/B in 1962), which had been developed as a response to the losses incurred in attacking heavily defended targets during the Korean War. Designed to be carried by carrier-borne light attack aircraft, not heavy land-based bombers, it was significantly smaller than its Soviet counterparts and had to be launched within visual range of the target because it employed joystick radio control. Early variants of the missile had a range of 8nm (12.5km), while the AGM-7C variant which entered service in 1964 had a range of 10nm (16km).

▶**SA-N-1 Goa missiles are loaded on to the launcher arms in preparation for firing during this Soviet exercise in the Baltic Fleet area. (TASS, 1982)**

The SA-N-1 Goa Missile

Soviet designation: M1 Volga-M
In service: 1962
Length: 6.1m(20ft)
Diameter: missile 0.37m(1ft 3in); booster 0.55m(1ft 10in)
Span: missile 1.2m(3ft 11in); booster 2.3m(7ft 7in)
Weight: 950kg(2,090lb)
Warhead: 60kg(132lb) HE
Propulsion: solid-fuel rocket with solid-fuel booster
Speed: Mach 3.5
Range: 18km(10nm)
Ceiling: 15,000m(50,000ft)
Min. altitude: 1,500m(5,000ft)
Guidance: radar/command
Fire control: Peel Group
Operating frequency: I-Band

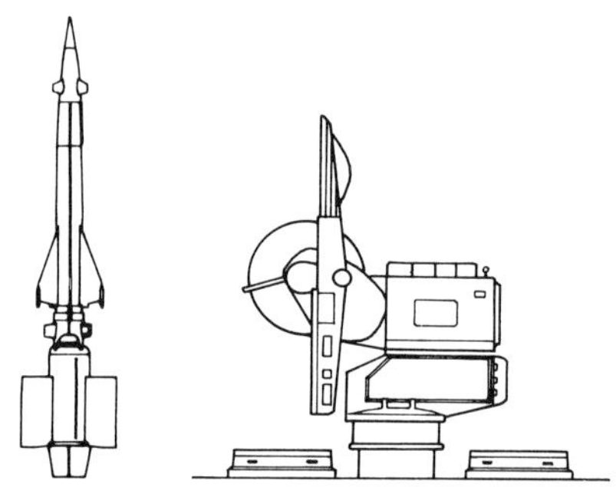

▶**The destroyer *Bravy* seen shortly after her conversion, with the Peel Group guidance radar concealed beneath a protective canvas cover. Note the after funnel, remodelled in order to deflect the blast from the missiles. The midships quadruple 45mm mountings and the forward bank of torpedo tubes were fitted only after the ship had completed trials with the SA-N-1 missile. (Skyfotos)**

Until the advent of the air-launched Harpoon missile in the early 1980s, Bullpup was the US Navy's only effective tactical anti-ship missile. Long-range stand-off weapons for both strategic and tactical use against land targets were proposed, but the Americans were unwilling to set up an elaborate over-the-horizon targeting network on the pattern of the one developed by the Soviets for the SS-N-3, preferring the small manned aircraft using visual targeting for anti-shipping attacks.

The comparatively short range of Bullpup meant that a surface-to-air missile of only modest performance was required. Whereas the US Navy was impelled towards ever-greater missile range in order to counter the threat of saturation strikes by land-based bombers armed with long-range missiles – later variants of the Talos ramjet-powered missile were credited with a theoretical range of 100nm (185km), and the long-range variant of Typhon projected as its successor with a range of 200nm (370km) – the Soviet Navy was content to adapt the land-based SA-3 Goa missile, which in its early version is reported to have had a maximum range of about 10nm (18km).

The SA-N-1 missile (Soviet designation: M1 Volga-M) was identical in configuration with its land-based counterpart, but was married to a twin-arm

stabilized launcher and a specially developed guidance radar (NATO designation: Peel Group). Stabilization of both the twin-arm launcher and the guidance radar was necessary because, like other Soviet land-based missiles, Goa employed radar/command guidance and a rolling, pitching platform would have made it difficult to 'gather' the missile into the beam of the tracking antennae.

The Peel Group radar combination comprises two pairs of solid 'orange peel' reflectors, each pair having one antenna in the vertical and the other antenna in the horizontal plane, the whole assembly being counterbalanced by prominent vanes projecting from the after side of the central mounting pillar. The pair of larger reflectors tracks the target, while the smaller pair tracks the two missiles, normally fired in a salvo, the combination of vertical and horizontal reflectors providing tracking in three dimensions. The angle of error between the missile and the target is measured and correction signals are transmitted to the missile via a radio command link.

The SA-N-1 guidance system is relatively self-contained, no separate height-finding radar being required, but this is achieved at the cost of an antenna that is significantly larger and heavier than Western ship-borne missile guidance radars of the period.

The SA-N-1 air defence system was trialled aboard a converted destroyer of the 'Kotlin' class, the *Bravy*. Construction of the latter, which had been laid down and launched at the 61 Kommuna Shipyard in 1956–7, was apparently suspended while revised plans were drawn up for her completion as a missile trials platform.

The fore part of the ship remained substantially unaltered except for the provision of a new air surveillance radar (NATO designation: Head Net A) to provide bearing data for the missile guidance radar. The after guns, however, were removed and replaced by a large deckhouse on which was mounted the prototype SA-N-1 launcher. The second funnel was redesigned to deflect the blast from the missiles, and the former lattice mainmast was replaced by a prominent tower mast at the top of which was located the Peel Group guidance radar.

Bravy was completed in late 1961 and conducted trials of the missile in the Black Sea during the early 1960s. In 1963–4 the ship underwent an extensive refit during which the original midships 45mm quadruple AA mountings and the forward bank of torpedo tubes were replaced, and two RBU-2500 anti-submarine rocket-launchers were fitted. She subsequently entered full operational service with the Black Sea Fleet, and served as the prototype for further 'Kotlin' surface-to-air missile conversions in the late 1960s (*see* Chapter 4).

The 'Kynda' Installation

The 'Kynda' was provided with a single twin-arm launcher, located between the forward bank of SS-N-3 missile tubes and the bow. The launcher enjoys excellent all-round arcs, but its proximity to the bows of the ship means that it is exposed to the worst of the seas. The installation is therefore protected by continuous shielding which also serves to deflect the missile exhaust from other near-by weapons systems.

Other navies have favoured the relatively sheltered quarterdeck location for their own surface-to-air missile installations, but in the 'Kynda' this arrangement appears to have been precluded by the lack of hull depth at the stern. Missile stowage for the longer Western missiles was generally horizontal, the missile rounds being run up at a shallow angle from a preparation room behind the launcher, whereas the six-metre Goa missile is stowed vertically and reloaded via doors directly beneath the launcher arms. Locating the launcher on the low quarterdeck of the 'Kynda' would presumably have interfered with the shafting arrangements.

The magazine capacity was originally estimated at 22 missile rounds, but the lower figure of 16 is now generally accepted. This is small by contemporary Western standards. Western missile ships of the period generally had a magazine capacity of 30–40 rounds per launcher and in some of the larger US Navy ships the figure was 60.

In part this may be explained by differences in operational philosophy, the 'Kyndas' being designed to undertake specific missions of relatively short duration, and in relative proximity to their support base. Their US Navy counterparts, on the other hand, were designed for prolonged operations in hostile waters, during which they might be subjected to air attacks over a period of days or even weeks.

A further factor is that the 'Kyndas' were probably not intended to operate alone, but would almost certainly have been accompanied by destroyers of the 'Kashin' class (q.v.), each of which was armed with a further two SA-N-1 missile systems. Given the 'modular' nature of the SA-N-1 system, the launcher and its reload magazine being installed as a unit of fixed volume, the relatively small capacity of the reload magazine was necessary if the system was to be fitted in ships of destroyer size. In the 'Kashin' two launchers, each with a capacity of 16 rounds, were considered preferable to a single launcher with double that capacity because two targets could be engaged

▶ A bow view of *Grozny* in the Black Sea, displaying much of the new electronic equipment added in the early 1970s. The Plinth Net missile control radars are on platforms projecting from the tower mainmast, with the four Guard Dog ESM radomes visible below them. Above them are four Top Hat ECM jammers, which were fitted from the outset. (Novosti, 1977)

simultaneously, and a malfunction in one or other of the launchers would have less serious consequences.

Missile Guidance

The air surveillance radars fitted at completion were the Head Net A model previously installed in the Rocket Ships of the 'Krupny' class and in *Bravy*. One antenna was fitted atop of each of the two tower masts. The duplication of air surveillance antennae (although in later ships these were not necessarily of the same model) was to become a standard feature of major Soviet surface units. The need for redundancy, often at significant cost in terms of topweight, appears to have been considered necessary because of the high failure rate of Soviet electronic equipment of the period and the difficulty of maintaining the equipment at sea.

In the early days of high-technology weapons systems it would have been difficult to provide crews (largely made up of short-term conscripts) with the skills necessary to operate them. Maintenance would have to be carried out almost exclusively at the support base because a ship the size of the 'Kynda' would have insufficient hull volume to carry the necessary spares, and underway support ships on the Western model were non-existent.

Head Net A was the latest in a series of E-Band radars developed by the Soviet Navy during the 1950s. The adoption of E-Band in preference to the lower-frequency sets generally favoured by Western navies had two primary advantages. The antenna could be smaller and lighter and was therefore better suited to installation on ships of destroyer size. Furthermore, the performance of the radar was markedly superior at lower altitudes, where the performance of the lower-frequency sets favoured by Western navies was adversely affected by interference from the sea, especially in adverse weather.

The disadvantage of E-Band was its relatively short range, Head Net A being credited with an effective range of only 60–70nm (110–130km) against a large aircraft, whereas contemporary Western air surveillance radars generally achieved detection ranges of 150–200nm (280–370km).

Presumably long range was considered less important by the Soviets, given that the SA-N-1 area defence missile is credited with a maximum range of only 10nm (18km). Moreover, the carrier-borne attack aircraft which presented the major threat to the 'Kynda' would approach their target at relatively low altitude compared with the land-based bombers which faced the Western carrier task forces. The superior performance of the E-Band radar against low-flying aircraft would therefore assist target designation to the missile

and gun fire control radars. The performance of the radar against surface ships and low-flying aircraft was further enhanced by stabilization of the antenna.

The single Peel Group fire control radar on the 'Kynda' is located on top of the bridge. The entire assembly can be trained in azimuth, and the paired 'orange peel' reflectors can be elevated independently to facilitate tracking of both target and missiles. Targets can be engaged well abaft the beam, but stern coverage is effectively blocked by the massive tower masts.

Gun Mountings

Twin 76.2mm dual-purpose gun mountings are superimposed on the low quarterdeck abaft the after bank of SS-N-3 launchers. These provide back-up for the SA-N-1 system together with coverage of the 'blind' stern arcs.

The mounting was originally intended for the projected 'Kotlin' successor, the 'Kruplin'. The decision to adopt the 76.2mm(3in) calibre for this weapon was undoubtedly influenced by similar developments in the US and British Navies, both of which during the post-war period opted for the heavier 3in weapon to replace the standard wartime 40mm AA gun. Both Navies developed twin enclosed mountings designed for a high rate of fire at high angles of elevation. The US model was credited with a theoretical 90 rounds per minute per gun; its Royal Navy counterpart achieving a more modest 60 rounds per minute. Both proved to be too complex and suffered frequent breakdowns.

For their own 76.2mm gun the Soviets appear to have settled for a more conservative 45 rounds per minute, and the mounting is probably more robust and reliable as a consequence. Certainly it has outlasted both its Western counterparts in service, and was being

The 76.2mm/60 D-P Gun System

Soviet designation: AK-276
In service: 1961
Angle of elevation: +85 degrees
Rate of fire: 45rpm per barrel

Projectile weight: 16kg(35lb) fixed round
Muzzle velocity: 900m/sec(2,950ft/sec)

Range: 11km(6nm); 7km(4nm) effective AA
Weight of mounting: 25 tons
Fire control: Owl Screech

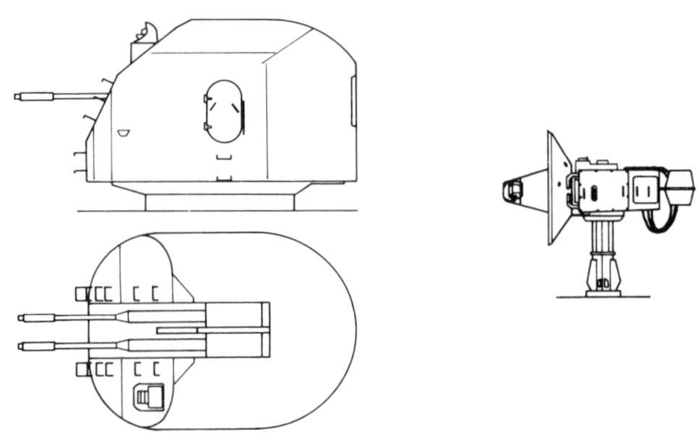

◀ **The after twin gun mountings and quadruple SS-N-3 launcher of *Admiral Golovko*. The helicopter landing pad is rarely used, and there are no hangar or maintenance facilities. (MoD, 1973)**

installed in Soviet surface ships into the early 1980s. The twin mounting is fully stabilized on the pattern of earlier Soviet 130mm and 100mm mountings fitted in post-war cruisers and destroyers, and allows for a maximum elevation of 85 degrees.

Fire control is provided by a single Owl Screech 'Tellerform' radar, mounted on a pedestal abaft the second funnel. This was the first installation of Owl Screech, which was derived from the smaller Hawk Screech radar used to control the quadruple 45mm and 57mm AA mountings on the 'Kotlin', 'Kildin' and 'Krupny'. The solid dish antenna is mounted between boxes housing the electronics, the entire assembly being rotated and elevated around a central mounting pillar. Stabilization is provided by prominent fins projecting from the rear of the electronics boxes.

Anti-Submarine Warfare

For the NATO navies the major preoccupation in the post-war era was the submarine, which presented a threat not only to offensive surface operations spearheaded by the carrier task forces, but also to the sea lines of communication between Europe and the United States. Western submarines did not present a comparable threat to Soviet surface operations.

It was envisaged by the Soviet Navy that the coastal zones would be patrolled by large numbers of small craft armed with 'blanket' weapons such as depth-charges and rocket-launchers, which were eminently well suited to anti-submarine operation in shallow waters, but that surface ships operating outside those zones would need only to cope with chance encounters. There was therefore little incentive to develop either high-performance low-frequency sonars or the stand-off missiles that were designed to take advantage of their increased detection ranges.

The Herkules keel-mounted sonar with which the 'Kynda' is believed to be fitted is a high-frequency model employing mechanical scanning and vacuum-tube electronics. Unlike the large, low-frequency sonars which began to enter service with the US Navy from the late 1950s, it is optimized for fire control rather than underwater surveillance and would therefore be more effective against a submarine which had already revealed its presence, possibly by a torpedo attack, than against one lying silently in wait, which it would probably fail to detect.

The primary anti-submarine weapon of the 'Kynda' is the RBU-6000 rocket-launcher, of which two are mounted side by side close to the bow. Anti-submarine rocket-launchers, which succeeded conventional depth-charge throwers modelled on Second World War types, first entered service with the Soviet Navy in the mid-1950s, but the RBU-6000 was the first to feature automatic reloading.

The launcher comprises twelve tubes in a horseshoe arrangement around a central mounting pillar. Each tube is approximately 1.6 metres long and fires a projectile with a diameter of 250mm. The tubes are elevated to the vertical for automatic reloading from a below-decks magazine, and are fired in a paired sequence.

Earlier Soviet rocket-launchers, the sixteen-barrel RBU-2500 fitted aboard Soviet destroyers from 1957/8, and the five-barrel RBU-1200 designed for smaller craft, had been manually loaded. However, the design of the RBU-6000 appears to have been heavily influenced by the 375mm rocket-launchers developed by Bofors during the early/mid-1950s. A four-barrel launcher, electro-hydraulically driven, with remotely controlled fuze-setting and automatic vertical reloading, had entered service with the Royal Netherlands and Swedish Navies in 1954–5, and had subsequently been exported to Japan and the Federal Republic of Germany. Even closer to the configuration of the RBU-6000 was the six-barrel launcher developed by Creusot-Loire to fire the standard range of Bofors anti-submarine rockets, which entered service with the French Navy in the late 1950s.

Reference sources generally credit the projectiles fired by the RBU-6000 with a range of 6,000m, a figure which derives from the Soviet type number for the launcher. This would constitute a dramatic improvement in performance over the rocket fired from the RBU-2500 which is credited with a range of only 2,500m (in itself no small achievement given the technology of the day). The launch tubes of the RBU-6000 appear to be of identical dimensions with those of the RBU-2500, and it is unlikely that the gas-ejection system with which the former is often credited would account for such a divergence in range capabilities.

Comparable Western systems of the period had significantly less range. That of the US Navy's Weapon Alfa, using the Mk 2 rocket, was under 900m; the M/50 rocket used by the early Bofors launchers had a maximum range of 830m; and the Terne III system developed by the Royal Norwegian Navy had a maximum range of 850m.

The Soviet projectiles are lighter and of smaller diameter than the Swedish and American rockets. They also sacrifice payload for propulsion because the pattern-firing made possible by the multiple-launch tubes requires each individual rocket to have a smaller lethal radius. Nevertheless, the figure of 6,000m seems unduly optimistic for the technology of the day, and is difficult to justify on the basis of Soviet sonar performance in the early 1960s. Throughout this period the RBU-6000 was referred to in NATO sources as the MBU-2500A, and it may be that the early rockets associated with the launcher had less range than the current models.

Torpedoes

As we have seen in the previous chapter, the Soviet Navy persisted with the surface-launched anti-shipping torpedo well into the 1950s, two quintuple banks of tubes being fitted in even the relatively advanced 'Kotlin'. In contrast, both the US and British Navies had abandoned the anti-shipping torpedo altogether in their latest destroyer designs and were removing at least one of the standard two banks of tubes from many of their war-built ships in order to compensate the added weight of new electronics. Of the major Western navies only the French Marine Nationale retained the anti-shipping torpedo in its post-war destroyers and these ships, like the 'Kotlins', were essentially derived from a pre-war design.

The standard Soviet anti-surface torpedo of the period was a 7.8-metre heavyweight with a 400kg warhead. The basic steam-turbine powered model, which received the Western designation Type 53–51, is broadly comparable to the British Mk 8 and the US Mk 14 torpedoes which proved so reliable in service during the war years. An HTP-powered variant, the Type 53–56, entered service in 1956 and an alternative 15kT nuclear warhead was available for the same torpedo by 1959.

The Type 53–51 torpedo was initially fired from an 8.5-metre torpedo tube designated M-48. The tubes themselves were hydraulically operated, the torpedoes being ejected by compressed air.[3] The M-48 was installed aboard the cruisers of the *Sverdlov* class, the destroyers of the 'Skory' and 'Kotlin' classes, the Guard Ships of the 'Kola' and 'Riga' classes and the fast patrol craft built during the 1950s.

In 1957 it was superseded by the M-57, a more advanced tube constructed of aluminium alloy and equipped with a heating system, which allowed torpedoes to be stored inside for up to three months without additional maintenance checks. The torpedo was ejected by a powder-charge. Both the 'Kildin' and the 'Krupny' were fitted with the new tube and it was this which equipped the 'Kynda' and its successors.

There can be little doubt that the M-57 tube was intended to fire the standard Soviet anti-surface torpedo of the period. However, the limited tactical value of this weapon, particularly when fired from a large modern surface combatant such as the 'Kynda', has led commentators to assume that the M-57 is a dual-purpose weapons tube, capable of firing anti-submarine torpedoes as well as anti-surface models.

The problem lies in identifying the anti-submarine torpedo which would have been fired from the tube when the 'Kynda' first entered service. A lightweight anti-submarine torpedo was being developed by the late 1950s for installation aboard a new generation of Guard Ships and Small Anti-Submarine Ships, but this was a small-diameter weapon fired from a short 400mm tube with a length of about 5m.

During the post-war period all the major Western navies showed a considerable interest in developing an effective anti-submarine homing torpedo capable of being fired from a surface escort. The major attraction of the torpedo as an anti-submarine weapon was its range, which greatly exceeded that of contemporary anti-submarine rockets and mortars and therefore more closely matched the improving ranges of the latest sonar equipment. The major difficulty encountered was that of eliminating the 'dead time' between the launch of the torpedo and its arrival at the predicted position of the submarine.

With the advent of nuclear propulsion, it would be a relatively simple matter for the submarine to escape the torpedo before it began its helical search pattern because the seeker in the torpedo homing head had a relatively short acquisition range. The solution proposed was wire guidance, but paying out a wire from a surface ship moving at 20–25 knots and being buffeted by heavy seas was never really practical; the wires broke, and the concept was quickly abandoned by most navies.

The anti-submarine torpedoes developed by the Western navies during the late 1940s and early 1950s were all full-diameter weapons. The British Mk 20(E) 'Bidder' and the US Navy's Mk 35 were 533mm(21in) torpedoes, while the French persisted with their pre-war 550mm diameter for the L3.

It has been suggested that the Soviet 400mm anti-submarine torpedo could be fired from the 533mm tubes of 'Kynda' provided that a tube liner were used.[4] This would certainly be feasible, but it is by no means clear that the Soviets had this in mind when the 'Kynda' and the earlier Rocket Ships were designed. The effective range of the 400mm torpedo is probably little greater than that of the RBU-6000 rocket-launcher. Moreover, the number of tubes provided – four in the 'Kildin', six in the 'Krupny' and 'Kynda' – appears to be inadequate, particularly if some of the tubes were to house anti-ship weapons.

The Guard Ships of the 'Petya' class had a bank of five tubes dedicated exclusively to the 400mm torpedo in addition to their RBU rocket-launchers, and the later

'Petya II' variant had its torpedo armament doubled. It is of interest to note that the 'Kynda' successor, the 'Kresta', also had its torpedo armament uprated to two quintuple banks of tubes, making it possible to obtain a more satisfactory 'mix' of anti-surface and anti-submarine weapons. Contemporary Western escorts armed with anti-submarine torpedoes had either large numbers of tubes (cf., the British Type 12 frigates) or a smaller number of reloadable tubes (cf., the French E52 frigates); there is no evidence of reload magazines on any of the Soviet ships.

More plausible is the possibility that 'Kynda' was intended to carry a mix of 533mm torpedoes some of which would have the nuclear warhead introduced in the late 1950s. These would be more useful than a torpedo with a conventional HE warhead if the cruiser were to be cornered by an enemy force of surface warships, as they could be fired at maximum range with potentially devastating effect.

In theory they could also be used against hostile submarines because the lethal radius of the nuclear warhead would compensate the inaccuracies inherent in preset guidance. The rationale for the US Navy's submarine-launched Mk 45 ASTOR torpedo which entered service in 1959, was similar, although it had the advantage of being allied to superior sonar technology. It is questionable whether the acquisition range of the sonar installed in the 'Kynda' would have been sufficient to permit the firing of a nuclear torpedo without the risk of considerable damage to the launch-ship.

Electronic Countermeasures

The 'Kynda' was the first Soviet warship to feature a comprehensive electronic countermeasures outfit. Early post-war types were fitted only with standard high/medium-frequency direction-finding loops for the interception of enemy radio transmissions. Otherwise the Soviets relied almost exclusively on 'active' means of detection such as radar or visual reconnaissance.

Small radomes for the detection of enemy radar transmissions first appeared on the 'Krupny', being located on the after side of the short lattice mainmast. The 'Kynda' ECM suite, however, was much more extensive and was clearly designed to cover a wider range of frequencies.

On the forward side of the mainmast were projecting platforms for a large double radome (NATO desig-

▼ *Grozny* was further modified in 1979–80, and now has a new deckhouse abeam the forefunnel for four AK-630 anti-missile gatlings. The Bass Tilt fire control radar has replaced the former Guard Dog ESM radomes, which have been removed. (US Navy, 1985)

nation: Bell Clout) and a smaller variant (NATO designation: Bell Bash) respectively. Each antenna comprised a larger radome mounted above the platform with a smaller radome for higher frequencies beneath.

Above and below these two platforms, further platforms projecting from the sides of the mainmast supported four identical radomes which received the NATO designation Top Hat. Grouped around the upper mainmast platform were three further unidentified ESM antennae together with a High Pole IFF mast and the HF/DF loop.

The extent and variety of the equipment, almost all of which was new to this class, made it clear that the Soviets intended to take electronic warfare very seriously. They appear to have perceived at an early stage that radar interception and deception measures would become crucial to the survival of the surface ship in a modern maritime environment.

This is not to say that the Soviets did not experience problems in developing and manufacturing effective EW sensors. The evidence for this is provided by the substantial platforms located on the lower foremast and the lower mainmast, which remained empty long after the ships entered service, and in the case of one unit, *Varyag*, remained unoccupied until the ship underwent a major refit in 1980–1.

Hull-Form and Propulsion

The Soviets had not previously built a ship of this size, intermediate between the 6-inch gun cruisers of the *Sverdlov* class and the destroyers of the 'Kotlin' class. The designers could therefore begin with a clean sheet and opted for a hull-form which not only differed markedly from previous Soviet types but which also departed from established Western practice.

The standard hull-form for US Navy and Royal Navy warships of the period featured a relatively flat bottom, with a soft bilge radius and vertical sides. Flare was employed only in the forward section. In contrast, the new Soviet hull-form was characterized by straight deadrise on each side of the keel and a hard bilge radius, with flared sides to the hull and a knuckle some 2.5m below the weather deck (*see* accompanying drawing). A relatively broad waterplane aft, terminating in a transom stern, provided more intact stability than its Western counterpart.

The Soviet hull-form has a greater wetted surface area than the standard Western hull-form, and therefore more frictional resistance at all speeds. However, it

▲A close-up of the midships section, showing the full extent of the modifications to *Grozny*. The 'back-up' optical fire control position for the AK-630 gatlings can be seen between the gun mountings, and there is a new deckhouse between the forefunnel and the tower mainmast. Modern 'cocooned' life rafts have replaced the old-style floats on the sides of the superstructures. (US Navy, 1985)

has less wave-making resistance at higher speeds, so performance at full power is superior to comparable Western types. The broad waterplane also provides more usable hull volume for a given length.[5]

The hull-form adopted for the 'Kynda' was clearly regarded as a success by the Soviet Navy because it has formed the basis for the hull-form of all subsequent types of major surface warship.

The 'Kynda' also introduced the turbo-pressurized boiler to the Soviet Navy. The attraction of the new plant was presumably the combination of high power and a relatively compact machinery layout. The requirement for high speed in the 'Kynda' demanded machinery capable of producing a total horsepower close to that of *Sverdlov* but which could nevertheless be accommodated in what was essentially an enlarged destroyer hull.

The machinery installation adopted appears to have been particularly successful and was retained for the 'Kresta' and 'Moskva' classes which followed. The later 'Kievs' also appear to have a 'doubled-up' 'Kynda' propulsion plant, and the destroyers of the 'Sovremenny' class which are currently building at the Zhdanov Shipyard have essentially the same machinery installation as the 'Kresta'.

Stability Problems

There are indications that trials with the first unit of the 'Kynda' class, *Grozny*, revealed serious problems which required rectification in the shipyard before the ship was cleared for full operational service. The second ship, *Admiral Fokin*, also appears to have received the necessary modifications after completion. Construction of the third and fourth ships, however, was delayed by six months, presumably to enable the modifications to be incorporated while building.

Instability was almost certainly at the root of the problems. Clearly too much had been attempted on the displacement, and the novelty of the Rocket Cruiser type, together with the uncertainties involved in calculating the weight of equipments that were under development even as the ship was being built, proved too much for the designers. The bulky launchers and their reloads, the twin tower masts with their plethora of primitive air surveillance and missile guidance antennae were compensated neither by adequate hull size nor by the weights lower down in the ship.

The solution was probably a combination of self-compensating fuel tanks and permanent ballast with a consequent increase in displacement and some loss of speed. These remedial measures appear to have been relatively successful because by the late 1970s the Soviets were happy to install new deckhouses and radars in these ships without removing any equipment as compensation.

Nevertheless, the problems experienced appear to have been sufficient to affect the deployment of the 'Kyndas'. Contrary to what one might have expected, given the intended mission of the Rocket Cruisers, 'Kyndas' have never deployed to the Northern Fleet area where any inherent topweight problems would be exaggerated by icing. Two ships, *Admiral Fokin* and *Varyag*, have served with the Pacific Fleet since the late 1960s. The other two units were transferred to the relatively calm waters of the Black Sea, frequently deploying to the Mediterranean as part of the Fifth Eskadra. *Grozny* later returned to the Baltic Fleet, serving as flagship during the mid-1980s.

Later Modifications

By the late 1960s *Admiral Golovko* had been fitted with two electro-optical systems designated Tee Plinth, and these were probably installed in *Admiral Fokin* at about the same time. In both cases the Tee Plinth sensors were mounted on the platforms at the base of the foremast. Tee Plinth, and its successor, Tin Man, are reported to provide closed-circuit TV facilities to the ship's operations centre located between decks.[6] They effectively replace human lookouts under NBC close-down conditions.

Subsequently, in 1972, *Grozny* appeared with a new sensor outfit on the lower tower mast platforms. It may be that this was the full sensor outfit that the ships were originally designed to carry because it included the first installation of Plinth Net, a parabolic lattice antenna, on the prominent mainmast platforms. The precise function of Plinth Net remains unclear, but it had by this time been installed in the Rocket Cruisers of the 'Kresta I' class, and as the 'Kresta' and 'Kynda' are the only Soviet ships fitted with the antenna, it was clearly related to the SS-N-3 Shaddock anti-ship missile. Plinth Net may be associated with the missile data

A Comparison between Soviet and Western Hull-Forms

Left: the classic Western hull-form, with its flat bottom, soft bilge radius and vertical sides. Right: the Soviet hull-form introduced by the 'Kynda', characterized by straight deadrise on each side of the keel, a hard bilge radius, and flared sides to the hull with a prominent knuckle.

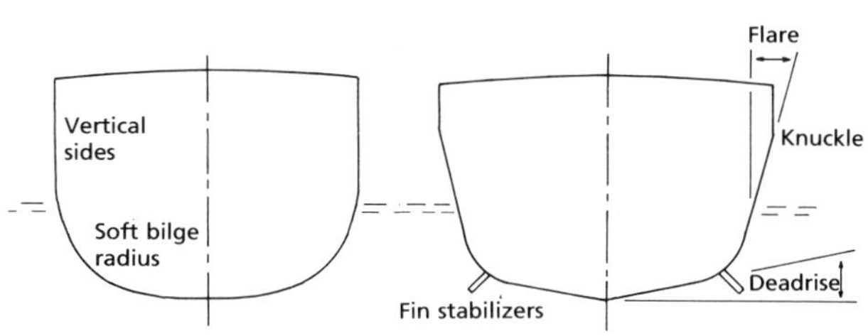

Admiral Fokin/Varyag: Profiles of Midships Section

Admiral Fokin as refitted in the late 1970s, with Head Net C substituted for the second Head Net A antenna. She has also received two Plinth Net data link antennae for her SS-N-3 missiles, and two Tee Plinth electro-optical systems. The original float-style life-rafts have been replaced by cocooned inflatables.

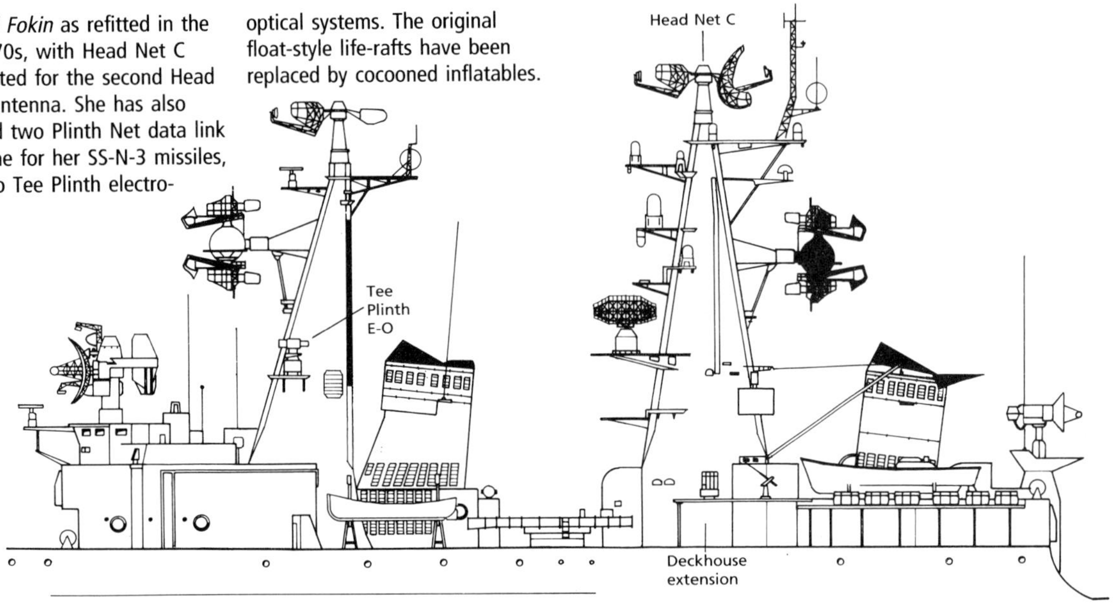

Varyag as refitted in 1980–1. Both Head Net A antennae have been replaced by the V-beam Head Net C. A new deckhouse has been constructed abeam the tower foremast for AK-630 anti-missile gatlings. The Bass Tilt fire control radars have been mounted in the positions formerly occupied by the Tee Plinth E/O sensors, which have been re-sited in the bridge wings. A second new deckhouse has been built between the forefunnel and the tower mainmast. *Grozny*, which was refitted at about the same time, is similar except that she retains both Head Net A antenna and Plinth Net, which has never been fitted in *Varyag*.

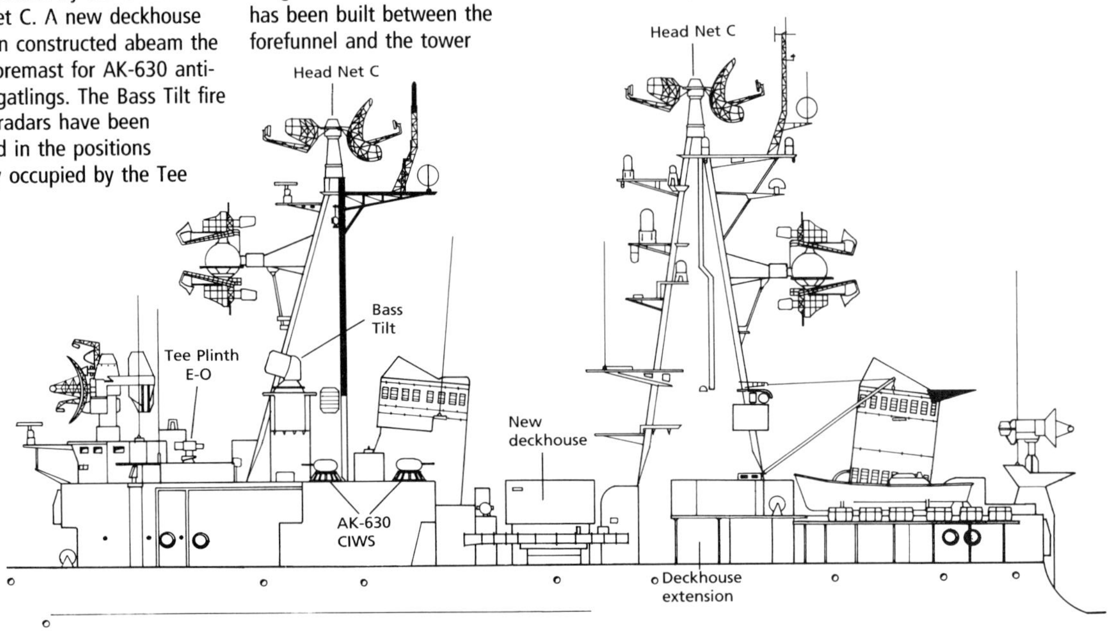

link, but a more plausible explanation lies in the limited tracking capacity of the Scoop Pair tracking radar which, even allowing for the independent rotation of its twin antennae, would be capable of tracking only two missiles simultaneously. Plinth Net may be employed to track and to communicate with missiles in the high-altitude cruise phase while Scoop Pair is used to gather and track a second pair of missiles closer to the ship.

Besides Plinth Net, *Grozny* was fitted at this time with two large ECM radomes which received the NATO designation Guard Dog. The only other Soviet ships to carry this system are two destroyers of the contemporary 'Kashin' class, and the NATO code-name implies that it is a larger counterpart to the Watch Dog ESM radome installed on Soviet destroyers and Guard Ships from about 1957. The radomes were removed from *Grozny* when she was refitted at the end of the 1970s although they remain aboard the destroyers.

During the late 1970s *Admiral Fokin* underwent a major refit in which Plinth Net was fitted, and the second Head Net A air surveillance radar was replaced by Head Net C, a 3-D back-to-back variant first trialled on a destroyer of the 'Krupny' class and fitted in later units of the 'Kashin' class from the mid-1960s.

This refit was followed by even more extensive refits to *Grozny* and *Varyag*. *Grozny*, refitted in 1979–80, retained both Head Net A antennae but had a large two-storey deckhouse built out to the sides of the ship abeam the first funnel to accommodate four AK-630 anti-missile guns (*see* 'Kresta II'). The Bass Tilt fire control radars for the latter were accommodated on pedestals abeam the forward tower mast, and a new deckhouse was constructed between the first funnel and the mainmast. Plinth Net was retained, but all four Guard Dog radomes were removed.

Varyag, refitted in the Pacific in 1980–1, was similarly modified. In this ship, however, both Head Net A air surveillance antennae were replaced by Head Net C. Plinth Net was not fitted, but *Varyag* received two Tee Plinth optical sensors which were installed in the bridge wings.

It has been reported that the remaining two units of the class have also received the AK-630 anti-missile gun, the last unit, *Admiral Golovko*, being taken in hand for modifications in 1985–6.

The piecemeal nature of Soviet refit procedures suggest that warship refits are not subject to the same degree of planning that characterizes Soviet construction. On the contrary, updating seems to be dependent both on the availability of shipyard labour and on the

▼As modified, the Pacific-based *Varyag* displays a number of important differences from *Grozny*. She never received the Plinth Net missile tracking radars. In contrast, both Head Net A air surveillance radars have been replaced by the later Head Net C V-beam model. (US Navy, 1987)

'Kashin' Class: Profile and Plan

The drawing shows *Obraztsovy* as she appeared in the mid-1970s.

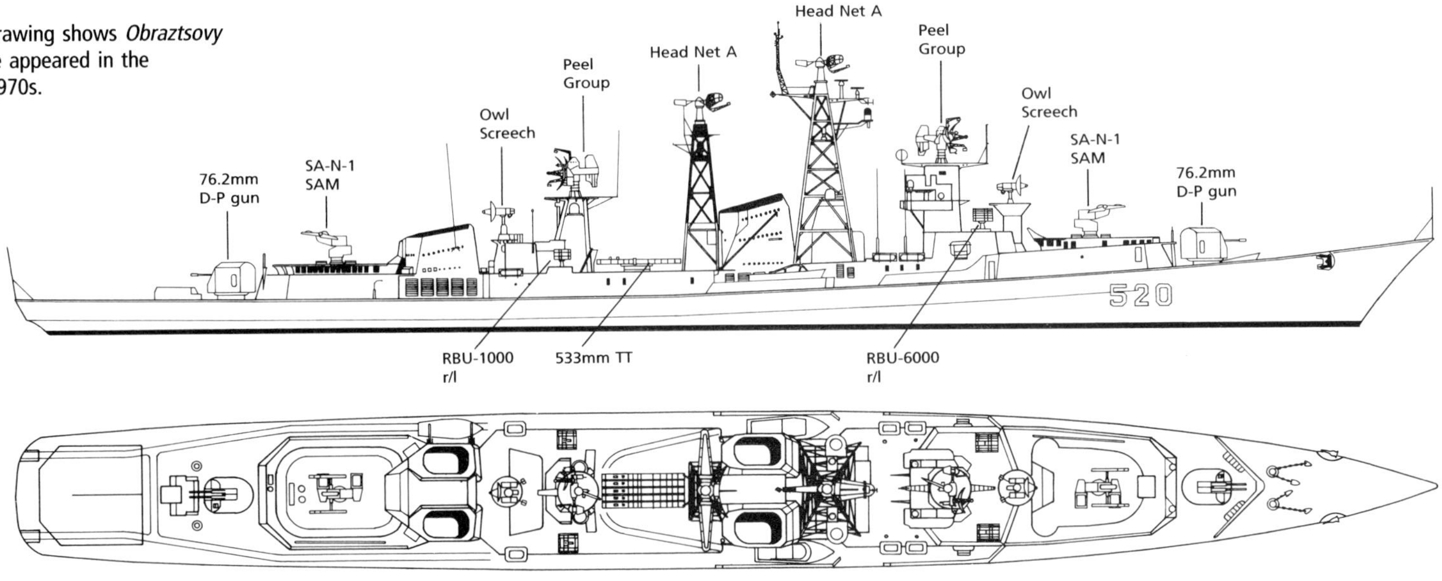

availability of particular items of equipment surplus to the construction programme.

Varyag paid off in 1990 and it appears likely that the other units will follow within the next two or three years.

The 'Kashin'

At about the same time that the first unit of the 'Kynda' class was laid down at the Zhdanov Shipyard in Leningrad, construction of a new type of destroyer was begun at the 61 Kommuna Shipyard in the Black Sea.

The 'Kashin' shared a number of common weapons systems with the 'Kynda', but in conception was closer to earlier Soviet destroyers and was initially typed *Eskadrenny Minonosets.*

Unlike the 'Kynda', which featured a cruiser hull with a short quarterdeck, the 'Kashin' has the same flush-decked hull as the 'Kotlin' and her derivatives, with marked sheer and flare forward. Comparisons with earlier Soviet destroyers extend to the symmetrical layout of the weapons forward and aft, and the retention of a quintuple bank of 533mm torpedo tubes on the centre-line.

In terms of its overall capabilities, too, the 'Kashin' is derived from the 'Kotlin'. The stabilized twin 130mm mountings of the latter have been superseded by twin-arm launchers for the SA-N-1 missile, while the twin 76.2mm dual-purpose mountings not only replace the quadruple 57mm mountings of the 'Kotlin', but have a limited anti-surface capability to compensate the loss of the 130mm guns. The 'Kashin' also features the mine rails of earlier Soviet destroyer types.

Apart from the technological advances which are evident from the installation of surface-to-air missiles and the new generation of electronics equipment, the major improvement in capabilities is in the area of anti-submarine warfare, with a full suite of automatically reloading anti-submarine rocket-launchers, allied in later units of the class to a sonar operating at medium frequency.

Everything about these large, high-performance destroyers suggests that they were designed not only for general tasks, but specifically to provide anti-air and anti-submarine escort for the Rocket Cruisers of the 'Kynda' class. In contrast to the 'Kotlin' derivatives of the 'Kildin' and 'Krupny' classes, they have no anti-ship missiles, even though later modifications of the design prove that it would have been possible to accommodate launchers for the horizon-range SS-N-2 Styx missile had such a feature been considered desirable.

The anti-surface capabilities of the 'Kashin' as originally designed were limited to the 76.2mm dual-purpose guns and the possibility of using the SA-N-1 missile, which employs radar/command guidance, in the anti-ship mode. The Rocket Cruisers would provide the long-range anti-surface strike capability, allowing the 'Kashins' to concentrate on defending the force

'Kashin' Class

Soviet Designation: Bol'shoy Protivolodochny Korabl' (BPK) – Large Anti-Submarine Ship

Construction

Name	Builder	In service
Komsomolets Ukrainy	61 Kommuna Shipyard	Feb 1962
Soobrazitelny	61 Kommuna Shipyard	Sept 1963
Ognevoy	Zhdanov Shipyard	Aug 1964
Otvazhny	61 Kommuna Shipyard	1964
Provorny	61 Kommuna Shipyard	1965
Obraztsovy	Zhdanov Shipyard	July 1965
Odarenny	Zhdanov Shipyard	Sept 1965
Stroyny	61 Kommuna Shipyard	1966
Slavny	Zhdanov Shipyard	June 1966
Steregushchy	Zhdanov Shipyard	Oct 1966
Krasny Kavkaz	61 Kommuna Shipyard	1967
Reshitelny	61 Kommuna Shipyard	Jan 1968
Smyshlenny	61 Kommuna Shipyard	1968
Strogy	61 Kommuna Shipyard	Aug 1968
Smetlivy	61 Kommuna Shipyard	Sept 1969
Smely	61 Kommuna Shipyard	1970
Krasny Krim	61 Kommuna Shipyard	Sept 1970
Sposobny	61 Kommuna Shipyard	Aug 1971
Skory	61 Kommuna Shipyard	Aug 1972
*Sderzhanny**	61 Kommuna Shipyard	1973

*Completed to modified design (see 'Kashin Mod.')

Displacement (tons)
Standard: 3,750
Full Load: 4,750

Dimensions
Length: 144m(472ft) overall
Beam: 15.8m(52ft)
Draught: 4.8m(16ft)

Propulsion
2-shaft COGAG; 4 gas turbines each 24,000hp; 96,000hp=36 knots max.

Complement
280

Weapons
ASW: 2 RBU-6000 rocket-launchers; 2 RBU-1000 rocket-launchers
AAW: 2 twin-arm SA-N-1 launchers (32 Goa missiles); 2 twin 76.2mm(3in) D-P guns
ASuW: 1 quintuple bank of 533mm(21in) torpedo tubes; rails for 20 mines

Sensors
Surveillance radars: 2 Head Net A, 2 Don-2 (early units); Big Net, Head Net C, 2 Don-Kay (later units)
Fire control: 2 Peel Group (SA-N-1); 2 Owl Screech (76.2mm)
Sonars: keel-mounted Herkules (Wolf Paw) HF (early units); Bull Nose MF (later units)
Electronic warfare: 1 Bell Slam; 6 Guard Dog (*Obrazitelny* and *Soobrazitelny* only); 2–4 Top Hat, 2 unidentified

against air and submarine attack. The higher construction rate of the 'Kashin', an average of two ships completed per year as compared with only one per year for their Rocket Cruiser counterparts, suggests that the typical anti-surface strike force initially envisaged might comprise a single 'Kynda' accompanied by two 'Kashins'.

Weapons Installation

The installation of a 'double-ended SA-N-1 area defence system in a relatively shallow hull clearly presented a problem to the ship's designers. The solution adopted was to mount both launchers on a continuous superstructure deck which runs for three-quarters of the ship's length. This had the effect of raising the after missile magazine clear of the after gearing and propeller shafts. The twin 76.2mm dual-purpose mountings are located forward and aft of the launchers, at upper-deck level.

The layout of the midships section was dictated by the massive paired uptakes for the gas-turbine propulsion machinery. The Peel Group guidance radars for the missiles are mounted on short tower structures forward and aft, with the Owl Screech fire control radars for the guns on pedestals beneath them. The forward tower, which is more substantial than the after tower, incorporates the navigation bridge. Twin Head Net A air surveillance antennae are mounted at the top of tall lattice masts stepped forward and aft of the forward pair of gas-turbine uptakes. The lattice masts also carry the electronic warfare sensors.

The quintuple bank of 533mm torpedo tubes is located on the superstructure deck between the mainmast and the after fire control radar tower.

Anti-Submarine Warfare

The 'Kashin' has a significantly enhanced anti-submarine suite as compared with the 'Kynda'. Besides the two trainable RBU-6000 rocket-launchers mounted at the base of the bridge tower, there are two RBU-1000 launchers amidships abeam the after radar tower.

The RBU-1000 appears to be an automatically reloading development of the RBU-600 rocket-launcher which began to be fitted in modified 'Kotlins' in about 1960. Both variants have six barrels with a diameter of 300mm (the RBU-6000 has barrels of 250mm diameter). In the RBU-600 the barrels are arranged as two vertical rows of three set close together, whereas the barrels of the RBU-1000 are more widely spaced, and the upper barrel in each group is set slightly inboard of the other two.

The RBU-1000 fires a heavier projectile over a shorter distance than the RBU-6000. Western reference

sources generally credit the projectiles with a maximum range of 1,000m in keeping with the type number of the launcher, and the 90kg rocket is reported to have a 55kg warhead. In contrast to the RBU-6000, which is generally located in a forward-firing position close to the bow, the customary position for the RBU-1000 is abeam the after superstructure.

Presumably the RBU-6000 is intended to combat submarines detected while manœuvring for an attack, or fast-running submarines endeavouring to escape, whereas the RBU-1000 is intended to destroy a submarine that has revealed itself at close range and has gone deep to evade its pursuers overhead. In effect the RBU-1000 replaces the BMB-2 depth-charge throwers which equipped the 'Kotlin' (modified 'Kotlins' had been replaced by the manually reloading RBU-600 in the early 1960s).

This would explain why the 'Kynda' mounts only the RBU-6000 launcher even though the RBU-1000 was available by the time the first unit was completed. While the RBU-6000 was useful to combat submarines on forward bearings manœuvring for a torpedo attack, the 'Kynda' would not be expected to hunt down submarines that had gone deep, but would move on, leaving an accompanying destroyers to pursue the contact.

Gas-Turbine Propulsion

Quite the most innovative feature of the 'Kashin' design was the adoption of an all-gas-turbine propulsion plant.

Up to this time the history of naval propulsion machinery in the twentieth century had been dominated by the steam turbine, powered by boilers operating at ever-higher steam pressures. The only other prime mover to make any impact was the diesel engine which the German Navy attempted to develop as an alternative for its commerce-raiding battleships. The diesel, however, while attractive in terms of fuel consumption, was limited in power output and therefore had to be combined with other units in multiple installations. It also proved to be difficult to maintain.

High-pressure steam plants created their own maintenance problems, both for the German Navy which installed them in the heavy cruisers built in the late 1930s, and for the US Navy, which adopted similar propulsion machinery during the post-war period. The European navies were generally less enthusiastic about high steam pressures and persisted with steam propulsion machinery of more conservative design, which was more volume-intensive but inherently more reliable.

▲The 'Kashin' was in many respects a 'Kotlin with missiles'. Note the symmetrical arrangement of the missile-launchers and guns. The SA-N-1 launchers had to be located on the deckhouses because there would otherwise have been insufficient hull depth to accommodate the vertically stowed missiles. This is *Obraztsovy*, one of the early group built at Zhdanov in the Baltic. (Skyfotos, 1967)

By the mid-1950s, however, advances in technology had opened up the possibility of obtaining high power ratings without a corresponding decline in reliability. The solution lay in adapting to naval use the jet engine which had made such a considerable impact on aircraft performance during the post-war period.

Whereas steam turbines operate most economically and reliably at low power, aero-engines are most fuel-efficient when operating at maximum thrust. A number of West European navies therefore began to look at the possibility of combining steam turbines or diesels, with their large cruising radius, with aero- or industrially derived gas turbines which would provide the necessary boost for high-performance operation.

The Royal Navy adopted a combined steam and gas (COSAG) installation for the destroyers of the 'County' class and the frigates of the 'Tribal' class, which were both contemporaries of the 'Kashin'. The 'Counties' had a 15,000shp steam turbine allied to two AEI G6 7,500hp gas turbines on each of two shafts; the 'Tribals' had a 12,500shp steam turbine and one G6 gas turbine for boost on a single shaft.

At about the same time the Federal German Navy was completing six frigates of the 'Köln' class, in which a Brown-Boveri 12,000hp gas turbine was paired with two MAN 3,000bhp diesels on each of two shafts in a combined diesel and gas (CODAG) installation.

By comparison with these somewhat tentative and experimental hybrid installations, the Soviet decision to adopt an all-gas-turbine plant for the 'Kashin' was both bold and imaginative. Moreover, the gas turbine which is the basic unit for the 'Kashin' propulsion machinery had twice the power output of the Brown-Boveri unit installed in the West German frigates, and three times that of the British G6 turbine. Western Intelligence sources were so taken aback by this development that for many years the 'Kashins' were credited with eight, or even twelve turbines.

In adopting such a radical machinery installation for the 'Kashin' the Soviet designers were making conscious decisions regarding performance and maintainability. An all-gas-turbine plant would result in quick response and high sustained speed at some cost to the ship's operating radius. This in itself provides an insight into Soviet tactical priorities; the 'Kashins' were designed not for world-wide 'policing' operations, but for short periods of intensive combat in zones within, or relatively close to, Soviet sea-space.

The large 24,000hp gas turbines of the 'Kashin' were not particularly well suited to patrol operations at cruise speed. Like other gas turbines they operated most efficiently at maximum power rating, and at lower power outputs were heavy consumers of fuel. The Soviets appear to have been well aware of these limi-

▼The layout of the midships section is dominated by the massive paired uptakes for the gas-turbine propulsion machinery. Note the quintuple bank of tubes on the centre-line, trained to port in this overhead view. (US Navy, 1984)

tations; contemporary Guard Ships of the 'Petya' class, which were designed for patrol operations in Soviet waters, had a CODAG plant similar in conception to the hybrid plants of the West European frigates and destroyers.

However, the Soviets appear to have been reluctant to develop the complex gearing which would have been necessary in order to combine prime movers with different characteristics (this was avoided in the 'Petya' by having the single gas turbine drive a third shaft). An all-gas-turbine plant had the attraction of being inherently simple and reliable in mechanical terms and this would help to reduce both manning and maintenance levels.

The West European navies would later reach much the same conclusions as a result of their experience with the early hybrid plants, and later installations would favour COGOG or CODOG arrangements in which either the high-power 'boost' turbine or the cruise turbine/diesel would drive the shaft.

In order to economize on fuel when cruising at 14–15 knots the 'Kashins' probably operate on a single turbine and a single shaft, with the after shaft trailing. This is less effective than single-shaft operations with the latest Western gas-turbine powered warships because the propeller blades on 'Kashin' are fixed, whereas contemporary Western vessels have controllable-pitch propellers which can be feathered to reduce drag. However, it may be that there is a mechanical cross-connection which enables both shafts to be powered by a single turbine. Such an arrangement would require that the shafts be parallel so as to avoid the use of bevel gearing.[7] There is evidence that this may be the case because the 'Kashins', unlike Soviet steam-powered warships of the period, have no propeller guards.

It is not clear what arrangements have been adopted to provide reverse power in these or later Soviet gas-turbine powered ships. In modern Western destroyers and frigates the pitch of the propellers is simply reversed, but the fixed-pitch propellers of the 'Kashins' preclude this option. Reversible couplings have been used both by the Royal Navy and the Italian Navy, at the cost of additional complexity in the gearing arrangements, but another possible alternative is a special reverse power turbine connected to the main reduction turbine bull gear – an arrangement which is apparently employed in Soviet gas-turbine powered merchant vessels.[8]

Machinery Layout

The four large gas turbines of the 'Kashin' are arranged as two groups of two, the exhaust uptakes being led up into broad paired funnels which are canted outboard. The wide longitudinal spacing between the paired funnels suggests that there are at least two and possibly three watertight compartments between the forward and after turbine rooms.

Both sets of air intakes are located forward of the paired funnels, suggesting that all four turbines face aft, driving two shafts of equal length through separate gearboxes. The main auxiliary machinery spaces are probably located between the forward and after turbine rooms, with a second generator room forward beneath the bridge tower.

This arrangement is very different from those adopted in later Western gas-turbine powered warships which generally have more flexible layouts facilitated by the adoption of controllable-pitch propellers. The all-gas-turbine frigates and destroyers of the western European navies generally have their separate boost and cruise turbines mounted 'back-to-back' with the gearing between. This has the advantage of grouping the exhaust trunking for all four turbines together within a single funnel, thereby economizing on valuable centre-line space.

Larger Western all-gas-turbine warships, such as the British 'Invincible' and the US 'Spruance' class, generally have the 'unit' machinery layout which is less

▶This overhead view of the Pacific-based *Strogy* displays to good effect the paired uptakes for the four large gas turbines. The uptakes are inclined outboard to keep the corrosive exhaust gases clear of the delicate electronic equipment located between them, and are of double-walled construction to provide natural cooling. (RAAF)

▼Close-up of the RBU-1000 rocket-launcher which, like the RBU-6000, can be both trained and elevated. The rockets are fired in sequence and the launcher reloads automatically from beneath with the tubes in the vertical position. (US Navy, 1978)

vulnerable to action damage. In this arrangement the turbines are grouped in pairs and disposed *en echelon*, each pair of turbines driving one of two unequal shafts through a single gear-box. In the US Navy's 'Spruance' this arrangement is reflected in the layout of the funnels, the fore-funnel being offset to port and the after one to starboard.

The 'Kashin' funnels, like those of later Soviet gas-turbine ships, are of double-walled construction with numerous openings to provide a natural cooling effect on the hot exhaust gases. Canting the funnels outboard also helps to keep the corrosive exhaust gases clear of the delicate electronics. The Soviets appear to have been so concerned about the effects of these gases on the radar and ECM antennae mounted on the tall lattice masts that the fore-funnels of the first unit completed were approximately 1.5 metres shorter than those of later ships. Following sea trials, the height of the funnels had to be increased, presumably because problems were experienced with the exhaust gases drifting across the decks.

Electronic Countermeasures

Photographs of destroyers of the 'Kashin' class in their early configuration are comparatively rare, but it appears that the planned electronic countermeasures outfit was similar to that of the 'Kynda'.

Early units of the class had a single Bell Slam antenna mounted on a platform projecting from the forward side of the lattice foremast. Pairs of Top Hat jammers were mounted above the bridge wings and on platforms projecting from the sides of the mainmast respectively, and there were two small unidentified antennae similar to those of the 'Kynda' on the sides of the foremast. At a lower level, on the sides of the lattice foremast and on the sides of the after radar tower, there were more substantial platforms apparently intended, as in the 'Kynda', to support four Guard Dog ESM radomes.

Obraztsovy, serving with the Baltic Fleet, appears to have been the first unit to receive Guard Dog, four radomes being fitted on the lower platforms with a further two located on additional platforms high on the sides of the foremast. The upper two radomes were later removed, but her ECM fit was otherwise unmodified.

The only other unit to be fitted with Guard Dog was *Soobrazitelny*, serving with the Pacific Fleet. However, she was given a more experimental installation in which Guard Dog competed for space with the new Tee Plinth and Tilt Pot electro-optical sensors which would be fitted in later units of the 'Kashin' class. Three Tee Plinth E/O sensors were installed around the lattice foremast, two of which were mounted on the lower platforms originally intended for Guard Dog, with the third located immediately behind Bell Slam on the upper foremast platform. Two diminutive Tilt Pot E/O sensors were mounted on the platforms projecting from the sides of the after radar tower. This would be the standard arrangement for later units of the class.

The three pairs of Guard Dog ESM radomes therefore had to be accommodated in a different fashion from those of *Obraztsovy*. One pair was located on platforms on the upper sides of the foremast; the second pair on new platforms projecting from the after corners of the mainmast, immediately above the torpedo tubes; and the third pair on the sides of the after radar tower, below the platforms carrying Tilt Pot. The latter feature is of particular interest as it is the only example of the 'horizontal' mounting of Guard Dog and appears to anticipate the installation of Side Globe in *Moskva*, the 'Krestas' and later Soviet cruiser-size warships. *Soobrazitelny* was almost certainly operated as a trials ship for all of these systems during the mid-1960s, as the range of equipment fitted was not repeated on any other unit of the class.

Later Modifications

The 'Kashin' programme continued well into the 1970s which suggests that the Soviets regarded the design as relatively successful. Early construction was limited to one ship per year at the 61 Kommuna Shipyard, but parallel construction at the Zhdanov Shipyard began in the early 1960s, and with the somewhat abrupt curtailment of the 'Kynda' programme production at the latter was briefly boosted to two 'Kashins' per year, before the programme was phased out altogether at Zhdanov in favour of other projects (notably the 'Kresta'). By this time, however, 61 Kommuna had increased its own production rate to an average of two ships per year, thereby compensating the cessation of construction at Zhdanov.

This was a period which saw the introduction into the Soviet Navy of large quantities of new equipment so it followed naturally that modifications would be made both to units under construction and to those already in service. As has already been noted in respect of the 'Kynda', these modifications lacked uniformity and were frequently dictated by the availability of certain items of equipment when ships underwent scheduled refits, with the result that individual ships subsequently deployed with minor differences in their electronics outfits.

The first ship to be completed with a modified radar outfit was *Stroyny*, completed by the 61 Kommuna shipyard in 1966. In place of the paired Head Net A air surveillance antennae of earlier ships, she had a new back-to-back variant designated Head Net C at the top of the foremast, and a larger antenna designated Big Net on the mainmast.

Head Net C is a 'V-beam' radar. Two Head Net A antennae are mounted back to back with the second tilted at an angle of 30 degrees to the horizontal. This effectively doubles scanning rate and dispenses with the need for the counterbalancing vanes which are a distinctive feature of the Head Net A antenna. Tilting the second antenna has the effect of displacing the fan-shaped elevation beam from the vertical, giving the Head Net C a height-finding capability previously

ECM Arrangements

Upper drawing: the standard 'Kashin' arrangement, with a single Bell Slam ESM antenna complemented by two Top Hat jammers in the bridge wings and two smaller Bell-type radomes on the upper platforms of the foremast.

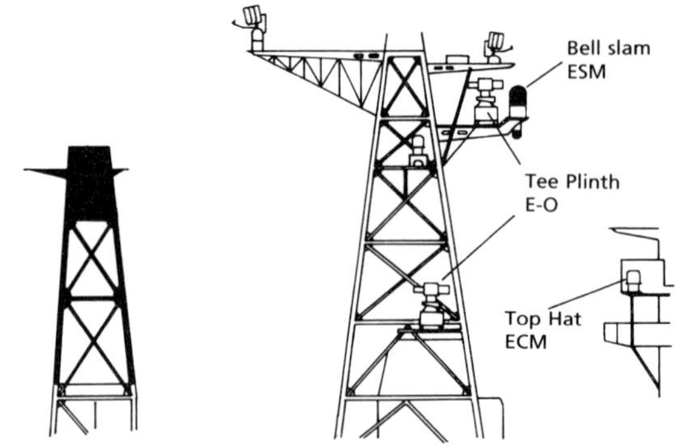

Middle drawing: *Obraztsovy*, with her unusual outfit of Guard Dog ESM radomes.

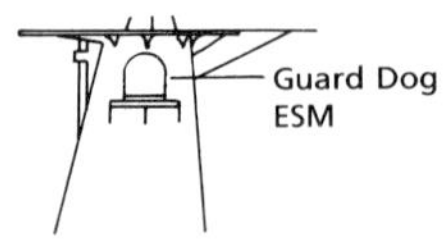

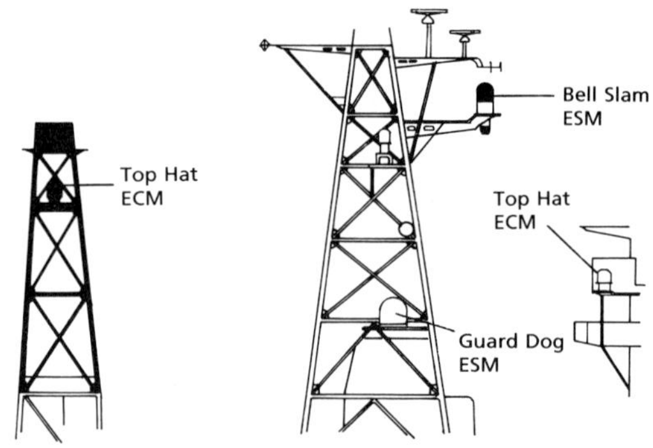

Lower drawing: arrangements on the Modified Kashin *Stroyny*, with the original ECM equipment replaced by the standard 'Krivak' outfit of Bell Shroud and Bell Squat.

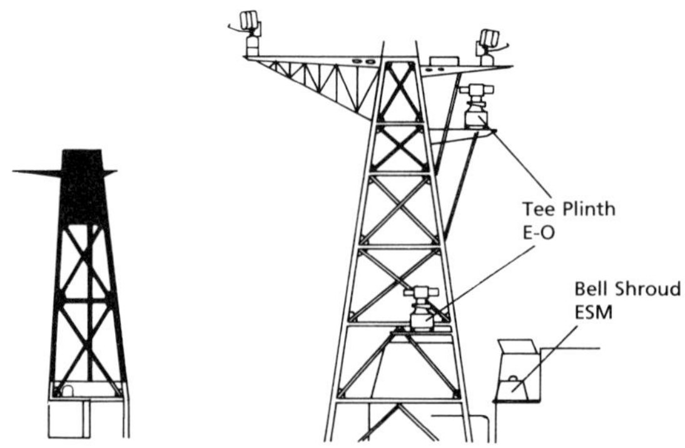

▲An early view of *Soobrazitelny*, with her unusual combination of Guard Dog ESM radomes and Tee Plinth/Tilt Pot electro-optical devices. Note the after pair of Guard Dog radomes, which are mounted horizontally on the sides of the after radar tower. (Novosti, 1967)

▶ The Pacific Fleet destroyer *Strogy*, seen here at Ethiopian Navy Days in 1972. She has the Head Net C V-beam antenna at the top of the foremast, and the distinctive Big Net air surveillance radar at the top of the mainmast. (MoD, 1972)

lacking in Soviet air surveillance radars. The vertical beam performs a conventional search function, and the operator measures altitude on a selected target by marking it and setting range gates on the tilted beam.[9] The technique is simple, but by no means as effective as the electronic frequency scanning techniques introduced into the US Navy in the early 1960s.

Big Net operates in the lower-frequency C-band and therefore has better performance against high-flying aircraft than the Head Net series. Maximum detection range is estimated at 100nm (170km), as compared with only 60–70nm (110–130km) for the latter. The parabolic antenna is of open lattice construction and has prominent stabilization vanes projecting from its base. Big Net was not a new radar, development having begun in 1948; it entered service in about 1957 when it began to be retro-fitted to cruisers of the *Sverdlov* class. Presumably its large size precluded its installation in the 'Kynda', but topweight clearly presented less of a problem in the 'Kashin'.

Early units of the class generally remained unmodified and no attempt was made to retro-fit them with the Head Net C/Big Net air surveillance combination. The sole exception was *Odarenny* which received two Head Net C radars in place of Head Net A during the 1980s.

At about the same time as the introduction of Big Net and Head Net C a new medium-frequency bow sonar (NATO designation: Bull Nose) entered service, probably on this class. It would subsequently be fitted in the 'Kanin' ASW destroyer conversions and a new generation of purpose-built anti-submarine ships (*see* Chapter 4).

Ships completed from 1968 to 1969 also featured new radars for surface surveillance and navigation. Early units were generally fitted with two Don-2 navigation radars, of which one was mounted on a narrow platform projecting from the forward radar tower and the other at the top of the foremast. Later ships had two Don-Kay radars for which the antennae were mounted at the forward and after ends of the upper foremast platform. This arrangement provided better coverage

▶Close-up of the midships section of *Obraztsovy*, showing the unusual ECM fit. Guard Dog ESM radomes are located on platforms at the base of the lattice foremast and on the sides of the after radar tower. There is a Bell Slam double radome on the platform projecting from the foremast, and Top Hat jammers are located above the bridge wings and on the sides of the lattice mainmast. There are similar unidentified radomes on the upper sides of the foremast. The ship retains her original paired Don-2 navigation radars in this 1976 view. (C&S Taylor)

▲*Strogy* under way in the Pacific. Note the SA-N-1 launchers which train to starboard and to port, respectively, in order to reload.

◀An unusual view of four Big Net 'Kashins' in formation, probably taken in the Black Sea. Each is operating a Ka-25 Hormone anti-submarine helicopter. The 'Kashin', like the 'Kynda', has a landing pad marked out on the quarterdeck but no hangar or maintenance facilities. These are undoubtedly land-based helicopters embarked for an exercise in close proximity to Soviet bases. (Novosti, 1977)

of the stern arcs and presumably assisted helicopter control when the ships operated in conjunction with rotary-wing land-based units. *Smetlivy* had her Don-Kay radars replaced by Palm Frond during the early 1980s.

The electronic countermeasures outfit of the later ships remained essentially unchanged, although the second pair of Top Hat ECM antennae was suppressed and Guard Dog was never fitted. In place of the latter there were the Tee Plinth and Tilt Pot electro-optical sensors trialled by *Soobrazitelny*.

The 'Modified Kashin'

Forward deployment, which became an important feature of Soviet peacetime operations from the late 1960s, brought the Soviet Navy into frequent contact with Western naval forces. This was especially true of the Mediterranean, where Soviet ships were often tasked with shadowing the two carrier battle groups of the US Sixth Fleet.

The Rocket Cruisers and missile submarines which made up the anti-carrier strike groups were ill-suited to the close shadowing role, while the oversized missile patrol boats of the 'Nanuchka' class (the only Soviet missile boats to deploy out of area) were relatively slow, short on endurance and uncomfortable in any sort of a sea. Their diesel engines also proved to be unreliable and subject to frequent breakdowns.

The 'Kashins', a large proportion of which served with the Black Sea Fleet, had the high speed, acceleration and endurance necessary for such operations, but were armed exclusively with air defence and anti-submarine weapons. These considerations appear to have prompted a modification programme for the 'Kashin' which would involve the installation of four fixed-angle

Modified 'Kashin' Class

Soviet Designation: Bol'shoy Raketny Korabl' (BRK) – Large Rocket Ship

Construction

Name	Builder	Converted
Sderzhanny	61 Kommuna Shipyard	1973*
Ognevoy	Zhdanov Shipyard	1973
Smyshlenny	61 Kommuna Shipyard	1974
Smely	61 Kommuna Shipyard	1974
Slavny	Zhdanov Shipyard	1975
Stroyny	61 Kommuna Shipyard	1980

Displacement (tons)
Standard: 3,950
Full Load: 4,950

Dimensions
Length: 146m(472ft) overall
Beam: 15.8m(52ft)
Draught: 4.8m(16ft)

Propulsion
2-shaft COGAG; 4 gas turbines each 24,000hp; 96,000hp=34 knots max.

Complement
280

Weapons
ASuW: 4 single SS-N-2C launchers
ASW: 2 RBU-6000 rocket-launchers; 1 quintuple bank of 533mm(21in) torpedo tubes
AAW: 2 twin-arm SA-N-1 launchers (32 Goa missiles); 2 twin 76.2mm(3in) D-P guns; 4 AK-630 gatlings

Sensors
Surveillance radars: Big Net, Head Net C 3-D (*Ognevoy* initially retained 2 Head Net A); 2 Don-Kay
Fire control: 2 Peel Group (SA-N-1); 2 Owl Screech (76.2mm); 2 Bass Tilt (AK-630)
Sonars: keel-mounted Bull Horn MF; Mare Tail MF VDS
Electronic warfare: 2 Bell Shroud, 2 Bell Squat; 4 16-tube chaff-launchers

Modified 'Kashin' Class: Profile View

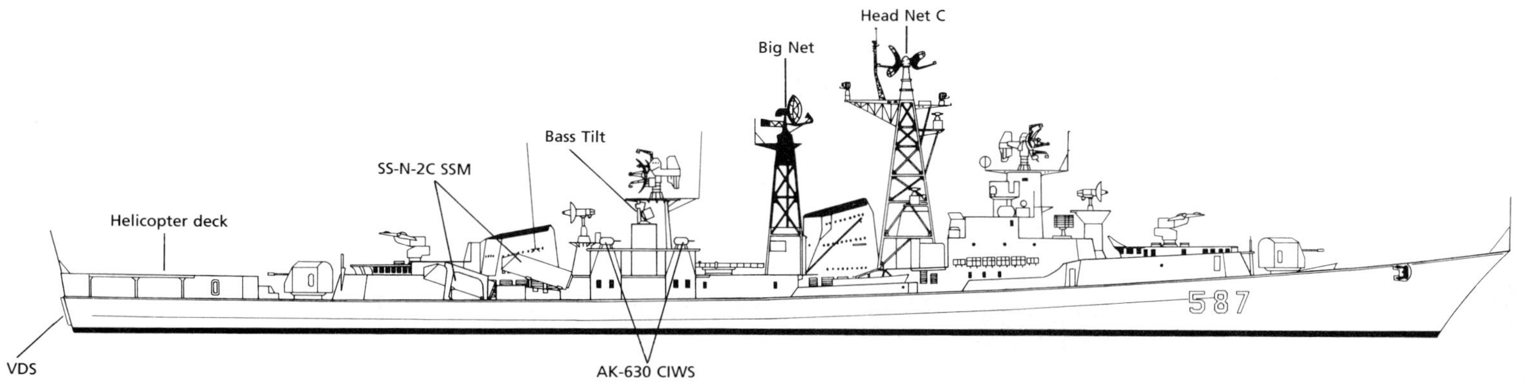

◀The launch tubes for the aft-firing SS-N-2C anti-ship missiles are particularly prominent in this stern view of *Stroyny*. Note also the raised helicopter deck above the stern and the hinged door for the Mare Tail variable depth sonar. (MoD, 1981)

▼All modified 'Kashins' have had their ECM outfit upgraded, but there were initial problems with the supply of equipment. The 'Kashin-Mod' seen here has empty platforms for the Bell Shroud ESM antenna at the after end of the bridge structure. (MoD, 1976)

▲**This close-up of *Stroyny* shows clearly the AK-630 anti-missile gatlings and their Bass Tilt fire control radars abeam the after radar tower. The outer door for the variable depth sonar is hinged upwards. (MoD, 1981)**

launch tubes for the horizon-range SS-N-2C missile. The SS-N-2C (initially thought by NATO to be a new missile and accordingly typed SS-N-11) was a variant of the SS-N-2A/B Styx which had been fitted in some 300 missile boats of the 'Komar' and 'Osa' classes during the 1960s. The new model had a maximum range estimated at 45nm (80km) and featured a combination of active radar and infra-red terminal homing.

The four missile launch tubes were installed on angled ramps abeam the after pair of funnels. Contrary to normal Soviet practice, the missiles are fired over the stern of the ship. The construction during the 1980s of three similar vessels for India, all of which were armed with forward-firing missile launch tubes, suggests that this was a conscious decision on the part of the Soviet Navy.

The tactic apparently envisaged was to operate these ships within visual range of the carrier and to monitor her position. In the event of hostilities the destroyer would 'call down fire' from other anti-carrier units shadowing the task force at stand-off range, and would attempt to get off a pre-emptive strike of her own before making her escape. The quick response of the gas turbines of the 'Kashin' made it well suited to the evasive manœuvres that would be required.

Further modifications to these ships included the provision of a variable depth sonar in a remodelled stern, with a raised helicopter platform on the VDS housing. The variable depth sonar is reported to be the same Mare Tail medium-frequency model fitted in *Moskva* and the 'Krivak'. Accommodating the VDS housing and flight deck involved lengthening the stern section by approximately two metres. Even so, the after 76.2mm gun mounting now has severely constricted arcs, particularly at lower angles of elevation.

To provide improved protection against air- and surface-launched missiles, which would be essential if the ships were to operate close to enemy carrier battle groups, the RBU-1000 anti-submarine rocket-launchers were removed and in their place there are now four AK-630 anti-missile 'gatlings' (*see* 'Kresta II') controlled by two Bass Tilt radars. Optical 'back-up' fire control positions for the AK-630 gatlings are located at the base of the mainmast.

The electronic countermeasures outfit has also been upgraded. The original mixed suite of Bell Slam and Top Hat has been removed altogether and has been replaced by two Bell Shroud and two Bell Squat antennae. The new ECM outfit is therefore identical with that of the 'Krivak' although, as with the latter class, there appear to have been problems in matching production with planned installations. Three of the first five 'Kashin' conversions emerged from reconstruction with empty platforms. A standard outfit of three Tee Plinth electro-optical sensors is carried.

The first unit to undergo conversion, *Sderzhanny*, is reported to have been modified during construction at the 61 Kommuna Shipyard; she was completed in 1973. Four other units quickly followed and there was then a gap of a full six years before the last unit, *Stroyny*, was taken in hand in 1979–80.

At about the same time, three of the four ships of the 'Kildin' class were taken in hand at Nikolayev for a similar conversion, which involved removal of the obsolete SS-N-1 anti-ship missile system and its replacement by a combination of fixed-angle launchers for the SS-N-2C missile and twin 76.2mm dual-purpose gun mountings. The SS-N-2C launch tubes were disposed in similar fashion to those of the modified 'Kashins', which suggests that the conversion was prompted by the same tactical considerations.

It was at first thought that the entire 'Kashin' class would undergo conversion, but with the exception of *Stroyny* all SS-N-2C conversions (including the 'Kildins') had been completed by 1975. It is of interest to note that all the converted ships were initially typed *Bol'shoy Raketny Korabl'* (Large Rocket Ships), but that the modified 'Kashins' were promptly reclassified *Bol'shoy Protivolodochny Korabl'* (Large Anti-Submarine Ships), in line with unmodified units of the class. This may indicate that changes in policy were responsible for the curtailment of the programme, and that the tactical concept which produced the Large Rocket Ship conversions quickly lost favour. However, the 'balanced armament' of the modified 'Kashin' certainly appears to have influenced the design of the destroyers of the 'Sovremenny' class, for which draft plans were being drawn up in the mid-1970s, and which in a number of respects can be regarded as the natural successor to the 'Kashin'.

In 1987 *Smely* was transferred to the Polish Navy to replace the former 'Kotlin SAM' missile destroyer *Warszawa*. Two years later the class began to decommission. By 1991 four ships had been stricken: *Odarenny*, *Soobrazitelny* and the modified *Ognevoy* and *Stroyny*. Other ships will undoubtedly follow over the next few years.

The SS-N-2C Missile

Soviet designation: P-20/21
In service: 1967
Length: 6.5m(21ft 4in)
Diameter: 0.78m(2ft 7in)
Span: 2.8m(9ft 2in)
Weight: 2,500kg(5,510lb)
Warhead: 500kg(1,100lb) HE
Propulsion: liquid-fuel rocket with solid-fuel booster
Speed: Mach 0.9
Range: 80km(45nm)
Guidance: inertial/active radar or IR homing
Fire control: none

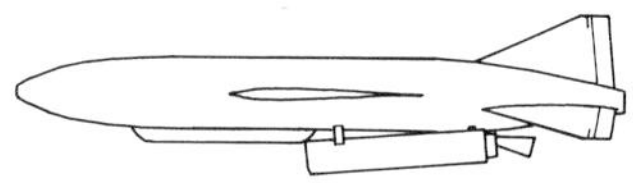

◀**The modified 'Kildin'-class destroyer *Prozorlivy*. The 'Kildins' were reconstructed in similar fashion to the 'Kashin-Mod', with four aft-firing launchers for SS-N-2C anti-ship missiles abeam the after funnel. (MoD, 1980)**

3. Anti-Surface and Anti-Submarine

IT IS GENERALLY ACKNOWLEDGED that the year 1959–60 marked a watershed in the development of the modern Soviet Navy. It was in that year that the second major post-Stalin review of defence policy took place, and the effects were far-reaching.

The decision which made the greatest impact on all the existing branches of the Soviet armed forces was the creation of a new branch, the Strategic Rocket Forces, which would take on full and total responsibility for nuclear missiles with a range in excess of 1,000km.

This development effectively put into reverse the strategic missile programmes of the other services and in particular that of the Soviet Navy, which had just embarked on a major construction programme based on three new types of submarine: the diesel-powered ballistic missile submarines (SSB) of the 'Golf' class; their nuclear-powered counterparts (SSBN) of the 'Hotel' class; and the nuclear-powered cruise missile submarines (SSGN) of the 'Echo' class.

Deprived of a strategic nuclear strike role, the Soviet Navy would be forced back on to its traditional 'Defence of the Homeland' mission. The strategic missile submarine programme had to be abandoned; Soviet naval missions, and the balance between the surface ships, submarines and aircraft which would be needed to fulfil those missions, would have to be completely rethought.

The Soviet surface fleet, for which there was no projected strategic nuclear mission, was not directly affected by the decision to set up the Strategic Rocket Forces, but the knock-on effects of this policy decision were considerable. The nuclear-powered submarines of the 'Echo' class, which could no longer be employed for strategic nuclear strike, were redesigned for the tactical anti-surface mission and armed with eight SS-N-3A missiles. The hitting power of just one of these submarines was therefore equivalent to that of a 'Kynda', while parallel production at Severodvinsk and Komsomolsk would boost construction to between five and six boats per year by the mid-1960s.

The 'Echo II' was complemented by the 'Juliett', a diesel-powered boat armed with four SS-N-3A missiles. These submarines were being completed at a steady rate of two per year beginning in 1962.

The new cruise missile submarine programmes undoubtedly had a major impact on the Soviet rocket cruiser programme. The submarines were not only designed essentially for the same mission, but were relatively simple and inexpensive to build. There were undoubtedly some elements in the Soviet Navy who argued that they were also potentially more effective.

There was concern as to how well the 'Kynda' might perform in a high-threat environment. Much of the sophistication and expense of the design was related to the need to protect the ship against aerial and submarine attack. A submarine, on the other hand, could reach an attacking position without revealing its presence and could dive deep if threatened by hostile counter-strikes.

The problems experienced during sea trials with the first of the 'Kyndas' did little to enhance the reputation of the design, and resulted in the completion of the last two units being delayed by about six months so that the necessary modifications could be incorporated. This resulted in a construction period of approximately 42 months, as compared with only 24–30 months for the nuclear-powered 'Echo' and substantially less for its diesel-powered counterpart, the 'Juliett'.

The 'Kynda' programme, which according to MccGuire[1] was originally to have totalled twelve ships, was therefore drastically curtailed. The last unit was laid down in September 1961 and construction at the Zhdanov Shipyard was temporarily shifted to the 'Kashin'. It would be two years before further rocket cruisers would be laid down, and these would be to a completely new design.

The Anti-Submarine Mission

The second major consequence of the 1959–60 defence policy review was a fundamental re-assessment of the threat posed by NATO forces.

The carrier task forces of the US Navy would continue to present a major threat to Soviet territory for the foreseeable future. By the late 1950s the 'super-carrier' construction programme was in full swing, with four ships of the *Forrestal* class in service and a further three (including the nuclear-powered *Enterprise*) nearing completion.

▲The Anti-Submarine Cruisers of the *Moskva* class were the Soviet Navy's first air-capable ships. They were built to counter the threat to the territorial USSR posed by the Polaris submarine. (MoD, 1981)

The A-3D Skywarrior carrier-borne, long-range, nuclear strike bomber was already in service and a jet-powered successor, the A-3J (later A-5) Vigilante, had entered production. A new all-weather, low-level strike aircraft with a combat strike radius of 1,000nm, the A-6 Intruder, was projected to enter service in the early 1960s.

There was, however, a threat to Soviet territory from another quarter. In 1957 the US Navy decided to deploy a major new nuclear strike weapon, the Polaris missile. The latter would be fired from nuclear-powered submarines with sixteen launch tubes, of which the first was due to be completed in 1960. The Polaris A-1 missile had a maximum range of 1,200nm (i.e., comparable to the combat radius of the A-3 Skywarrior), but the A-2 was projected to have a range of 1,500nm and this would be increased to 2,500nm in the A-3 variant which entered service in 1964.

Anti-submarine warfare had received a relatively low priority in the Soviet Navy during the immediate post-war period. The Soviet Union, unlike the NATO allies, did not have a large merchant fleet to protect, nor was it dependent on sea lines of communication for the transportation of troops, food and raw materials. Its small sea-going surface fleet was intended to react to threatened incursions into Soviet waters by Western

▲The hybrid carrier/cruiser layout of the *Moskvas* is readily apparent in this view of *Leningrad*. The 'cruiser' weapons systems and their associated fire control radars rise in steps from the forecastle, while the after part of the ship is given over exclusively to helicopter operations. Note the marked overhang of the flight deck, and the four Ka-25 Hormone A helicopters lined up for take-off. (RDN, 1981)

carrier task forces and amphibious fleets, and was therefore designed for only limited exposure to NATO submarines in low-threat areas.

Soviet sonars continued to be derived from late-war Allied high-frequency models. They were mechanically scanned and had a strictly limited effective range, being intended for fire control rather than for long-range detection. Soviet anti-submarine firepower consisted exclusively of short-range 'blanket' weapons such as multiple rocket-launchers and depth-charges. Despite rapid advances in Soviet rocket technology, there was no attempt to develop long-range precision weapons comparable to the Western ASROC, Malafon and Ikara. This line of development was not pursued for two reasons: current sonars were incapable of providing the necessary underwater fire control; and missiles of this size would effectively displace anti-ship and air defence weapons, which for the Soviet Navy had far greater priority.

The 1959 Defence Review established anti-submarine warfare as a major Soviet Navy mission. A 'crash' programme to develop the necessary underwater sensors and long-range weaponry was begun and new ships were designed to hunt down the Polaris submarines. This programme would eventually deliver powerful low-frequency sonars, variable depth sonars, nuclear-tipped anti-submarine missiles and anti-submarine helicopters equipped with dunking sonars and armed with homing torpedoes. It would also be responsible for the helicopter carriers of the *Moskva* class and the Il-38 May maritime patrol aircraft.

The creation of an anti-submarine mission alongside the existing anti-surface mission was to result in revisions in the classification system for Soviet surface forces. Henceforth there would be two major strands in the naval construction programmes: ships designated *raketny* ('rocket') with a primary anti-surface mission; and ships designated *protivolodochny* ('anti-submarine'). Each of these two categories would be subdivided into ships of various sizes: *kreyser* ('cruiser') for major surface warships intended as the focal point of anti-surface or anti-submarine operations; *bol'shoy korabl'* ('large ship') for surface warships designed for out-of-area operations; *maly korabl'* ('small ship') for minor units intended to operate within the Soviet maritime defence zones; and *kater* ('cutter') for fast patrol boats.

The two units of the *Moskva* class, the first of which was completed in 1967, would be the first of a series of Anti-Submarine Cruisers (*Protivolodochny Kreyser*). In the interim, however, the 'Kashins', which had initially received the traditional Soviet 'destroyer' classification *Eskadrenny Minonosets*, were re-designated '*Bol'shoy Protivolodochny Korabl'*' (Large Anti-Submarine Ship). This would become an important category in the Soviet Navy during the 1970s as a result of interim conversion programmes and major new construction programmes dating from the mid-1960s.

Moskva Class: Profile and Plan plus Port-Side Superstructure

The full profile and plan drawings show *Leningrad* as she appeared in 1970. The inset shows the port side of the superstructure of *Moskva* as first completed. Note the original solid triangular counter-balancing vanes on the Top Sail 3-D radar antenna, and the asymmetrical layout of the engine room vents and ECM arrays.

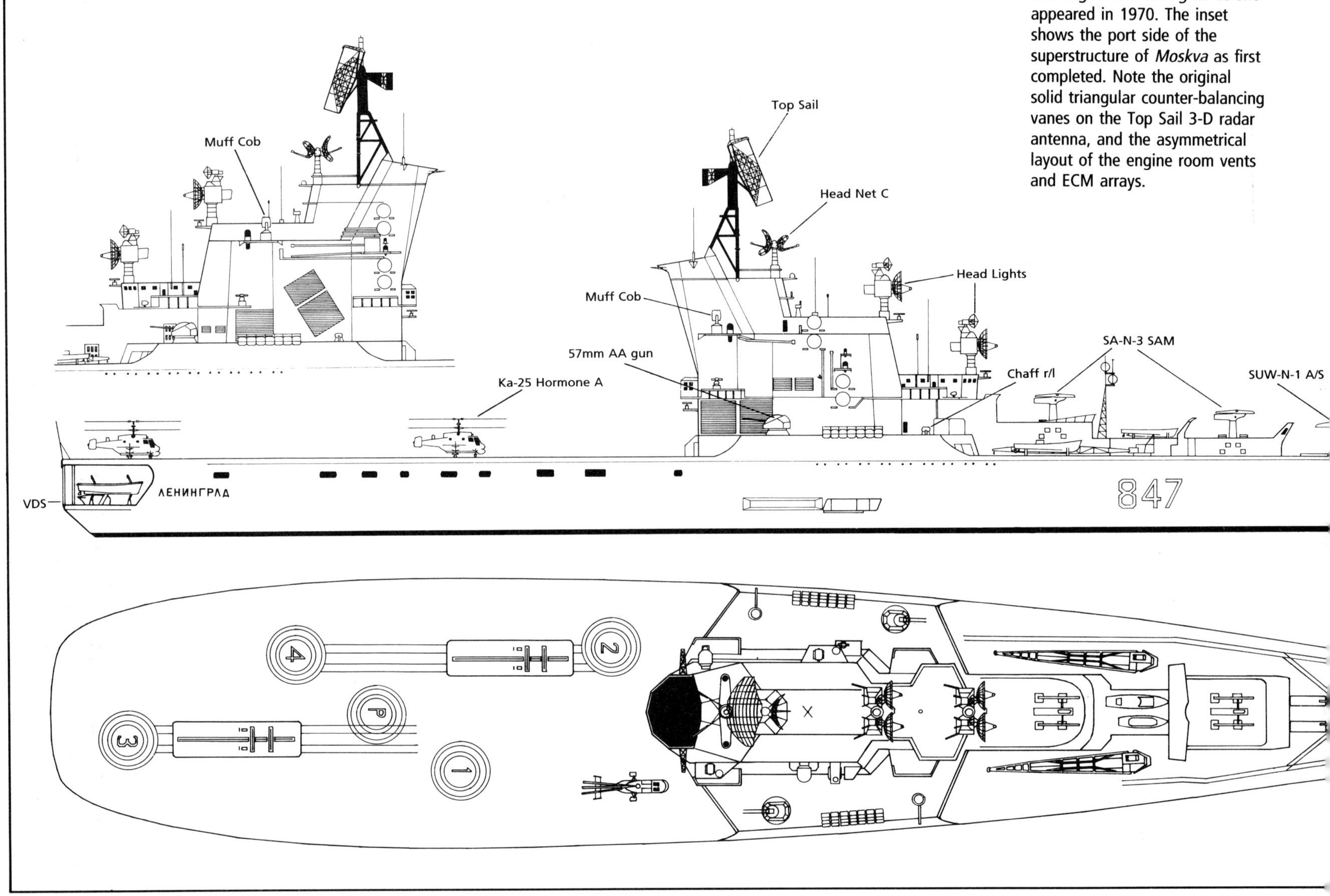

The *Moskva* Class

The initial design work for the *Moskva* class must have begun shortly after the 1959 Defence Review. The name-ship was laid down at the Black Sea Shipyard in 1962 and her sister *Leningrad* followed her on to the slipway some two years later.

The choice of the Black Sea Shipyard (formerly the Marti Shipyard) is interesting in a number of respects. The yard had been engaged exclusively in mercantile construction since Krushchev's termination of the *Sverdlov* programme in the mid-1950s. However, it was one of the traditional 'big ship' construction yards of the Soviet Navy and was one of the few Soviet shipyards that had building ways of sufficient size to accommodate the 190-metre hull of *Moskva*.

Nevertheless, the Baltic Shipyard (formerly the Ordzhinikidze Shipyard) would have been equally capable of constructing ships of this size, and the selection of a shipyard located in the southern USSR may therefore have some strategic significance. The major political centres of the Soviet Union were either inland or located on land-locked seas. Polaris missiles fired

▶Close-up of the flight deck, displaying the fifth (central) helicopter landing spot added during the 1970s. At the forward end of the flight deck can be seen the small double hanger located between the boiler uptakes. This is often

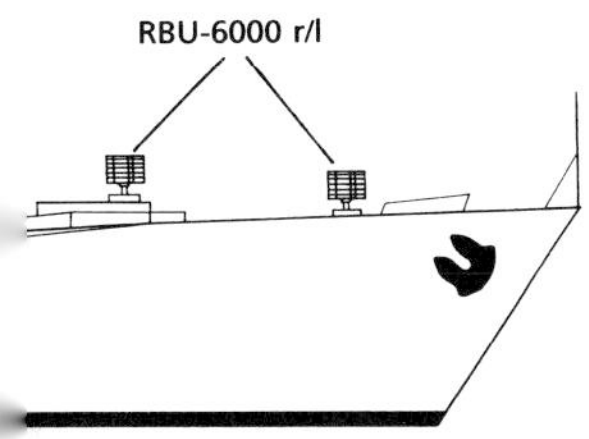

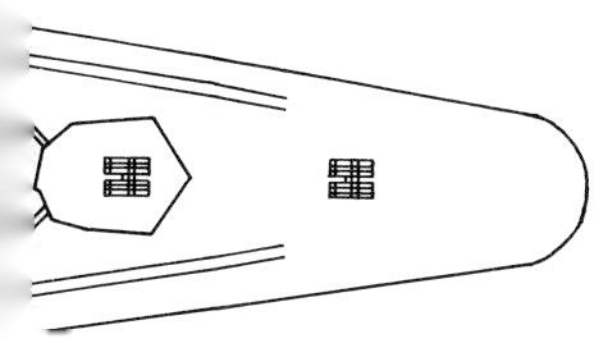

used as a garage for the deck tractors, but in this view accommodates a Ka-25 Hormone. Note the winches and towing apparatus for the Mare Tail variable depth sonar in the stern cutaway. (MoD, 1981)

from northern waters would therefore have to cross Scandinavia in order to strike at Moscow and Leningrad and would have insufficient range to strike the major cities of the southern USSR.

The eastern Mediterranean, on the other hand, provided some of the best launch points against Soviet military and civilian targets, hence the forward basing of a squadron of the new US ballistic missile submarines at Rota, in southern Spain. Access to the eastern Mediterranean from the base at Rota involved the transit of narrow 'choke points' in the central Mediterranean, which offered possibilities for the monitoring of SSBN movements to and from their patrol stations.

The early 1960s saw persistent efforts by the Soviets to secure port facilities in Egypt from President Nasser after the loss of their foothold in Albania in 1961. These facilities were conceded only after the defeat of their Arab client in the 1967 Middle East War, but from that moment on the Soviet Union had access not only to the naval base of Port Said but also to airfields in Egypt and Libya. The massive injections of military aid to rebuild the Egyptian armies after 1967 were an indication of the price the Soviets were prepared to pay in order to secure these concessions.

Hunting the SSBN

It is no easy matter to hunt down an underwater target whose primary aim is concealment. Anti-submarine warfare had been a major area of concern for the Western navies ever since the first 'Battle of the Atlantic' in the early years of the century, and continued to occupy much of the attention of the NATO alliance,

Moskva Class

Soviet Designation: Protivolodochny Kreyser (PKR) – Anti-Submarine Cruiser

Construction

Name	Builder	Laid down	Launched	In service
Moskva	Black Sea Shipyard	1962	1964	July 1967
Leningrad	Black Sea Shipyard	1964	1966	1968

Displacement (tons)
Standard: 15,500
Full Load: 19,200

Dimensions
Length: 190m(623ft) overall
Beam: 34m(112ft) flight deck; 26m(85ft) waterline
Draught: 7.6m(25ft)

Propulsion
2-shaft geared steam turbines; 4 turbo-pressurized boilers; 100,000shp=30 knots max.

Complement
850 (incl. air group)

Weapons
Aircraft: 14 Ka-25 Hormone A anti-submarine helicopters; 1 Ka-25 Hormone C SAR helicopter
ASW: 1 twin-arm SuW-N-1 launcher (18 FRAS-1 missiles); 2 RBU-6000 rocket-launchers; 2 quintuple banks of 533mm(21in) torpedo tubes
AAW: 2 twin-arm SA-N-3 launchers (44 Goblet missiles); 2 twin 57mm guns

Sensors
Surveillance radars: Top Sail 3-D, Head Net C 3-D; 3 Don-2
Fire control: 2 Head Lights (SA-N-3); 2 Muff Cob (57mm)
Sonars: keel-mounted Moose Jaw LF; Mare Tail MF VDS
Electronic warfare: 8 Side Globe; 2 Bell Clout, 2 Bell Slam, 2 Bell Tap, 2 Bell ?; 2 Top Hat; 2 twin-barrelled chaff-launchers

▼An overhead of an Anti-Submarine Cruiser of the *Moskva* class. Note the contrast between the tapered bow and the broad flight deck. The hull-form does not appear to have been particularly successful. The *Moskva*s have been observed to trim by the bow and are reputed to be poor seaboats. This unit has the original flight deck markings, with four helicopter landing spots disposed forward and aft of the aircraft lifts. (US Navy, 1970)

dependent as it was on protecting the sea lines of communication between Europe and the United States.

The convoy system had been found to be the only effective defence against the submarine in both world wars. However, many naval officers viewed convoying as an unpleasant and expensive necessity which could be eliminated if only more effective means could be found of detecting and attacking enemy submarines in the open oceans before they had the opportunity to attack.

There had been periodic flirtations with the idea of the anti-submarine 'hunting group' during and between the world wars, but this concept had proved relatively ineffectual because of the limited technology of the day. With the advent of specialized ship-borne ASW aircraft equipped with their own means of detection, however, submarine hunting operations became a practical proposition. Aircraft could cover a much broader expanse of ocean more quickly and more effectively than surface ships. Moreover, the anti-submarine helicopter, which entered service with the major Western navies during the mid-1950s, could hover above a suspected contact to listen, and could therefore be equipped with its own 'dunking' sonar to localize the target.

Armed with the new technology, the US Navy converted a number of its war-built *Essex*-class carriers to the CVS (Anti-Submarine Support Carrier) role, equipped with the low-frequency SQS-23 sonar and operating a combination of S-2 Tracker aircraft and HSS-1 Seabat helicopters, to serve at the centre of hunter-killer (HUK) task groups. Many of the smaller NATO navies created similar submarine-hunting groups based around the British war-built light fleet carriers.

These developments clearly influenced Soviet thinking when confronted with the problem of how to counter the Polaris submarine. Moreover, there was no alternative to offensive anti-submarine warfare for the Soviets. If the HUKs failed in their primary mission of sub-hunting, the West could still fall back on the convoy system, and the anti-submarine support carriers would prove invaluable in providing a protective envelope for the merchantmen. However, the convoy system had proved to be effective against the U-boats precisely because the latter had to converge on the convoy in order to sink ships. Ballistic missile submarines would not reveal themselves in the same way; indeed, concealment was an essential feature of their mission. They would therefore have to be hunted and found before they could be attacked.

Without a tradition, indeed any great experience of anti-submarine warfare, the Soviet Navy was careful to study all the new developments and select those which held out the promise of an effective solution to the specific problem of countering the Polaris submarine.

The NATO hunter-killer groups had been devised for mid-ocean ASW in the North Atlantic, and their effectiveness would depend increasingly on inputs from the SOSUS broad-area passive detection system based on a network of hydrophones laid on the seabed. SOSUS had been first tested in 1954, and was rapidly extended over the course of the next decade. The Soviets had no comparable system and the geostrategic situation of the Soviet Union made the development of such a system unattractive. The continental shelf which surrounded the waters of the North Atlantic, and on which the SOSUS hydrophones were located, was 'owned' by NATO. Moreover, the waters of the Mediterranean, which were crowded and relatively shallow, were inherently unsuited to broad-area passive acoustic systems, even if the political problem of locating military surveillance installations on foreign territory could be resolved.

In designing the *Moskvas*, the Soviet Navy was therefore influenced by developments in the major Mediterranean navies, those of France and Italy. France, with continuing colonial commitments worldwide, had preferred to use the small war-built carriers it had acquired from Britain and the United States for power projection; Italy had never possessed such a ship. Both developed helicopter-carrying cruiser hybrids for Mediterranean anti-submarine operations in the late 1950s.

The Italian *Andrea Doria* and *Caio Duilio*, which were laid down in May 1958, had a 30m by 16m flight deck and a large above-decks hangar aft, designed to accommodate three SH-3 Sea King helicopters. Designed around an enlarged destroyer hull, these ships were otherwise of conventional cruiser configuration with an SQS-23 low-frequency sonar and a US Terrier air defence missile system forward.

The French *Jeanne d'Arc*, laid down two years later, was a larger vessel designed to serve in peacetime as a training ship, and in wartime to be employed either as a helicopter assault ship or for anti-submarine warfare. She had a full 62m by 21m flight deck aft with a hangar beneath it, and a wartime capacity of eight large Super Frelon anti-submarine helicopters. A conventional cruiser superstructure forward accommodated all command and control spaces together with the electronics and most of the air defence weapons. As with the Italian vessels, these were originally to have included an area defence missile system (the French Masurca) but this plan was later abandoned.

The French Navy was also in the vanguard when it came to the development of the variable depth sonar. A

◀**In building helicopter carrier/cruisers, the Soviet Union was following a precedent set by the two major Mediterranean powers, France and Italy. The French *Jeanne d'Arc* (above), laid down in 1960, was designed for a dual role as an amphibious assault ship or an anti-submarine helicopter carrier. The *Vittorio Veneto*, a contemporary of the *Moskvas*, was the culmination of a series of Italian helicopter cruisers which began with *Andrea Doria*, laid down in 1958. Both types have their hangar beneath the helicopter flight deck. Note the Terrier air defence missile system of *Veneto* (*Jeanne d'Arc* was originally intended to carry the French Masurca, which was comparable in performance). (Marine Nationale/Italian Navy)**

prototype was installed in a modified *escorteur d'escadre*, *La Galissonnière*, in 1962 and subsequently underwent extensive trials in the Mediterranean. The variable depth sonar was housed in a 'fish' streamed over the stern of the vessel. It could therefore be lowered beneath the thermal layers which during certain seasons trapped the emissions of conventional hull-mounted sonars in the surface duct, thereby preventing the detection of submarines operating at depth.

A Hybrid Configuration

In adopting a hybrid carrier/cruiser configuration for the *Moskvas*, the Soviet Navy was therefore following a pattern already established in the major Mediterranean navies.

However, the *Moskvas* were to be significantly larger than their French and Italian counterparts. Full load displacement is about 19,200 tons as compared with 12,400 tons for *Jeanne d'Arc* and a mere 6,500 tons for the Italian ships. The profile of the Soviet vessels is dominated by a massive central superstructure block which divides the ship neatly into two distinct sections. The conventional forward section houses the major weapons systems and their related magazines; rocket-launchers and missile-launchers rise in graduated steps towards the central superstructure block on which the surveillance and fire control radars are mounted, also in steps. The superstructure ends abruptly in a large pyramid-shaped uptake for the steam-turbine exhausts.

The after section is broad and flat, an uninterrupted expanse of flight deck with the hangar beneath. It terminates in an open stern, the flight deck extending above it and supported on stanchions – an arrangement reminiscent of Japanese carriers of the Second World War. The open stern accommodates two whalers on davits, with the variable depth sonar handling gear between them.

The flight deck itself measures approximately 80m by 34m and overhangs the after part of the hull, the side-plates being gently curved outwards to meet it. Beneath the flight deck is the hangar which extends from just forward of the stern right up to the superstructure, at which point it must necessarily be terminated by the funnel uptake trunking and the magazine hoists serving the starboard gun mounting. The extent of the hangar is also indicated by the row of ventilation grilles located just beneath the flight deck; these probably serve to disperse any build-up of petrol and engine exhaust fumes from the hangar area. The dimensions of the hangar are probably about 64m by 22m, with aviation repair workshops located to the sides. This would permit a maximum of four Ka-25 helicopters to be stowed abreast except where the hangar is interrupted by the lift mechanisms.

Hangar and flight deck are served by two long narrow lifts, each about 16.5m by 4.5m, arranged *en echelon*, the forward lift offset to port and the after lift to starboard. The lift surfaces are of bare steel, contrasting with the remainder of the flight deck which has a brownish cladding. They have fixed rails and attachments to enable the helicopters to be securely locked on to them.

Inside the massive funnel structure at the forward end of the flight deck is a sizeable covered bay which can be closed off by two hinged folding doors. This appears to serve as a 'garage' for the deck tractors, but height is sufficient for two helicopters to be accommodated side by side, thereby presenting a viable alternative to strike-down in the hangar during intensive flying operations.

Initially, four helicopter spots were provided: one forward and aft of each of the two lifts. In about 1970 a fifth, central, spot was added; unlike the other spots it is not numbered, but is marked with the Cyrillic letter 'R'. The helicopter spots are covered by square mesh mats during helicopter operations to prevent the helicopter slipping on the deck.

Just above the bay doors at the forward end of the flight deck is a catwalk surmounted by a Don-2 navigation radar which, besides covering the stern arcs of the ship, presumably serves for helicopter control. Above it is a second catwalk at the centre of which is located the flight control cabin. A less familiar aid to helicopter operations – but one that has since become standard in the Soviet Navy – is the provision forward of the flight deck of stanchions from which striped windsocks are suspended.

Helicopters

The Soviet anti-submarine cruisers were designed from the outset to operate the Kamov Ka-25 Hormone helicopter which would enter service in the same year as *Moskva*. The Ka-25 Hormone was developed via the tiny Ka-15 Hen (in service 1958) and the pre-production Ka-20 Harp (first seen in 1961), and features the classic Kamov combination of twin contra-rotating rotors and a tailplane comprising a horizontal stabilizer with three vertical fins. This arrangement results in good lift characteristics and a compact fuselage, although the twin superimposed rotors require greater hangar clearance than comparable Western types.

The Ka-25 is significantly smaller than the standard NATO carrier-borne helicopter of the period, the Sea King, and has little more than half the endurance of the latter, but the relatively capacious fuselage together with the favourable lift characteristics of the design, ensures a comparable payload. A dunking sonar and magnetic anomaly detector are standard fit, and there is a surface search radar (NATO designation: Puff Ball) housed in a prominent radome beneath the helicopter's nose. A sizeable weapons bay can accommodate homing torpedoes, sonobuoys or depth-charges.

The two lifts fitted in the *Moskvas* have slots and guide-rails to accommodate the Hormone as part of the securing mechanism. The lifts are deliberately sized for the Hormone, and it was clearly not envisaged that other types would be operated from the ships. When in 1974 *Leningrad* embarked two Mi-8 minesweeping helicopters for minesweeping operations in the Suez

◀**A Ka-25 Hormone A anti-submarine helicopter in flight. Note the under-nose radome for the Puff Ball surface surveillance radar, and the distinctive Odd Rods IFF antenna mounted directly above it. (US Navy, 1974)**

Canal, the helicopters had to remain on deck because, with a fuselage 18.3m long, they were too large to be struck down. The standard Soviet land-based anti-submarine helicopter, the Mi-14 Haze, is a derivative of the Mi-8 and cannot therefore be operated by the anti-submarine cruisers.

Trials with the Yak-38 Forger A, which were conducted aboard *Moskva* in 1973, almost certainly involved land-based aircraft. A specially built platform, presumably covered with heat-resistant material, was used for the trials and was placed over the after lift. The standard Forger A tactical fighter would have been an extremely tight fit for the remaining lift, and the Forger B two-seat training version would have been too long to be accommodated.

In contrast, the lifts of Western carriers are generally sized to accommodate a variety of aircraft and this feature has frequently made possible a change of role. The disadvantage of the Soviet approach is that it imposes constraints on the dimensions of future aircraft. Significantly the Hormone replacement, the Ka-27 Helix, is similar in size and general configuration to its predecessor, its parameters being fixed by existing shipboard installations.

The air group operated by the *Moskvas* in peacetime is reported to total fourteen Ka-25s including one or two SAR/utility variants. There is limited space for parking helicopters on the flight deck; helicopter capa-

The Ka-25 Hormone A ASW Helicopter

In service: 1966
Length: 9.75m(32ft) fuselage
Width: 3.76m(12ft 4in)
Height: 5.4m(17ft 8in)
Weight: 4,765kg(10,500lb) empty; 7,500kg(16,550lb) max. take-off
Engines: two Glushenkov GTD-3F turboshafts, each 617kW(900shp); later aircraft have GTD-3BM, 738kW(990shp)
Speed: 193km/h(104kts) cruising; 209km/h(113kts) max.
Ceiling: 3,350m(11,000ft)
Operational radius: 175km(95nm); endurance 1.5–3hr
Armament: two 400mm A/S torpedoes, nuclear/conventional depth-bombs in internal weapons bay; max. weapons load 1,000kg(2,200lb)
Sensors: Puff Ball chin surface surveillance radar; Oka-2 dunking sonar; towed MAD boom; sonobuoys from internal weapons bay; Odd Rods IFF

Moskva Class: Inboard Plan

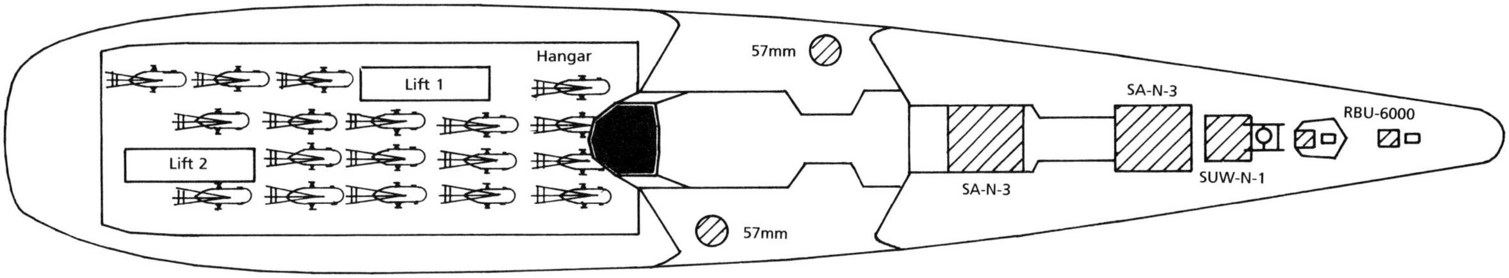

city is therefore limited by the size of the hangar. Estimates of the hangar floor space available suggest that a maximum of eighteen helicopters could be accommodated.

Anti-Submarine Missiles

The Ka-25 Hormone was equipped for independent submarine-hunting operations around the *Moskvas*. It could therefore conduct anti-submarine patrols in recognized choke-points or be vectored out to investigate a long-range contract. Closer in, however, there was a need for a new type of stand-off weapon to match the improved detection ranges promised by the large, low-frequency sonar then under development. Such a weapon would have to be able to 'kill' a fast-manœuvring nuclear-powered submarine.

The Western navies had followed two lines of development: the first, represented by the American ASROC (in service 1961), was that of a simple ballistic rocket carrying either a homing torpedo or a nuclear depth-charge; the second, represented by the French Malafon and the Australian Ikara (both in service 1964), was that of a cruise missile of aeroplane configuration capable of mid-course guidance with a homing torpedo payload.

Despite their prowess in the field of guided missiles, the Soviets chose the ballistic rocket option for the *Moskvas*, although a missile similar in conception to Malafon and Ikara, the SS-N-14, would appear three years later aboard the Large Anti-Submarine Ships of the 'Kresta II' class. The FRAS-1 (Free Rocket Anti-Submarine) missile, which uses a solid-fuel propellant, is reported to be derived from the Soviet Army's FROG-7, although the missile body is significantly shorter and more compact to facilitate shipboard installation. It is fired from a twin-arm launcher designated SUW-N-1.

The Soviet decision to opt for a ballistic rocket with inertial guidance appears to have been influenced by a number of factors. Since the intended target was a Polaris submarine, a nuclear warhead was presumably deemed acceptable since the weapon would be used as a pre-emptive counter to an imminent nuclear strike. Fitting a nuclear warhead in turn obviated the requirement for mid-course guidance because its destructive radius would compensate the 'dead time' between missile launch and arrival in the target area. The US Navy's ASROC, which had an alternative conventional homing torpedo payload, had a range of only 6nm (11km), while the FRAS-1 has a maximum range estimated at 16nm (30km).

Clearly, not even the new Soviet low-frequency sonar, which was probably designed for convergence-zone performance, could provide effective fire control out to 30km. However, the missiles could be launched on the basis of target data supplied by the Ka-25 Hormone, and the traditional Soviet emphasis on 'collaborative' tactics meant that target data might also be supplied by another surface vessel operating in the area.

The SUW-N-1 Launcher with its FRAS-1 Missile

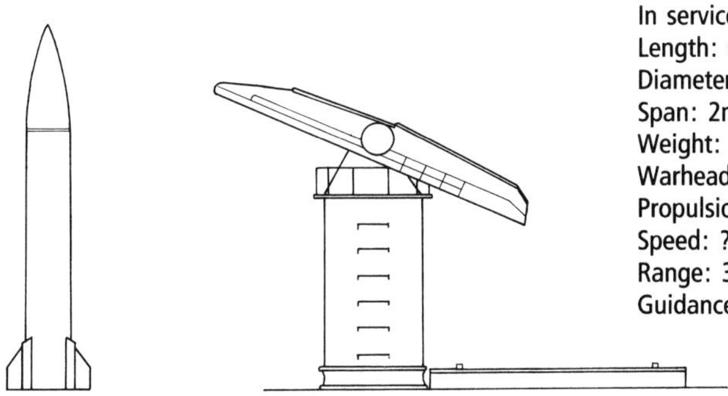

In service: 1967
Length: 6.1m(20ft)
Diameter: 0.75m(2ft 5in)
Span: 2m(6ft 6in)
Weight: 2,000kg?(4,400lb)
Warhead: nuclear
Propulsion: solid-fuel rocket
Speed: ?
Range: 30km(16nm)
Guidance: inertial

A further factor in the decision to opt for a ballistic rocket may have been the relative simplicity of the stowage arrangements which make for an enhanced magazine capacity. Cruise missiles of the period generally had to be fired from launchers that incorporated a ramp, and had to be stowed on horizontal racks with complex reloading mechanisms. A ballistic rocket, on the other hand, could be stowed in a rotating magazine ring and fired from a launcher similar to those used for surface-to-air missiles.

The reload hatches for the SUW-N-1 launcher are approximately 4.5m by 1.25m and are located immediately abaft the twin-arm launcher, suggesting that the missiles are stowed horizontally in one or two magazine rings with a capacity of perhaps 16–20 missiles, reloads being lifted from the top of the ring (or rings) and run up on to the launcher arms at an angle of about 45 degrees.

The low-frequency hull sonar (NATO designation: Moose Jaw) developed for the *Moskvas* is reported to be housed within a large, retractable dome abaft the bows of the ship. In theory a sonar of this size and operating at this frequency would be comparable to the US Navy's SQS-26 which had entered service during the early/mid-1960s and which was designed to give improved detection ranges out to 18,000m. But even the SQS-26, which had been developed via other earlier models, initially encountered severe reliability problems, and it was only when the sonar was matched with the advanced processors which became available during the 1970s that it reached its full potential. The sonar installed in the *Moskvas* presumably uses the relatively primitive vacuum tube and hybrid electronics observed in other Soviet military systems of the period, and this must limit performance.

The variable depth sonar (NATO designation: Mare Tail) trialled on the *Moskvas* quickly proved its worth, and its apparent success led to the installation of similar models on virtually all Soviet major surface combatants during the 1970s. It operated at a higher frequency than the hull sonar, however, and the raw data from the two sonars could not therefore be co-ordinated in the same way as the 'paired' bow-mounted and variable depth sonars favoured by the French Navy. The two sonars installed in *Moskva* operate independently and the data is monitored at separate consoles.

As completed *Moskva* and *Leningrad* had the customary anti-submarine rocket-launchers and torpedo tubes in addition to their helicopters and missiles, thereby reaffirming the traditional Soviet emphasis on redundancy and defence in depth. The two RBU-6000 rocket-launchers were sited one above the other close to the bow. The quintuple banks of torpedo tubes were concealed behind sliding hull doors amidships, immediately abaft the stowage bay for the accommodation ladders. They were removed during the mid-1970s.

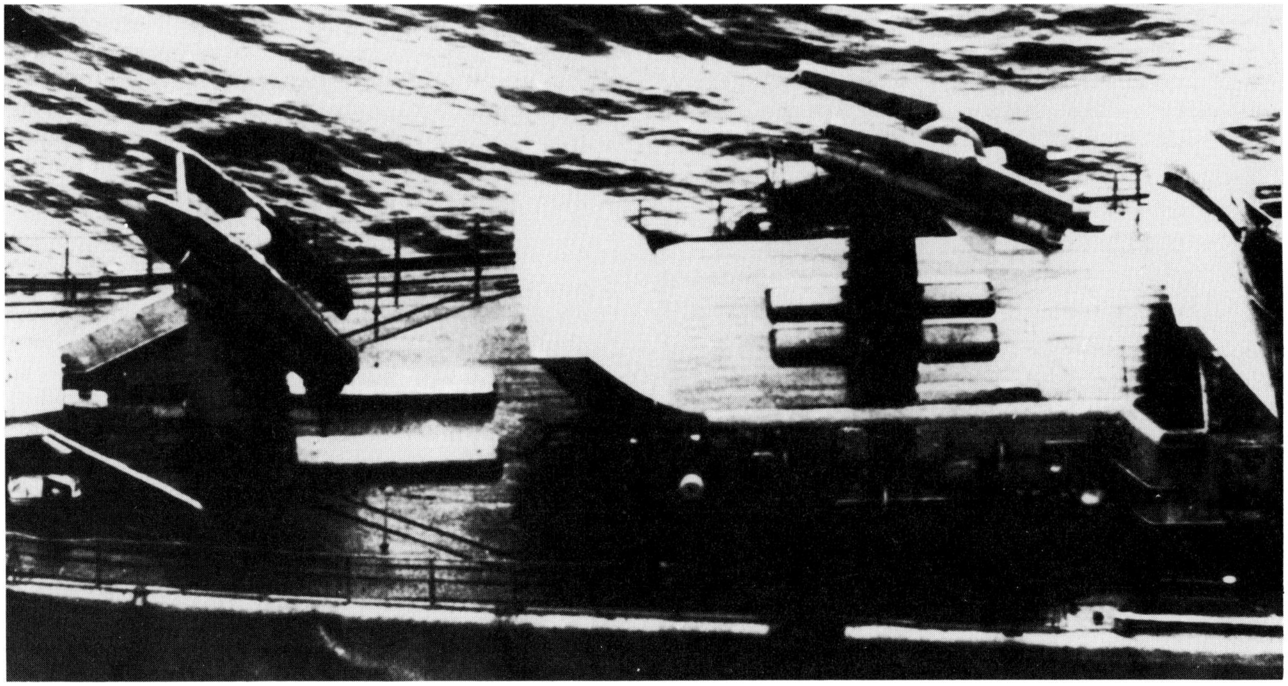

◀The SUW-N-1 anti-submarine missile-launcher (left of picture) and one of the two SA-N-3 air defence missile-launchers (right) aboard an Anti-Submarine Cruiser of the *Moskva* class. The FRAS-1 missile, which is a simple ballistic rocket, closely resembles the US ASROC in its general configuration, although it is larger and has substantially greater range. (US Navy, 1977)

Air Defence

The innovatory nature of the appearance and ASW systems of the *Moskva* class was matched by a new air defence system designated SA-N-3 Goblet. The missile was first thought to be a derivative of the land-based SA-6 Gainful, but this theory was subsequently found to be in error. The SA-N-3 was developed specifically to Soviet Navy requirements, and to this day remains the only Soviet surface-to-air missile to have been developed exclusively for naval use. Development is reported to have been first proposed to Khrushchev in 1961, approval being given shortly afterwards.[2]

The need to develop a missile with superior performance to Goa was presumably dictated by the tactical mission of the *Moskvas*, which would be expected to undertake submarine-hunting operations in areas dominated by the carrier air power of the US Sixth Fleet without the benefit of land-based fighter cover. The large size of the anti-submarine cruisers would make it possible to accommodate heavier, more capable electronics systems and missile magazines of greater capacity, thereby imposing fewer constraints on the design bureau responsible for the development of the new missile. In view of the independent sub-hunting operations envisaged for the ships, a dual anti-air/anti-ship capability was probably built into the missile,[3] thereby eliminating the need of dedicated surface-to-surface missiles for self-defence.

The 'Goblet' missile has a length of just over six metres. It has a stubby cylindrical body with a conical nose, and unusually large cropped-delta cruciform wings with smaller cruciform tail surfaces indexed in line. The large delta wings were presumably adopted to give longer range and superior stability in the surface-to-surface mode, and fold for stowage so as to maximize magazine capacity, which is estimated at 22 missiles per launcher.

The smaller cruciform tail surfaces are used for steering, but are unusual in that the ailerons are set into the leading edge of the fins. At the tips of the fins are fixed antennae which are thought to be interferometers for the missile's proximity fuze. Between the tail fins on each side of the missile body are two stub-fins tipped by streamlined fairings. These are probably rearward-pointing receiving antennae to pick up command signals emitted by the fire control radar. There are also four prominent strip antennae on the fuselage sides which may also be related to the proximity fuzing.

Unlike Goa, Goblet is a single-stage weapon with no booster, probably employing a dual-thrust solid-fuel rocket motor.[4] The maximum effective range of the early SA-N-3A Mod O variant carried by the *Moskvas* is nevertheless estimated at 30km (16nm), indicating significant advances in propellant technology.

Goblet is fired from a twin-arm launcher. Two systems are fitted in the *Moskvas*, both of which are located forward between the SUW-N-1 anti-submarine missile-launcher and the bridge. The launchers are superimposed to eliminate mutual interference on forward arcs, and the deckhouses on which they are mounted have prominent blast deflectors at their forward end.

The SA-N-3 launcher is not stabilized, which indicates that the problems of missile acquisition experienced with early Soviet PVO-derived naval SAMs had by this time been resolved. The provision of four square reload hatches led to initial speculation that there were separate belt feed systems for two different types of missile, one of which might be optimized for surface-to-surface engagements. It now seems more likely, however, that the four-hatch layout was adopted to ensure a rapid rate of fire with the launcher trained on either beam (in conventional two-hatch systems the launcher has to be re-aligned with the ship's axis in order to reload).

If that is indeed the case, rapid reloading must have been achieved only at the expense of increased complexity of the magazine arrangements with a corresponding reduction in capacity. This appears to be borne out by later developments. In the 'Kara' and 'Kiev' classes the Soviets would revert to the conventional two-hatch arrangement, and magazine capacity was increased from 22 to 36.

Command guidance was retained for the SA-N-3 system. NATO navies had by this time abandoned command guidance in favour of semi-active guidance which promised improved accuracy at longer ranges, but the Soviet Navy was faced not with the threat of

The SA-N-3 Goblet Missile

Soviet designation: unknown
In service: 1967
Length: 6.1m(20ft)
Diameter: 0.6m(2ft)
Span: 1.4m(4ft 7in)
Weight: 550kg(1,210lb)
Warhead: 80kg(175lb) HE fragmentation
Propulsion: dual-thrust, solid-fuel rocket
Speed: Mach 2.5
Range: SA-N-3A 30km(16nm); SA-N-3B 55km(30nm)
Ceiling: 25,000m(80,000ft)
Min. altitude: 100m(330ft)
Guidance: radar/command
Fire control: Head Lights
Operating frequency: G/H Band

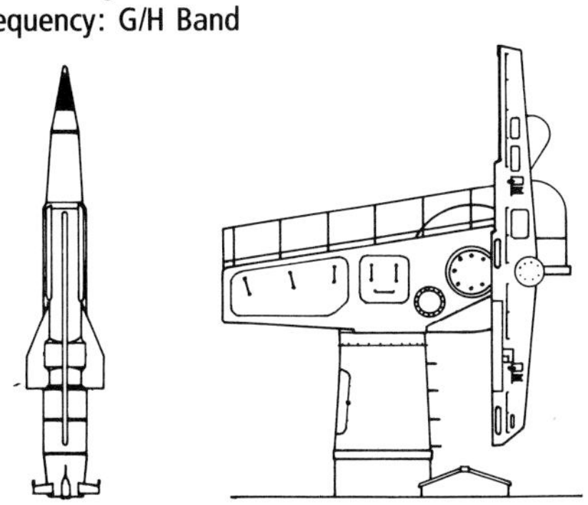

high-flying land-based bombers armed with long-range stand-off weapons, but by relatively small carrier-borne attack aircraft flying at medium or low altitude and equipped with short-range anti-ship missiles. The range at which aircraft needed to be engaged was of less significance than the ability to handle multiple targets quickly and efficiently, hence the Soviet preference for double-ended SAM systems employing twin-arm launchers. Moreover, command guidance was inherently well suited to the employment of the missile in its secondary role against surface ships.

The Head Lights radar which provides guidance for the Goblet missile is even bulkier and heavier than its immediate predecessor, Peel Group. It comprises two pairs of dishes of open mesh construction disposed symmetrically about a central pillar. The lower two dishes, which have a diameter of just under four metres, are mounted on the front face of large boxes which presumably house their electronics. The upper pair, with a diameter of about 1.8m, are mounted on top of the boxes on a small lattice assembly. Between the two lower dishes is a small solid dish which is presumably the command link. The entire assembly is suspended from the top of the mounting pillar on a yoke structure of tubular steel construction, and can be rotated and elevated. Stabilization is evident from the prominent balancing vanes projecting from the rear of the electronics boxes.

It seems probable that the larger dishes track the target while the smaller ones, which can be elevated independently, track the missiles. The solid dish transmits command signals to the missiles for course correction. Like its immediate predecessor, Peel Group, the Head Lights fire control radar appears to be designed for single-target engagement, with a standard firing policy of two missiles per target.

The other major element in the SA-N-3 fire control system is the massive Top Sail tracking radar, the first Soviet three-dimensional radar to employ frequency-scanning techniques. The rapid and efficient engagement of air targets within the effective range of the Goblet missile clearly required a tracking radar capable of giving range, bearing and height data on a number of targets at several times that distance.

Top Sail operates in the C/D-band (i.e., well below that of contemporary US Navy 3-D radars) and employs a large curved lozenge-shaped reflector with cropped corners. The reflector is illuminated by a waveguide wrapped around a tube running parallel to its axis and supported top and bottom. The heavy framework of the main antenna body, which is tilted at 20 degrees to the vertical, is counter-balanced by prominent V-shaped vanes. Originally these were of solid triangular configuration, but in 1973–4 the Top Sail antenna of *Moskva* was modified to resemble the installation of the 'Kresta II' and 'Kara' classes, with smaller vanes suspended from a lattice structure of tubular steel. This modification was a useful recognition feature until it was extended to her sister *Leningrad* in the late 1970s.

On the *Moskvas*, as on later surface units fitted with the radar, Top Sail occupies a prominent position at the top of a heavy lattice mast immediately forward of the funnel. A Head Net C V-beam air surveillance/tracking antenna is located directly beneath it and provides a degree of redundancy and back-up.

The electronic apparatus associated with the SA-N-3 system contributes in no small way to the impressive appearance of these ships. However, both Top Sail and Head Lights are extraordinarily heavy and bulky and testify to deficiencies in Soviet antenna design and metal technology. They represent a 'brute force' approach to technological progress whereby improved performance takes precedence over other considerations. Topweight clearly presents less of a problem in a ship the size of *Moskva*, but when the Soviets attempted to extend installation of the SA-N-3 system to smaller units, considerable ingenuity was required to ensure an adequate margin of stability. The production lines for the SA-N-1 Goa missile would therefore remain open for some years yet, and ships below cruiser size would continue to be fitted with the latter system long after it was regarded as obsolescent in Western military circles.

The only other active air defence weapons fitted in the *Moskvas* are two twin 57mm AA gun mountings,

▶An excellent close-up of the electronics arrays of *Moskva*, which are mounted on a block superstructure which rises in steps. From right to left: the two Head Lights guidance radars for the SA-N-3 Goblet missile, the Head Net C V-beam air/surface surveillance radar, and the massive Top Sail 3-D target tracking radar, with the original triangular balancing vanes. The extensive ECM arrays are located on the sides of the superstructure. Occupying pride of place are the four Side Globe broad-band jammers which are complemented by smaller Bell Clout, Bell Slam and Bell Tap ESM radomes. The Muff Cob fire control radar for the twin 57mm AA guns is located on top of the platform projecting from the side of the funnel, with a Tee Plinth electro-optical device beneath. This arrangement is reversed on the port side of the superstructure. (US Navy, 1968)

The 57mm/80 AA Gun System

Soviet designation: AK-257
In service: 1963
Angle of elevation: +90/−10 degrees
Rate of fire: 120rpm per barrel
Projectile weight: 5kg(11lb) fixed round
Muzzle velocity: 1,000m/sec(3,280ft/sec)
Range: 9km(5nm)
Weight of mounting: 25 tons
Fire control; Muff Cob

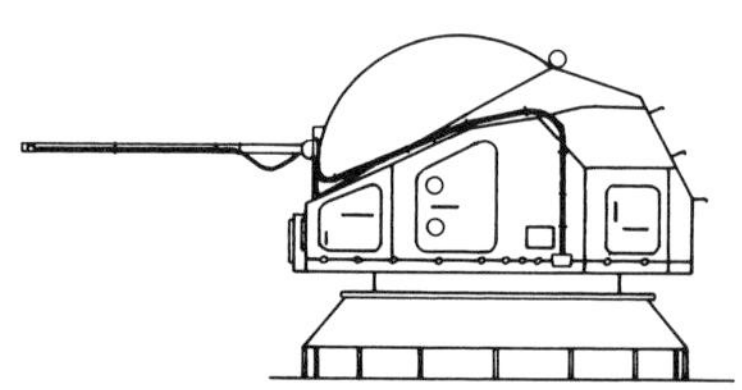

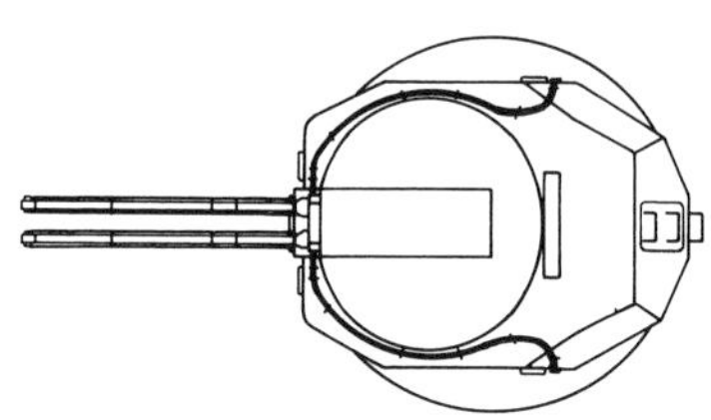

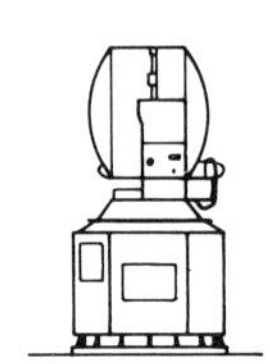

disposed *en echelon* on each side of the superstructure block at 01 level. The 57mm/80cal gun mounting, which had first entered service on the Small Anti-Submarine Ships of the 'Poti' class and was also installed on the contemporary Rocket Cruisers of the 'Kresta' class, effectively superseded the twin 76.2mm mounting fitted in 'Kynda' and 'Kashin'. It combines fully automatic operation with an exceptionally high rate of fire – estimated at 110–120rpm per barrel.

The fire control radar is Muff Cob, one of a series of Soviet fire control radars with a 'searchlight' configuration. It can be distinguished from the smaller Drum Tilt and Bass Tilt control radars by the trunnion yoke which enables the antenna 'drum' to be elevated as well as trained. Muff Cob also has an on-mount optical tracking sensor which elevates with the main antenna. Each of the 57mm mountings on the *Moskvas* has its own Muff Cob fire control radar.

Electronic Countermeasures

Moskva was given the most complete electronic countermeasures outfit yet seen on a major Soviet surface warship; clearly, the development of effective electronic warfare equipment was being accorded high priority in the Soviet Navy. Soviet cruisers, unlike the large Western task force escorts of the day, had no carrier-borne interceptors to protect them from attack from the air, but relied entirely on their own weapons and sensors for self-defence. Surface-to-air missiles might take their toll of enemy aircraft, but jamming and deception promised the only effective counter to small air-launched anti-ship missiles in the mid-1960s.

The ECM arrays of the *Moskvas* are dominated by the distinctive Side Globe radome which made its first appearance in this class. Side Globe presumably functions as a broad-band jammer, creating a high level of noise throughout the frequency spectrum. Eight Side

Globe radomes are fitted – four per side – and these are complemented by pairs of smaller 'Bell' radomes: Bell Clout, Bell Slam and Bell Tap. The antennae are disposed asymmetrically on each side of the superstructure.

The asymmetrical arrangement of the ECM arrays on the *Moskvas* is an indirect consequence of the layout of the boiler rooms which are concentrated amidships, beneath and immediately forward of the massive single funnel. The boiler rooms appear to be *en echelon*, as the prominent air vents for the starboard boiler room are located at the after end of the superstructure, whereas the port-side vents are more central, and this is reflected in the disposition of the respective ECM arrays. The port-side Side Globe radomes are mounted in pairs one above the other on the side of the funnel, together with the smaller Bell Slam and Bell Tap antennae, which are mounted on platforms between them. Bell Clout and the Muff Cob fire control radar for the twin 57mm anti-aircraft gun mounting are mounted close together at the forward end of the superstructure. This layout is reversed on the starboard side (*see* drawing).

The waveguides for these antennae are encased in protective conduits which run along the outside of the superstructure. They appear to lead to a point high in the main superstructure block which presumably houses an EW operations room.

A major innovation introduced by the *Moskvas* was the provision of 'active' electronic countermeasures in the form of a trainable twin-barrelled launcher for chaff rockets similar in conception to the US Navy's Mk 33 Chaffroc. The launchers are disposed *en echelon* at 01 deck level, a layout which reflects that of the 57mm gun mountings.

The electronic warfare suite of the *Moskvas* was to set a standard for major Soviet surface warships for many years to come. The 'Krestas' and early 'Karas' would have an almost identical outfit, while major elements of the system would also be a feature of *Kiev* and *Minsk*, *Kirov* and the *Slavas*. Not until the early 1980s would a new generation of EW sensors emerge.

Twin-Barrelled Chaff Rocket-Launcher

Replenishment

The *Moskvas* were the first modern Soviet warships to be designed for out-of-area deployment and would therefore have to be replenished with weapons and stores when operating outside the confined waters of the Black Sea. The Soviet Navy's capability for under way replenishment was virtually non-existent at this time, nor was there any great hope of an improvement in this situation; the army-dominated Soviet High Command would undoubtedly have baulked at the creation of any infrastructure which might support an ocean-going fleet in the Western mould.

When Soviet surface units began to deploy to the Mediterranean during the late 1960s the Soviet Navy therefore established a network of anchorages in the eastern half of that sea. Ships riding at anchor could then be replenished by a variety of non-specialist mercantile vessels.

To facilitate the transfer of stores and munitions the *Moskvas* were fitted with two heavy-lift cranes, and an elaborate network of rails encompassed the forecastle deck. The cranes can transfer missile reloads either directly to the reload hatches or to the rail system, and can handle the ships' boats. The rail system itself extends forward as far as the RBU-6000 anti-submarine rocket-launchers, and the port-side rails extend all the way aft to the first superstructure deck which probably incorporates a lift for general stores and provisions.

Service History

The deployment cycle of *Moskva* and *Leningrad* has generally involved service with the Black Sea Fleet or the Mediterranean Fifth *Eskadra* for one ship while the other undergoes refit and work-up in the Black Sea. Deployments are frequently made in company with other anti-submarine units. It is possible that the Soviets may have originally intended that the ships deploy to the Indian Ocean via the Suez Canal, but this option was no longer available after the closure of the latter in 1967 and Soviet forces in the Indian Ocean had to be supported from the Pacific Fleet. It is of interest to note that *Leningrad* headed the Soviet force deployed in 1974 to assist in mine clearance operations in the Suez Canal, embarking two specially configured Mi-8 Hip helicopters for this mission. Useful experience in helicopter minesweeping techniques was gained although it is reported that the Soviet part in the operation was not a conspicuous success.

The possibility that one or both of these ships might be bottled up in the Black Sea at the outbreak of hostilities has not been lost on the Soviets. *Moskva* is reported to have conducted helicopter operations in

▶This close-up of the superstructure and forecastle deck of *Moskva*, at anchor off Kithira Island, Greece, shows the elaborate arrangements made for alongside replenishment in this class. The ship's boats are handled by two large cranes which can also handle a variety of solid stores and munitions. A railway track facilitates the transfer of stores and munitions around the forecastle deck. (US Navy, 1974)

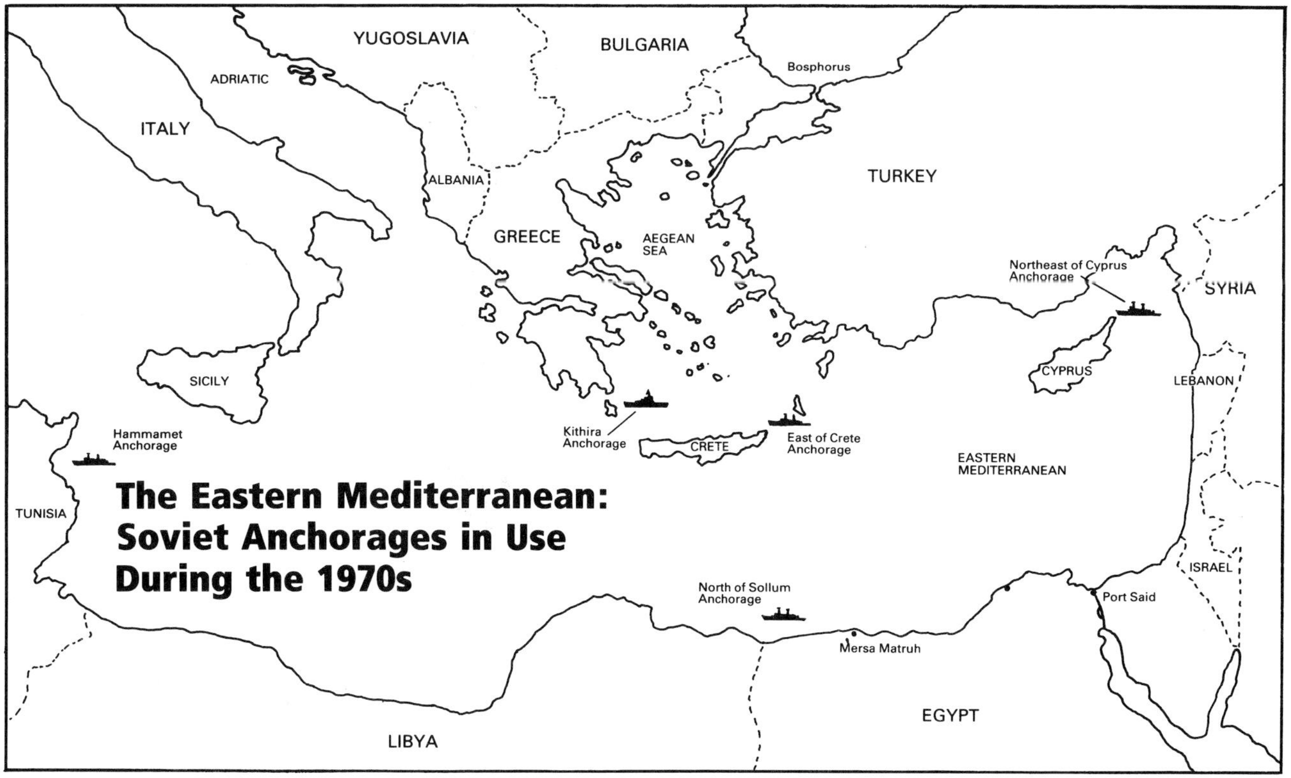

The Eastern Mediterranean: Soviet Anchorages in Use During the 1970s

◀ **This overhead view of *Leningrad* dates from 1974, when the ship took part in mine clearance operations in the Suez Canal. The helicopters on deck are Mi-8 Hip minesweeping helicopters which proved too large to be struck down via the lifts, and therefore had to remain on deck throughout the operation. (MoD, 1974)**

support of amphibious landings in the Black Sea shortly after completion, and in the summer of 1981 *Leningrad* took part in a similar but larger-scale exercise in the Baltic in company with *Kiev* and other major units.

Both ships have remained largely unmodified throughout their service careers except for the removal of their torpedo tubes during the late 1970s. In view of their importance it is perhaps surprising that no attempt has been made to retro-fit them with the AK-630 anti-missile gatlings which have been a feature of all major Soviet surface units since 1970. Nor is there as yet any evidence that the Ka-25 Hormone will be replaced by the improved Ka-27 Helix.

A New Rocket Cruiser

The first of the Rocket Cruisers of the 'Kresta I' class was laid down in September 1964, two years after work had begun on *Moskva* and three years after the fourth and last unit of the 'Kynda' class, *Varyag*, had been laid down at the same Zhdanov Shipyard.

Clearly the anti-carrier mission continued to occupy a prominent place in the Soviet surface ship construction programme even after the 1959 Defence Review. With the exception of small surface units intended to operate exclusively in Soviet waters, the only vessels with a primary anti-submarine mission under construction during the early and mid-1960s were *Moskva* and *Leningrad*. The 'Kashins' were to be reclassified as Large Anti-Submarine Ships (BPKs), but the new designation would not be accompanied by any change in the weapon/sensor outfit originally specified, which featured a full complement of anti-submarine rocket-launchers but was otherwise weighted in favour of air defence.

The three-year hiatus between the laying down of *Varyag* and the first of the 'Krestas', *Admiral Zozulya*, is nevertheless significant in that it is indicative of a lower priority for Rocket Cruiser construction in the wake of the 1959 Review, which resulted in much of the weight of the anti-carrier programme being shifted to submarines armed with cruise missiles. It also suggests that major redesign work was necessary partly as a result of experience with the 'Kynda', but also as a consequence of a radical reappraisal of the Rocket Cruiser concept.

Many of the problems experienced with the 'Kynda' were related to the 'compactness' of the design, which resulted in stability problems and cramped accommodation. The 'Kresta' hull was there-

▶**The 'Kresta I', although in conception derived from the 'Kynda', was a very different ship both in terms of build and in the balance of her armament. The single heavy tower mast with its integrated funnel structure is flanked by short radar towers for the two Peel Group missile guidance radars. (US Navy, 1971)**

'Kresta I' Class: Profile and Plan

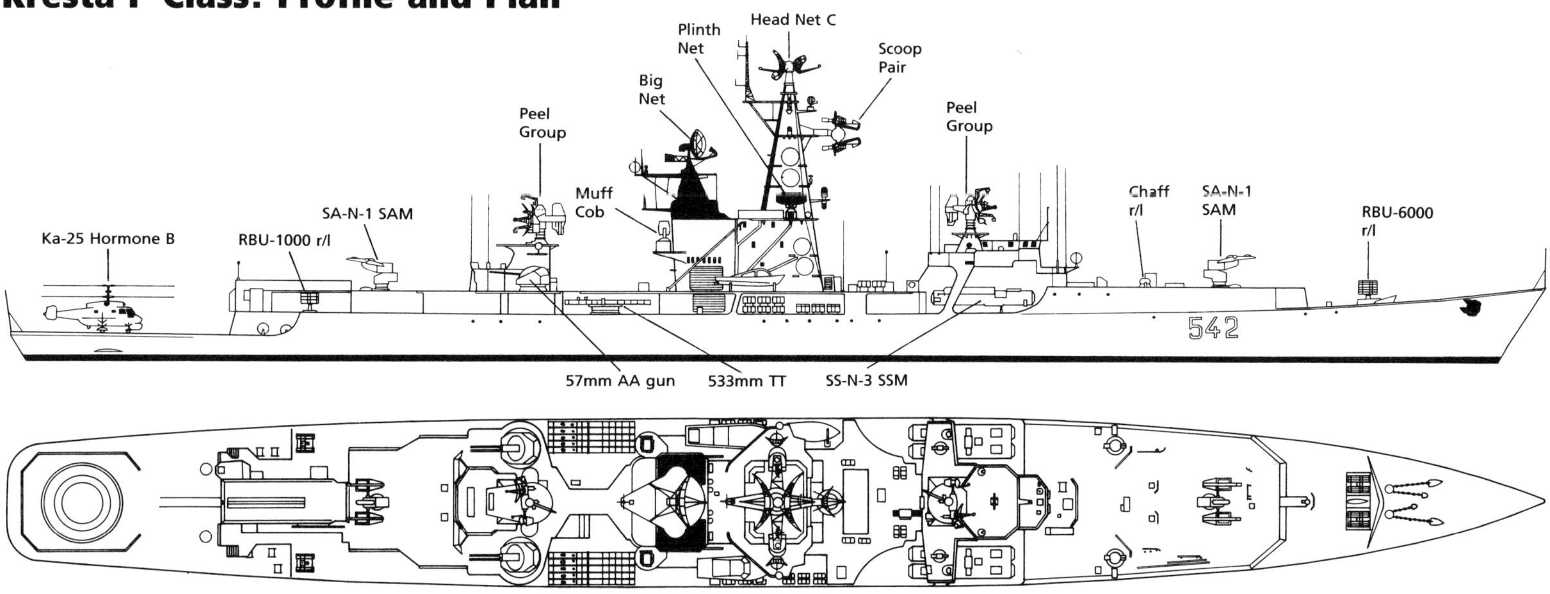

'Kresta I' Class

Soviet Designation: Raketny Kreyser (RKR) – Rocket Cruiser

Construction

Name	Builder	Laid down	Launched	In service
Admiral Zozulya	Zhdanov Shipyard	Sept 1964	Oct 1965	Mar 1967
Vladivostok	Zhdanov Shipyard	1965	Aug 1966	Jan 1968
Vitse-Admiral Drozd	Zhdanov Shipyard	1965	Jan 1967	Aug 1968
Sevastopol	Zhdanov Shipyard	1966	June 1967	July 1969

Displacement (tons)
Standard: 6,150
Full Load: 7,500

Dimensions
Length: 155.5m(510ft) overall
Beam: 17m(56ft)
Draught: 6m(20ft)

Propulsion
2-shaft geared steam turbines; 4 turbo-pressurized boilers; 100,000shp=34 knots max.

Complement
380

Weapons
ASuW: 2 twin-SS-N-3 launchers (4 Shaddock missiles); 1 Ka-25 Hormone B missile targeting helicopter; 2 quintuple banks of 533mm(21in) torpedo tubes
AAW: 2 twin-arm SA-N-1 launchers (32 Goa missiles); 2 twin 57mm guns
ASW: 2 RBU-6000 rocket-launchers; 2 RBU-1000 rocket-launchers

Sensors
Surveillance radars: Big Net, Head Net C 3-D; 1 Don-Kay, 1 Don-2
Fire control: 1 Scoop Pair (SS-N-3 guidance), 2 Plinth Net (SS-N-3 data link); 2 Peel Group (SA-N-1); 2 Muff Cob (57mm)
Sonars: keel-mounted Herkules (Wolf Paw) HF
Electronic warfare: 8 Side Globe; 1 Bell Clout, 2 Bell Slam, 2 Bell Tap; 2 twin-barrelled chaff-launchers

fore made significantly larger, with an additional 12.5m in overall length and a 1.3m increase in beam. The forecastle deck was carried aft almost to the stern, creating additional internal volume. The exhaust uptakes from both boiler rooms were grouped together into a single broad funnel amidships, thereby economizing on centre-line space and facilitating the accommodation of weapons-launchers forward and aft.

The most significant changes in configuration, however, were related to a dramatic shift in the balance of the weapons systems. The design of the 'Kynda' had been dominated by the massive quadruple launchers for the SS-N-3 Shaddock missile, which commanded the key centre-line positions forward and aft. In the 'Kresta' these positions were occupied by two SA-N-1 Goa surface-to-air missile systems, and the paired launchers for the SS-N-3 anti-ship missiles were relegated to the beam, being located directly beneath cantilevered bridge wings. Not only was the number of SS-N-3 launchers halved, but reloads were completely eliminated. Moreover, the launchers on the 'Kresta' could be elevated but unlike those of the 'Kynda' could not be trained in azimuth.

The reduction to four launchers brought with it a number of important design benefits. As all SS-N-3 launchers were forward-firing they required only a single Scoop Pair guidance radar installation. The second radar tower mast of the 'Kynda' could therefore be eliminated, with a consequent reduction in topweight. The designers still wished to accommodate two air surveillance antennae, but rather than persist with two tower masts adopted the expedient of splitting the funnel uptakes and mounting the second radar antenna on a raised platform between them. The single tower mast, which was even broader and more substantial than those of the 'Kynda', supported a Head Net C V-beam air surveillance antenna, the single Scoop Pair missile guidance radar installation (on the face), Plinth Net tracking/data link antennae for SS-N-3 to port and to starboard, and all the electronic countermeasures arrays.

Despite the cruiser hull and heavier superstructures, the design of the 'Kresta' clearly owed at least as much to the 'Kashin' as it did to the 'Kynda'. The twin-arm launchers for the SA-N-1 Goa missile were mounted on top of raised deckhouses which were integrated into a single continuous superstructure deck running three-quarters the length of the ship. The forward deckhouse extended to the ship's sides, thereby protecting the SS-N-3 launchers tucked beneath the bridge wings from the worst of the weather. The Peel Group guidance radars were sited on separate towers forward and aft, also in the manner of the 'Kashin'. As with the latter, the forward tower incorporated the navigation bridge, although this was an altogether more substantial structure than that of the destroyers.

Anti-submarine capabilities were on a par with those of the 'Kashin'. There was a keel-mounted Herkules hull sonar, twin RBU-6000 rocket-launchers close to the bow, and twin RBU-1000 heavy rocket-launchers aft. The torpedo tubes were in quintuple mountings, two sets being mounted amidships on the beam, whereas the 'Kashin' had only a single centre-line installation. This may reflect a mixed complement of anti-ship and anti-submarine torpedoes.

Electronics

The 'Kresta' was almost certainly designed from the outset to carry the more effective Big Net/Head Net C

▶The 'double-ended' layout of the 'Kresta' is reminiscent of the 'Kashin', with which it has a number of features in common. Note, however, the compactness of the central superstructure block which bristles with radar and ECM antennae, the Side Globe broad-band jammers being particularly prominent. Unlike the 'Kynda', the 'Kresta I' class was fitted with Plinth Net from the outset. (RDN)

▶The SS-N-3 launchers, which occupied the dominant centre-line positions on 'Kynda', are tucked beneath the cantilevered bridge wings of 'Kresta'. The paired launchers are elevated to fire, but cannot be trained. Note the Tee Plinth electro-optical sensors in the bridge wings; these provide an all-round view to the operations room under NBC close-down conditions. (US Navy, 1970)

radar combination which replaced the twin Head Net A installation of the 'Kynda' and early 'Kashins'. Big Net was primarily for long-range air surveillance. Its antenna was comparatively heavy and bulky, and performance close to the surface was less crucial so it did not need to be installed at mast-top height. The first 'Kashins' to receive the Big Net/Head Net C radar combination were completed in 1966 when the first of the 'Krestas' was fitting out at the Zhdanov Shipyard.

Other systems are similar to those of the contemporary Anti-Submarine Cruisers of the *Moskva* class. Four pairs of Side Globe broad-band jammers are mounted on the sides of the lower mast. There is a single Bell Clout antenna on a platform projecting from the forward face of the tower, complemented by pairs of smaller Bell Slam and Bell Tap antennae on the tower mast sides. Two twin-tube chaff rocket-launchers are fitted on the forward deckhouse immediately abaft the SA-N-1 missile-launcher. The electronic warfare centre is almost certainly housed within the mast structure itself.

The 'Kresta' also features the twin 57mm automatic gun mounting in place of the 76.2mm mounting adopted for the 'Kynda' and 'Kashin'. This has superior performance in the anti-aircraft role, although at some expense to anti-surface capability. The twin gun mountings are installed abeam the after Peel Group radar tower, with the Muff Cob fire control radars on platforms projecting from the after corners of the funnel structure. There are two Tee Plinth installations in the bridge wings and smaller Tilt Pot sensors on platforms projecting from the sides of the after radar tower.

Flight Deck and Hangar

For the first time in a conventional surface warship design, the Soviets incorporated a helicopter hangar and flying-off deck close to the stern. Both the 'Kynda' and 'Kashin' had small helicopter landing areas marked out above the stern, but these were suitable only for operating land-based helicopters and no maintenance facilities were embarked.

The helicopter which the 'Kresta I' was to operate was a variant of the Ka-25 Hormone (NATO designation: Hormone B), equipped with a Big Bulge-B radar for surface surveillance. The Hormone B is fitted for missile targeting and like its larger land-based counterpart, the Bear D reconnaissance bomber, can transmit data via a video data link back to the missile launch ship, thereby enabling mid-course corrections to be made.

The hangar clearance required by the Hormone, which has the distinctive Kamov coaxial rotor configuration, presents problems for the constructor because the hangar can easily obstruct other weapons

◀**The 'Kresta' was the first conventional Soviet surface warship to have a flight deck and hangar for its own helicopter. The double hinged doors of the hangar open outwards. Locating the flight deck on the low quarterdeck means that the hangar roof does not obstruct the after SA-N-1 missile-launcher. The unit depicted is *Admiral Zozulya*. (MoD, 1975)**

▶A Ka-25 Hormone B missile targeting helicopter in flight. Its Big Bulge B surface surveillance radar is the smaller counterpart to the radar carried by the Bear D land-based reconnaissance aircraft. Like the Bear, the Hormone B is fitted with a video data link which enables its own radar picture to be transmitted back to the launch ship. (US Navy, 1986)

systems, particularly in a 'double-ended' ship. The solution adopted for the 'Kresta I' was to retain the low quarterdeck of the 'Kynda' for use as a helicopter landing deck, so that the hangar roof does not interrupt the arcs of the after SA-N-1 launcher. This also facilitates movement of the helicopter between hangar and flight deck, at the expense of a very low helicopter operating platform. Operations in high sea states in northern waters appear to have convinced the Soviets that this solution results in an unacceptably low level of helicopter availability; later designs based on the 'Kresta' hull all have a raised flight deck, even at the expense of a complex 'split-floor' hangar incorporating a lift.

The Tactical Dimension

Although the major weapons systems of the 'Kresta I' are identical with those of the 'Kynda', the very different balance of the former type's armament suggests an important change in Soviet tactical priorities. The precise mode of employment initially envisaged for the 'Kresta' remains problematical.

The matter is not helped by confusion regarding the Soviet Navy's own classification of these vessels. By the time the first unit, *Admiral Zozulya*, was completed in March 1967 they were being referred to as Large Anti-Submarine Ships (BPK). Yet some ten years later they were reclassified as Rocket Cruisers (RKR).

The 'anti-submarine' classification is not entirely convincing. Despite the large size and cost of these ships, their anti-submarine weapons and sensors are on a par only with the 'Kashin', which itself only received the BPK classification retrospectively, being initially designed as a fleet destroyer with general-purpose capabilities.

The classification of the 'Kresta I' as a Large Anti-Submarine Ship is almost certainly related to a major policy change which took place in the mid-1960s and which will be discussed more fully in the following chapter. It should be noted that the first of a new anti-submarine variant of the 'Kresta', the 'Kresta II', was laid down at the Zhdanov Shipyard in late 1966, several months before *Admiral Zozulya* was completed. The initial classification of the 'Kresta I' as a BPK probably has more to do with the internal politics of the Soviet Navy than with any particular aspect of the original design. Once the construction programme begun in the late 1960s had produced the requisite quotas of blue-water anti-submarine vessels, and attention could again be given to the anti-ship mission, the 'Kresta I' reverted to what was almost certainly its original classification as designed, that of Rocket Cruiser.

Nevertheless, the differences in the balance of the armament of the 'Kynda' and 'Kresta I' clearly reflect changes in Soviet philosophy with regard to anti-

surface operations. Air defence capabilities show a two-fold increase, with the provision of a second SA-N-1 installation being complemented by more capable air surveillance and targeting radars. Anti-surface capabilities, on the other hand, have been halved and the total number of SS-N-3 Shaddock missiles carried is only one-quarter that of the 'Kynda'.

These changes suggest that the 'Kresta' was designed to operate farther outside the Soviet maritime defence zones than the 'Kynda'. The enhanced air defence capabilities would to some extent compensate the increased difficulty of providing air cover in areas far from Soviet bases, and would allow the new cruisers more autonomy in anti-carrier operations. Moreover, although long-range reconnaissance was still largely dependent on the lumbering Bear D bomber, there was now an integral over-the-horizon targeting capability in the form of the Hormone B helicopter.

Exercise 'OKEAN 70', involving more than 200 surface ships and submarines, was staged largely to demonstrate the ability of the Soviet Navy to make near-simultaneous co-ordinated strikes involving aircraft, surface ships and submarines at a number of points in the Atlantic and the Pacific. One of the new cruisers, *Vitse-Admiral Drozd*, operated in the western Atlantic out of Cuba in conjunction with Bear D aircraft operating from Havana. The extreme forward location in which *Drozd* operated in the course of this exercise would almost certainly have been untenable in time of war. Nevertheless, the exercise was indicative of Soviet attempts to test the feasibility of long-distance anti-carrier operations co-ordinated from land-based command positions.

Service Life

The four 'Kresta Is', completed at the Zhdanov Shipyard at the rate of one per year, joined the fleet from 1967 to 1970. Three were initially allocated to the Northern Fleet, while *Vladivostok* went to the Pacific. The significance of these deployments has to be seen in context with those of the 'Kynda' class. By the early 1970s the Northern Fleet had three 'Kresta Is', the Pacific Fleet had a single 'Kresta I' and two 'Kyndas', while the remaining pair of 'Kyndas' served with the Black Sea Fleet.

The deployment pattern is indicative of Soviet anti-carrier priorities. The Northern Fleet required ships with greater range and autonomy than the other fleets because NATO carriers operating in the southern reaches of the Norwegian Sea could be employed against a variety of Soviet military targets. In the Pacific and the Mediterranean, on the other hand, the range

▲ *Vitse-Admiral Drozd* following her mid-1970s refit in which the AK-630 anti-missile system was added. The Ka-25 Hormone B missile targeting helicopter, seen here being spotted for take-off, gave these ships a degree of autonomy which enabled them to undertake anti-surface operations outside Soviet sea-space. (MoD)

factor was of less significance, it being envisaged that the anti-carrier groups would pick up enemy carriers and amphibious forces only when they entered the areas in which Soviet surface forces were concentrated.

In the case of the Mediterranean Fifth *Eskadra*, the 'Kyndas' could more easily be supported by smaller missile-armed units such as the then-new Small Rocket Ships (*Maly Raketny Korabl'*) of the 'Nanuchka' class. With the experience of the 1973 Middle East conflict in mind, the Soviets would soon begin a missile conversion programme for destroyers of the 'Kashin' and 'Kildin' classes (*see* Chapter 2); many of the ships would be modified in Black Sea shipyards. The employment of these vessels for close shadowing of the carriers of the US Sixth Fleet would assist with the targeting of the long-range SS-N-3 missiles carried by major Soviet anti-surface units operating unseen beyond the horizon.

Over-the-horizon targeting for the 'Kresta Is' operating with the Soviet Northern Fleet would continue to rely on aircraft operating within radar range of the carrier task force. The provision of on-board helicopters to facilitate mid-course correction for the missiles was therefore an important step towards self-sufficiency.

It should be noted that from the early 1970s until a decade later, no Rocket Cruisers served with the Baltic Fleet. The NATO strategy of the day did not envisage risking carriers in these constricted and dangerous waters, so there was little point in retaining anti-carrier units with the Baltic Fleet. The latter was by now composed almost exclusively of light forces – fast attack craft armed with missiles and torpedoes, 'Guard Ships' and anti-submarine corvettes – and amphibious vessels, supported by a handful of older *Sverdlovs*, 'Kotlins' and 'Skorys' and a few modern missile-armed destroyers.

Because of its traditional naval association and its concentration of key shipyards and maintenance facilities, the Baltic also became the major work-up and training area for surface combatants that had recently entered service or had undergone refit. The Northern Fleet, in particular, continued to be largely maintained by the Baltic shipyards. Although this presented a neat and logical solution to a problem that was at once historical and geographical, it also begged the question of how effectively these units could be maintained in the Northern Fleet area in the event of hostilities, when NATO forces would attempt to seal off the narrow straits which separate Denmark from Norway.

Later Modifications

In 1973, following a lengthy deployment to the Northern Fleet, *Vitse-Admiral Drozd* returned to the Baltic to

Vitse-Admiral Drozd: Profile of Midships Section

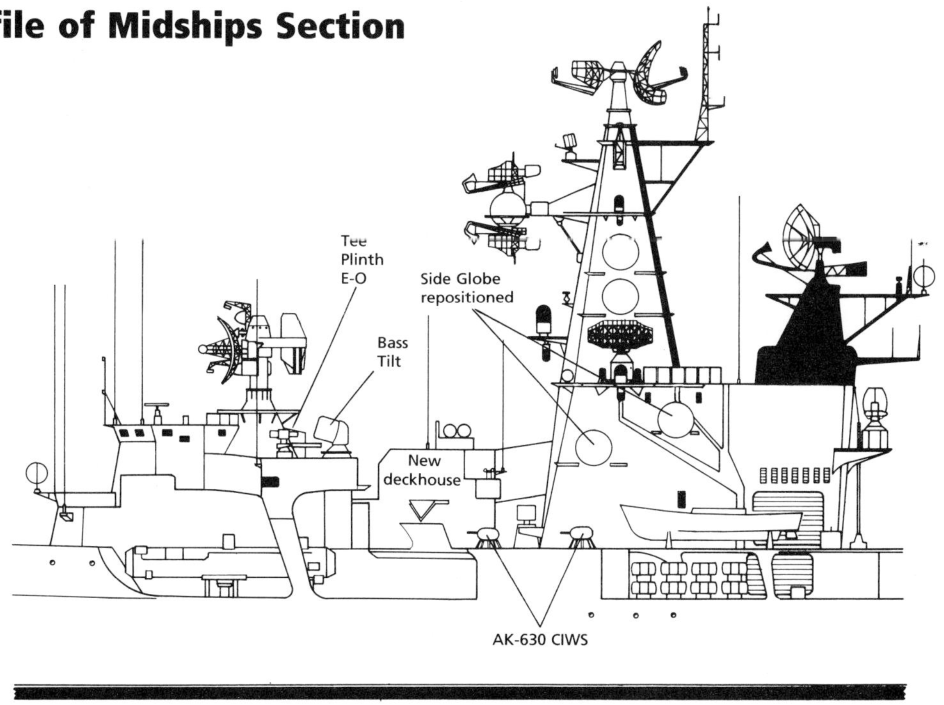

Vitse-Admiral Drozd as refitted 1973–5. A new deckhouse has been constructed out to the ship's sides abeam the tower mast for AK-630 anti-missile guns. The Bass Tilt fire control radars are located at the after end of an extended bridge structure, and there is an optical fire control 'bin' directly above the gun mountings themselves. Raised blast shields at the forward end of the deckhouse protect the guns from the rocket exhaust of the SS-N-3 anti-ship missiles. There is now a large two-storey deckhouse, presumably for additional accommodation or electronics, between the bridge structure and the tower mast.

undergo an extensive refit. She re-emerged in early 1975 with a number of important modifications.

Immediately forward of the central tower mast the first superstructure deck had been extended outboard to the sides of the ship to accommodate the magazines for two pairs of AK-630 gatlings – an arrangement similar to that of her half-sisters of the 'Kresta II' class (*see* Chapter 4). In order to accommodate the associated Bass Tilt fire control radars without at the same time displacing the Tee Plinth electro-optical sensors, the bridge wings had to be extended aft.

Between the bridge structure and the tower mast there is now a prominent deckhouse two decks high; this may house additional electronics, or possibly crew accommodation. Similar deckhouses are a feature both of the later units of the 'Kresta II' class and of the 'Kyndas' modified to receive the AK-630 anti-missile system. On the *Drozd* curved blast shields project from the sides of the deckhouse to protect the forward gatlings from the blast generated by the twin boosters of the SS-N-3 missiles. To accommodate these changes the port-side whaler and davits have been suppressed and the lower Side Globe ECM radomes re-sited (*see* drawing).

At the end of the 1970s *Sevastopol* underwent a major refit prior to joining the Pacific Fleet. She received the new two-storey deckhouse, but not the AK-630 gatlings. She paid off in 1990, the first of the class to do so.

Admiral Zozulya has been in refit at Kronshtadt since 1985 and may also be stricken, while *Vladivostok* has been inactive in the Pacific Fleet since 1988. Only *Vitse-Admiral Drozd* has been active throughout the 1980s, during which she has served as flagship of the Baltic Fleet. The anti-surface mission of these ships in the Northern and Pacific Fleets has been taken over by the powerful new Rocket Cruisers of the *Kirov* and *Slava* classes.

◀Overhead of *Vitse-Admiral Drozd* in 1986. The modifications made during the mid-1970s are clearly visible amidships. In view of the size and importance of these ships, it is perhaps surprising that no attempt was made to extend the installation of the AK-630 anti-missile system to the other three units of the class. (MoD, 1986)

4. The Large Anti-Submarine Ships

▼The 'Kresta II' was the first of a series of Large Anti-Submarine Ships (BPKs) that was to dominate the Soviet surface ship construction programme of the 1970s. The primary mission of these ships would be to support the deployment of Soviet submarines through the NATO anti-submarine barriers in the North Atlantic, and to defend the perimeter of the SSBN bastions set up in the early 1970s. (MoD, 1975)

WESTERN ANALYSIS of Soviet naval developments during the 1960s has generally seen the major anti-submarine construction programme embarked upon during the second half of that decade as being a natural progression from the decision to build the *Moskvas* and to initiate the development of a wide range of ASW weapons and sensors. A careful consideration of the available evidence, however, suggests that this analysis was in error and that the new ships belonged to a separate programme, underpinned by a different operational philosophy.

The waters are muddied by the relatively lengthy gestation period of the *Moskvas*, which were laid down in 1962 and 1964 respectively but emerged from the shipyards only towards the end of the decade. Indeed the second ship, *Leningrad*, entered service little more than a year before the first of the BPKs of the 'Kresta II' class, but completion dates are clearly less significant in this respect than the dates when keels were laid. Moreover, 'Kresta II' was a modification of an existing design whereas *Moskva* was the first vessel of her type in the Soviet Navy; keel-laying of *Moskva* must therefore have been preceded by a relatively protracted design period.

Moskva was laid down a full four years before the first of the 'Kresta IIs' and she and her sister-ship belong to the 1959–67 programme which resulted from Krushchev's defence review. The major anti-submarine construction (and conversion) programme begun during the late 1960s appears to have been decided on in about 1964–5, in the wake of Krushchev's fall from power. The weapons and sensors which would

be installed in the new ships were largely derived from those developed for the *Moskvas* but they were intended to be used for a different set of missions.

The Pro-Submarine Mission

The year 1964 also marked a watershed in the Soviet Navy's submarine construction programme because it was in that year that a new type of nuclear-powered ballistic missile submarine (SSBN), the 'Yankee', was sanctioned, and development of a new generation of long-range sea-launched ballistic missiles (SLBM) was begun. This would eventually culminate in the giant SS-N-8 missile, with a range in excess of 4,200nm, capable of striking at targets throughout North America from waters close to the Soviet Union, which entered service in 1973 aboard submarines of the 'Delta' class.

In the short term this would mean that Soviet submarines of all types would need to transit the NATO anti-submarine barrier strung across the Greenland/ Iceland/United Kingdom (GIUK) Gap, and would be vulnerable to long-range detection even in open-ocean areas because of the increasing capability of the SOSUS underwater surveillance system, based on networks of passive hydrophones allied to increasingly sophisticated processors. In the longer term the Soviet Navy could envisage the possibility of conducting SSBN patrols exclusively from Soviet waters where the submarines could be herded together and protected from hostile threats.

In his writings published in the mid-1970s Admiral Gorshkov would constantly stress the importance of supporting submarines with both aerial and surface forces, and would attribute the ultimate failure of the German Navy's U-boat campaign of the Second World War to its neglect of this axiom.[1] The timing of his words is particularly significant, the major anti-submarine surface ship programme embarked upon in the late 1960s being by that time in full swing.

The new ships would be designed for two distinct missions. The larger units would be equipped, like the *Moskvas*, for sub-hunting operations directed against the Polaris submarine, both in the eastern Mediterranean and the Norwegian Sea. This mission would decline in importance over the years, given the 2,500nm range of the Polaris A-3 missile and its successor, the C-3 Poseidon, and the lack of success in detecting and tracking the submarines. At best, Soviet anti-Polaris operations would achieve only minor revisions in Western SSBN patrol areas, and longer missile ranges meant that only a handful of submarines would need to patrol in areas where there was a risk of detection by Soviet anti-submarine units.

A mission of increasing importance would be the 'pro-submarine' mission, in which Soviet units would assist the deployment of Soviet submarines and protect them against Western anti-submarine forces. This mission would focus on two separate types of operation: the active support by mobile surface forces of submarines deploying into hostile waters; and the creation of defensive anti-submarine barriers on the perimeter of the Soviet defence zones in the Arctic and the north-west Pacific, thereby creating 'bastions' within which Soviet ballistic missile submarines could patrol in relative safety.[2] The northern bastion would be based on the Barents Sea; that in the Far East on the Sea of Okhotsk, enclosed to the south by Hokkaido and the Island of Sakhalin, and to the east by the disputed Kurile Islands.

The larger ships in the new programme, the design of which was based on enlarged destroyer hulls, would be built in the two shipyards currently responsible for the construction of blue-water surface ships: the Zhdanov Shipyard in the Baltic, and the 61 Kommuna Shipyard in the Black Sea. Zhdanov would use an existing type, the 'Kresta'-class Rocket Cruiser, as the basis of its new design. The 61 Kommuna Shipyard would develop its own design (the 'Kara'), similar in many respects to the Zhdanov design and featuring many of the same weapons systems, but incorporating gas-turbine propulsion technology derived from its experience as lead-yard for the destroyers of the 'Kashin' class.

However, the requirement for large numbers of surface ships to police the perimeters of the bastions spawned a new type of ship with anti-submarine capabilities on a par with the larger vessels, but with less emphasis on all-round air defence capabilities. In particular, these ships would lack an area defence missile system, defence against hostile aircraft being performed by a combination of radar-controlled guns and a new short-range missile system, while only passive measures would be provided to counter anti-ship missiles.

The concept of a 'high/low' mix was by no means new to Soviet military thinking. The major advantage of pursuing such a course at this particular juncture was that it allowed a rapid expansion in construction which would otherwise have been impossible. The Zhdanov and 61 Kommuna Shipyards would be working to full capacity to produce the larger surface ships, and any attempt to convert other major Soviet shipyards to naval construction would disrupt the large mercantile building programme then under way. Moreover, the machine tooling and worker skills in these shipyards were now geared to mercantile construction, and a further conversion programme would be prolonged and costly.

It was therefore decided to use the expertise accumulated in the smaller shipyards. The Yantar Shipyard, Kaliningrad, had been responsible for the entire series of Guard Ships (*Storozhevoi Korabl'*) since the Second World War. These included the post-war 'Kola' and 'Riga' classes, and the 'Petyas' and 'Mirkas' built throughout the 1960s. Yantar therefore became the lead-yard for the 'second rank' BPKs, which would effectively take on the Guard Ship function (and would eventually be reclassified as such), but which unlike their predecessors would be built primarily for anti-submarine warfare and would patrol the perimeters of Soviet sea-space rather than operating within those waters.

The other shipyard selected was that of Kamysh-Burun at Kerch on the Black Sea. Kamysh-Burun was a relatively new shipyard, but during the 1960s had been responsible for the construction of large numbers of Small Anti-Submarine Ships (*Maly Protivolodochny Korabl'*) of the 'Poti' class.

The 125-metre 'Krivak' was almost certainly the largest hull that could be comfortably built at either shipyard. The largest vessel previously built at Kaliningrad was the 'Kola' with an overall length of 96m; later Guard Ships had been shorter. The 'Potis', built at Kamysh-Burun, had a length of only 60m. Construction of the 'Krivak' may therefore have involved enlarging the existing slipways and upgrading other facilities at the shipyards.

Intermediate Conversions

The new programme was given high priority. It involved the parallel construction of three new types, two of which had similar capabilities, in four different shipyards. The modified 'Kresta II' could be pressed into service fairly rapidly, but the design process for the other two types would take somewhat longer, and ships were urgently needed to perform the new missions. By the mid-1960s only seven 'Kashins', now reclassified as Large Anti-Submarine Ships, had been completed, with a further ten due to enter service by the end of the decade. These were the only ships with modern hull sonars and a full outfit of automatically reloading anti-submarine rocket-launchers, and they would have to be distributed between four fleets. There were also eleven 'Kotlins' which had been modernized from 1960 to 1962 with RBU-2500 and RBU-600 manually loaded rocket-launchers, and which again had to be distributed between the four Soviet fleets.

In addition to the programme of new construction, the fruits of which would not be available for some years yet, the Soviets therefore resorted to the customary expedient of an intermediate programme of conversions for existing vessels. Some of the weapons systems employed to convert these ships may have become available as a result of the premature halt in the 'Kynda' construction programme.[3]

The first ships to undergo conversion were eight previously unmodified units of the 'Kotlin' class, the first of which was taken in hand probably in 1966–7. The after guns were removed and an SA-N-1 area defence missile system installed. The conversion was similar to that extended to the SA-N-1 trials ship *Bravy* in the early 1960s except that the funnel remained unmodified, and a Head Net C back-to-back air surveillance radar antenna replaced Head Net A. The midships 45mm gun mountings were suppressed and were later replaced in four ships by twin 30mm gun mountings controlled by Drum Tilt fire control radars; the forward 45mm mounting was replaced by a 57mm mounting which was removed from the 'Krupny' class (*see* below). A further important difference between these later conversions and *Bravy* was that they were fitted with two automatically reloading RBU-6000 rocket-launchers forward of the bridge in place of the manually reloaded RBU-2500.

Extensive as the conversion was, the 'Kotlin-SAM' retained the traditional EM classification, probably

▼At the lower end of the high/low mix of BPKs was the 'Krivak', equipped with the latest underwater weaponry and sensors but lacking the area air defence capabilities of her larger contemporaries. An elegant, compact design, the 'Krivak' was built in shipyards that had previously been responsible only for the construction of coastal types. (MoD, 1982)

because rebuilding did not extend to fitting improved submarine detection capabilities in the form of a modern hull sonar.

One or two ships were converted every year from 1966 onwards, the last conversion being completed in 1972. The Soviets do not appear to have regarded the SA-N-1 as in any way inadequate or obsolescent; it continued to be fitted in other converted destroyers up to 1977. Each 'Kotlin-SAM' represented the equivalent of half a 'Kashin' and therefore served as a useful supplement to the latter programme. The 'Kotlin' conversions also retained their forward 130mm twin gun mounting, which gave them a useful fire support capability which the more modern ships did not possess.

The other type to undergo conversion was the 'Krupny', which was rebuilt so extensively that it was subsequently reclassified 'Kanin' by NATO. The first ship was taken in hand at the Zhdanov Shipyard in 1967–8, to be followed by five other units, the last of this initial batch of conversions being completed in 1973. The two remaining unmodified units, *Gnevny* and *Gordy*, which were in service with the Pacific Fleet, were subsequently taken in hand at Vladivostok and their conversion had been completed by 1977.

The topsides of all ships were virtually stripped, both SS-N-1 launchers and their respective magazines being removed. They were given a remodelled bow which was sharply raked with the anchor hawsepipes well forward. This permitted the fitting of the new Bull Nose medium-frequency sonar due to be installed in the new BPKs. Two of the original four 57mm gun mountings were re-sited on superimposed deckhouses forward of the bridge. Arrangements at the after end of the ship were similar to those of the 'Kotlin-SAM', with a single SA-N-1 missile installation on a new deckhouse, and its Peel Group guidance radar at the top of a short tower mast forward of the second funnel. However, the greater length of the 'Krupny' also allowed the fitting of a raised platform extending from the after end of the deckhouse which could be used for operating either a land-based anti-submarine helicopter or for 'cross-decking' helicopters carried by other accompanying surface combatants.

Amidships, the triple side-mounted torpedo tubes of the 'Krupny' were replaced by quintuple mountings. Twin 30mm gun mountings controlled by Drum Tilt fire control radars would later be fitted in all ships. Even more significant was the replacement of the original hand-loaded RBU-2500 anti-submarine rocket-launchers by three automatically reloading RBU-6000, of which one was located close to the bow, while the other two replaced the midships 57mm gun mountings abeam the tower mast.

In the same way that the weapons systems for the 'Kotlin' conversions were made available as a result of the curtailment of the 'Kynda' programme, it seems

▼Eight destroyers of the 'Kotlin' class were taken in hand in 1966–7 for conversion to the 'Kotlin-SAM' configuration, the modifications being broadly based on the SA-N-1 trials ship *Bravy*. This is the former *Spravedlivy*, transferred to Poland in 1970 as *Warszawa*. She and her sister *Skromny* were fitted with the older-model RBU-2500 anti-submarine rocket-launcher, but the other six conversions received the automatically reloading RBU-6000. (C&S Taylor, 1975)

'Kanin' Class: Profile

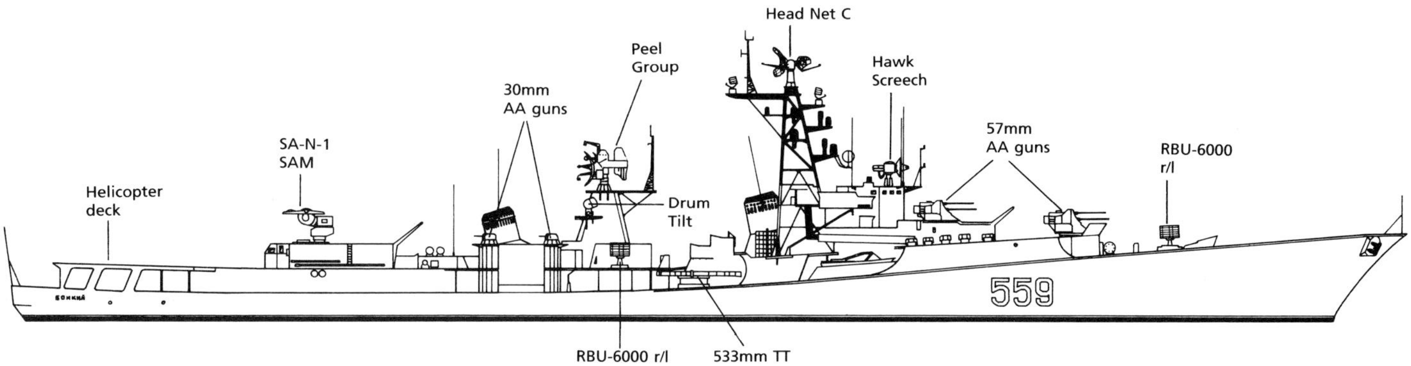

likely that the SA-N-1 missile systems fitted in the 'Kanin' were released by the decision to equip the 'Kresta II' BPKs with the more capable SA-N-3 system initially developed for the *Moskvas*.

Unlike the 'Kotlin-SAM' conversions, the 'Kanins' were reclassified as Large Anti-Submarine Ships (BPK) when they re-entered service. It therefore needs to be stressed that these ships were not simply re-designed as a vehicle for SA-N-1 missile systems which became available; they were modified primarily for anti-submarine warfare, and were given major new capabilities – a new hull sonar and a helicopter deck – which were lacking in the 'Kotlin-SAM'. Moreover, they carried an enhanced payload of anti-submarine weapons: three, rather than two, RBU-6000 rocket-launchers; and two quintuple banks of dual-capable torpedo tubes.

Fitting the SA-N-1 missile system was necessary to enable the ships to operate with the new BPKs in open-ocean areas. It is perhaps significant that whereas the 'Kotlin-SAM' conversions were subsequently distributed between the four fleet areas, the 'Kanins' were divided between the Northern Fleet (five ships) and the Pacific Fleet (three ships). They were to remain in service until the late 1980s when they were sold for scrap.

The 'Kresta II'

The decision to proceed with a new generation of purpose-built BPKs was probably taken in about 1965 when two (possibly three) Rocket Cruisers of the 'Kresta' class had already been laid down. Various alternative BPK designs were presumably considered including designs based on existing types.

The 'Kresta' proved eminently suitable for conversion to a BPK. There was adequate space and weight

'Kanin' Class

Soviet Designation: Bol'shoy Protivolodochny Korabl' (BPK) – Large Anti-Submarine Ship

Construction

Name	Converted	Refit dates
Gremyashchy	Zhdanov Shipyard	
Zhguchy	Zhdanov Shipyard	
Uporny	Zhdanov Shipyard	1968–74
Derzky	Zhdanov Shipyard	
Zorky	Zhdanov Shipyard	
Boyky	Zhdanov Shipyard	
Gordy	Vladivostok	1974–7
Gnevny	Vladivostok	

Displacement (tons)
Standard: 3,660
Full Load: 4,650

Dimensions
Length: 140.5m(470ft) overall
Beam: 14.6m(48ft)
Draught: 5.8m(19ft)

Propulsion
2-shaft geared steam turbines; 4 boilers; 72,000shp=34 knots max.

Complement
300

Weapons
ASW: 3 RBU-6000 rocket-launchers; 2 quintuple banks of 533mm(21in) torpedo tubes
AAW: 1 twin-arm SA-N-1 launcher (16 Goa missiles); 2 quadruple 57mm guns; four twin 30mm AA guns

Sensors
Surveillance radars: Head Net C 3-D, 2 Don-Kay
Fire control radars: 1 Peel Group (SA-N-1); 1 Hawk Screech (57mm); 2 Drum Tilt (30mm)
Sonars: bow-mounted Bull Nose MF
Electronic warfare: 2 Bell Slam, 2 Bell Tap, 4 Top Hat, 4 Bell series

◀ The 'Kotlin' conversions were closely followed by a more drastic rebuilding programme for the 'Krupny', which received not only the SA-N-1 area defence missile system, but a completely new outfit of anti-submarine weaponry and sensors. The bow was remodelled to accept the new Bull Nose MF bow sonar, which provided target data for three RBU-6000 rocket-launchers and anti-submarine homing torpedoes launched from quintuple banks of dual-capable tubes. As converted, the ships were reclassified by the Soviet Navy as Large Anti-Submarine Ships (BPKs) and received the NATO designation 'Kanin'. (US Navy, 1973)

for a 'double-ended' area defence SAM system, a 'containerized' missile system beneath the bridge wings, and a flight deck and hangar to support a helicopter, in addition to the customary close-range air defence and anti-submarine weapons. Moreover, all these systems could be accommodated with minimal structural changes.

With the abandonment of a long-range anti-ship missile system, it was possible to substitute the standard anti-submarine 'A' variant of the Ka-25 Hormone helicopter soon to enter service aboard the *Moskvas* for the Hormone B missile targeting variant intended for the new Rocket Cruisers.

The provision of a suitable stand-off anti-submarine missile system capable of countering fast-manœuvring nuclear-powered submarines proved more problematical. Such a system was already in the process of development for the *Moskva* class in the form of the FRAS-1 missile and its SUW-N-1 twin-arm launcher. However, installation of the latter aboard the 'Kresta' was possible only if the area air defence capability were halved because the SUW-N-1 launcher, which was reloadable and therefore required space for a magazine ring beneath it, would effectively replace the forward SAM launcher.

This does not appear to have been the only reason for rejecting the FRAS-1 missile as unsuitable for the new BPKs. The FRAS-1, initially intended for use against the Polaris submarine, was nuclear-tipped and employed inertial guidance. The concept dated from the 1959 Defence Review which focused primarily on the implications of all-out nuclear war. Clearly, against a Polaris submarine taking up position for a nuclear strike against the Soviet Union, a nuclear-tipped missile was both legitimate and, in purely tactical terms, likely to prove more effective than a comparable weapon with a payload of high-explosives.

The early 1960s were to see important changes in Western theories of nuclear war which could not fail to have an impact on Soviet thinking. In the context of a 'limited' nuclear war, in which a conflict might proceed by a series of graduated 'escalatory' steps, any weapon that could deliver only a nuclear payload would have its use restricted by political considerations. These would far outweigh the relative effectiveness of such a weapon from a purely tactical standpoint.

Moreover, the new BPKs were not intended exclusively for hunting SSBNs. The pro-submarine mission would involve them in operations against nuclear-powered attack submarines (SSNs) attempting to penetrate the Soviet bastions, and diesel-powered patrol submarines attempting to intercept Soviet ballistic missile and cruise missile submarines as they sortied into the Norwegian Sea. In the context of limited nuclear war, which might now be preceded by a prolonged conventional phase, these operations were at an altogether lower level of 'escalation' than direct attacks on Polaris submarines. They would therefore have to be performed using conventional weapons, at least in the early stages of a conflict. The FRAS-1 missile had insufficient flexibility and would therefore have to be complemented by a weapon with a conventional high-explosive payload.

Western equivalents of such a missile were already in service although there were two different strands of development. The US Navy had opted for a dual-capable missile in the form of ASROC which, like the FRAS-1, was a simple rocket with a ballistic trajectory and could carry alternative nuclear and conventional homing torpedo payloads. The combination of inertial guidance with a conventional payload was possible only because ASROC had a relatively short range (5.4nm) tailored to the performance of the SQS-23 sonar. The French Malafon and the Australian Ikara (subsequently adopted by the British Royal Navy), on the other hand, were designed to accept only a homing torpedo payload. They had greater range than ASROC and were susceptible to mid-course correction based on sonar updates. In theory this meant that the homing torpedo could be dropped closer to the true position of the submarine, thereby facilitating acquisition of the target by the acoustic homing head of the torpedo as it performed its underwater search.

Longer range for a ship-borne anti-submarine missile, however, was of value only if there were a corresponding improvement in sonar performance or if collaborative tactics, using sophisticated data links, were used. Malafon had a maximum range of 6.5nm – little better than ASROC – while even Ikara, which was designed to take account of future developments in sonar, was limited to 10nm.

The SS-N-14 Missile

The Soviet SS-N-14 evolved to meet the new requirement was similar in conception to the French and Australian missiles in that it could be guided while in flight. It was built around a specially designed homing torpedo (NATO designation: E45–70A), midway in size between the 'lightweight' 324mm torpedoes used in the US and Australian missiles and the larger 533mm L4 torpedo employed by the French Navy. With a diameter of 450mm, the E45–70A is just under 4m long and has electrical propulsion based on silver zinc batteries. Active/passive acoustic homing guides the torpedo to its target and the warhead is reported to be activated by impact or magnetic influence triggers.[4]

The missile itself is significantly larger than any of its Western counterparts and is fired from cylindrical launchers some 8m long and 1.75m in diameter. In the 'Kresta II' eight missiles are carried, housed in two quadruple 'box' launchers mounted at a fixed angle beneath the bridge wings.

The similarity between the launchers employed for the SS-N-14 missile and those used for Soviet anti-ship missiles such as the SS-N-3, the SS-N-2B and the SS-N-9 (in service 1969–70 on the Small Rocket Ships of the 'Nanuchka' class) initially led Western Intelligence to make an erroneous assessment of the nature of the missile and the capabilities of the new BPKs. The missile was originally typed SS-N-10 by NATO and stated to be a new horizon-range anti-ship missile. The error was no doubt prompted by the size of the launchers which clearly housed a missile considerably larger than any of the Western anti-submarine missiles, and by

'Kresta II' Class: Profile and Plan

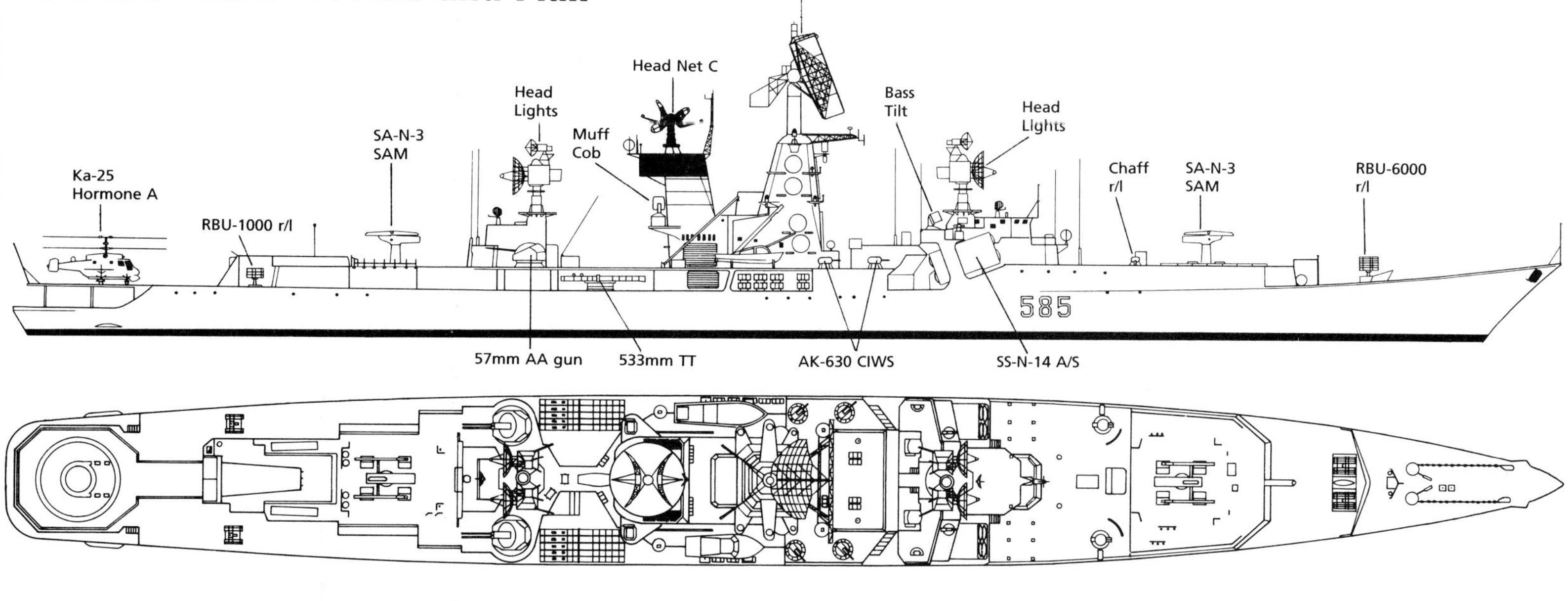

their location on the 'Kresta II', where they replaced the twin SS-N-3 launchers of the 'Kresta I' beneath the bridge wings.

The truth of the matter was revealed in about 1974 when a NATO helicopter swooped on a BPK with launcher doors open, and photographs clearly showed the head of a homing torpedo incorporated into the missile body. But it was to be a further three years before the correction began to filter into reference sources, by which time many commentators had decided that the 'Kresta II' was simply a natural development of the 'Kresta I', rather than a radically new type.

The comparatively large size of the SS-N-14 missile suggests a maximum range well in excess of its Western counterparts. The figure given in most reference sources is 30nm (55km), approximately three times that of Ikara and five times that of Malafon and ASROC. This must also have contributed to the initial confusion regarding the missile's configuration and performance parameters, it being by no means clear how the Soviets could exploit this range in an anti-submarine missile.

The maximum range of the SS-N-14's Western counterparts corresponded closely with the detection ranges of current or projected sonars, and Soviet sonar technology of the late 1960s was thought to be nowhere near as advanced. The Bull Nose bow sonar installed in the 'Kresta II', although of relatively modern design, continues to be classified as medium frequency in all open-source publications, and would certainly not have been in the same class as the US Navy's SQS-26 and its derivatives. The Soviets may have envisaged 'convergence zone' sonar operation as a future possibility and developed a missile which could take advantage of technological progress in acoustic detection, but the sonar actually installed in the 'Kresta II' would have been capable of targeting out to a maximum of perhaps 8,000–10,000m – i.e., close to the *minimum* range of the missile, which is generally estimated to be 4nm (7,400m).

It has therefore been assumed that 'collaborative' targeting would be used, using target data provided by other units operating in the same area. This would fit

'Kresta II' Class

Soviet Designation: Bol'shoy Protivolodochny Korabl' (BPK) – Large Anti-Submarine Ship

Construction

Name	Builder	Laid down	Launched	In service
Kronshtadt	Zhdanov Shipyard	June 1966	Feb 1968	Sept 1969
Admiral Isakov	Zhdanov Shipyard	June 1966	Dec 1968	Sept 1970
Admiral Nakhimov	Zhdanov Shipyard	May 1967	Apr 1969	Aug 1971
Admiral Makarov	Zhdanov Shipyard	Mar 1968	Apr 1970	June 1972
Marshal Voroshilov	Zhdanov Shipyard	Jan 1969	Nov 1970	May 1973
Admiral Oktyabrsky	Zhdanov Shipyard	May 1969	June 1971	Sept 1973
Admiral Isachenkov	Zhdanov Shipyard	Feb 1970	Mar 1972	Aug 1974
Marshal Timoshekno	Zhdanov Shipyard	June 1971	Oct 1973	Aug 1975
Vasily Chapaev	Zhdanov Shipyard	Feb 1972	Jan 1975	Sept 1976
Admiral Yumashev	Zhdanov Shipyard	Nov 1973	Oct 1976	Oct 1977

Displacement (tons)
Standard: 6,200
Full Load: 7,700

Dimensions
Length: 158.5m(521ft) overall
Beam: 17m(56ft)
Draught: 6m(20ft)

Propulsion
2-shaft geared steam turbines; 4 turbo-pressurized boilers; 100,000shp=34 knots max.

Complement
380

Weapons
ASW: 2 quadruple SS-N-14 launchers (8 Silex missiles); 1 Ka-25 Hormone A helicopter; 2 RBU-6000 rocket-launchers; 2 RBU-1000 rocket-launchers; 2 quintuple banks of 533mm(21in) torpedo tubes
AAW: 2 twin-arm SA-N-3 launchers (44 Goblet missiles); 2 twin 57mm guns; 4 AK-630 gatlings.

Sensors
Surveillance radars: Top Sail 3-D, Head Net C 3-D; 2 Don-Kay, 1 Don-2
Fire control: 2 Head Lights C (SA-N-3/SS-N-14); 2 Muff Cob (57mm); 2 Bass Tilt (AK-630)
Sonars: bow-mounted Bull Nose MF
Electronic warfare: 8 Side Globe; 1 Bell Clout, 2 Bell Slam, 2 Bell Tap, 2 Bell Crown; 2 twin-barrelled chaff-launchers

The SS-N-14 Silex Missile System

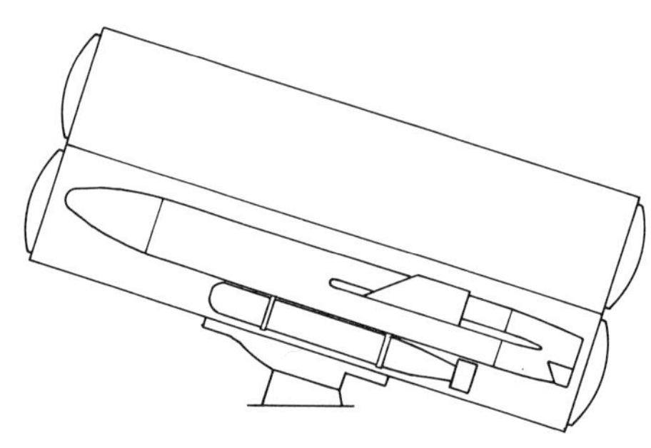

In service: 1970
Length: 7.6m(25ft)
Diameter: 0.85m(2ft 9in)
Span: 3m(9ft 10in)
Weight: 3,200kg(7,050lb) approx.
Payload: E45-70A homing torpedo/5kT nuclear
Propulsion: solid-fuel rocket
Speed: Mach 0.95
Range: 7–55km(4–30nm)
Guidance: command
Fire control: Eye Bowl or Head Lights B

well into the Soviet 'combined arms' philosophy, with surface ships, submarines and maritime patrol aircraft providing mutual support. Significantly, the first of a new generation of LRMP aircraft, the Il-38 May, would enter squadron service in 1969 – one year before the first of the 'Kresta IIs'. Equipment included sonobuoys which could be used to localize a fast-moving contact. The aircraft could then mount is own attack using depth-bombs and homing torpedoes, or could call down fire from friendly surface units armed with the long-range SS-N-14.

Clearly, the effectiveness of 'combined arms' anti-submarine warfare employing LRMP aircraft in conjunction with surface units is dependent on the availability of suitable airfields within the radius of the area of operations. It is therefore of interest to note that during the early 1970s the Soviet Navy operated Il-38 Mays out of Egyptian airfields over the eastern Mediterranean.

A much-neglected feature of the 'Kresta II' concept is the ability to use its single onboard helicopter to assist missile targeting at longer ranges. Although many NATO navies had adopted the manned helicopter for anti-submarine frigates built during the 1960s, these helicopters were smaller and very different in conception from the anti-submarine 'A' variant of the Soviet Ka-25. The British MATCH (MAnned Torpedo-Carrying Helicopter) concept, based on the diminutive Westland Wasp, viewed the helicopter simply as a weapons carrier. It was flown under control from the frigate out to the position of the submarine established by the ship's own sonar, and was then directed to release its homing torpedoes. The helicopter had no detection sensors of its own and relied exclusively on the frigate to provide updated sonar data on the target. It was in effect a re-usable alternative to the anti-submarine missile. Other Western navies, notably those of France and Italy, developed similar systems.

The Ka-25 Hormone differed in that it was equipped with its own detection apparatus in the form of an

▶ ***Admiral Makarov*, the fourth ship of the 'Kresta II' class, as she appeared in 1974. The overall layout of the 'Kresta I' has been retained, despite major changes in her primary weapons systems. (US Navy)**

Oka-2 dunking sonar and a towed magnetic anomaly detector boom, and could dispense sonobuoys from the under-fuselage weapons bay. This made the helicopter more independent of the ship; it could localize and classify long-distance contacts. In theory it could also make the initial contact, although this was less likely given the high frequency at which its dunking sonar operated. However, it could also respond to contacts established by other vessels operating in the area. It was therefore able to provide targeting data for vessels perhaps 20–30nm distant. This would enable any surface ship in the area to deploy its SS-N-14 missiles over their full range without the necessity for direct sonar contact with the target.

Air Defence

Many of the other physical changes in the 'Kresta II' design relate to the adoption of the SA-N-3 air defence system in place of the SA-N-1 system installed in the 'Kresta I'.

The Goblet missile was markedly superior in every respect to Goa and the decision to adopt it as the area defence weapon of the new BPKs is unsurprising. However, the Goblet missile guidance system, which was dependent on the massive Top Sail antenna for multi-target tracking in three dimensions, had been designed primarily for installation in the *Moskvas*, which were much larger ships. Indeed, there has been speculation to the effect that *Moskva* and *Leningrad* were to have been the first ships of a larger class, numbering at least four units, and that cancellation of the third and fourth units released the SA-N-3 systems which were subsequently fitted in the 'Kresta IIs'.[5]

What made installation possible was the abandonment of the SS-N-3 anti-ship missile, together with the bulky Scoop Pair guidance radar which provided tracking and mid-course correction. In the 'Kresta I' this radar had been mounted high on the face of the central tower mast, and was largely responsible for the large size of the latter. The SS-N-14 missile did not require a dedicated fire control radar and could use the same Head Lights guidance radar which provided fire control for Goblet with a minimum of modification. This enabled the tower mast of the 'Kresta II' to be cut down in size to compensate the added weight of Top Sail. The Head Net C air surveillance and tracking radar was relocated on the platform between the funnel uptakes in place of the Big Net antenna of the 'Kresta I'.

Anti-Missile Guns

The 'Kresta II' featured an electronic warfare suite comparable to that of the *Moskva* and 'Kresta I' classes, and was fitted with twin-tube chaff rocket-launchers

◀ A bow view of *Admiral Oktyabrsky* taken in the English Channel in 1974. Note the quadruple box launchers for the SS-N-14 anti-submarine missile beneath the cantilevered bridge wings. Just abaft the forward SA-N-3 missile-launcher, protected by blast shields, are twin-barrelled chaff-launchers for anti-missile defence. (Skyfotos)

for active countermeasures. A new departure, however, was the provision of a close-in anti-missile system based on small-calibre, multi-barrel guns.

The AK-630 gun mounting (NATO designation: ADMG-630) is approximately 2m in diameter and features a six-barrelled gatling gun with a rate of fire estimated at 3,000rpm. The barrels, which are encased in a cylindrical sleeve, are of 30mm calibre and have a continuous belt feed system which picks up ammunition rounds from a magazine located in the deckhouse on which the gun is mounted. The upper casing of the gun housing, which is hinged at its after end to facilitate maintenance, is probably made of GRP or light alloy.

The multi-barrel gun can be elevated to 90 degrees and can be depressed to compensate movements of the ship. Operation is fully automatic. Fire control is provided by a drum-shaped FC radar designated Bass Tilt, which is mounted separately and not on the mounting itself, as on later Western systems. This inevitably slows the reaction of the system and poses problems of alignment which would not exist with an on-mount radar, but it enables a single FC radar to control a pair of gun mountings.

Two pairs of AK-630 gun mountings are fitted in 'Kresta II', controlled by Bass Tilt radars to port and starboard. The gun mountings are installed on deckhouses extending outboard from the first superstructure deck at the base of the tower mast. The Bass Tilt radars are mounted on pedestals at the after end of the bridge structure, the Tee Plinth electro-optical sensors being relocated at the forward end of the bridge wings. As a result of either development or production problems, the first four units of the class entered service without Bass Tilt and the radars have never been fitted.

Immediately inboard of the AK-630 gatlings is a single-storey deckhouse, at the after end of which are circular tubs for two Kolonka optical sights, to be used for back-up in a heavy ECM environment or in the event of failure of the Bass Tilt radars. The last three ships were completed with a two-storey deckhouse – a feature also of the modified 'Kyndas' and 'Kresta Is' (*see* Chapter 3). This presumably houses either additional accommodation or additional electronics.

The AK-630 Anti-Missile Gun System

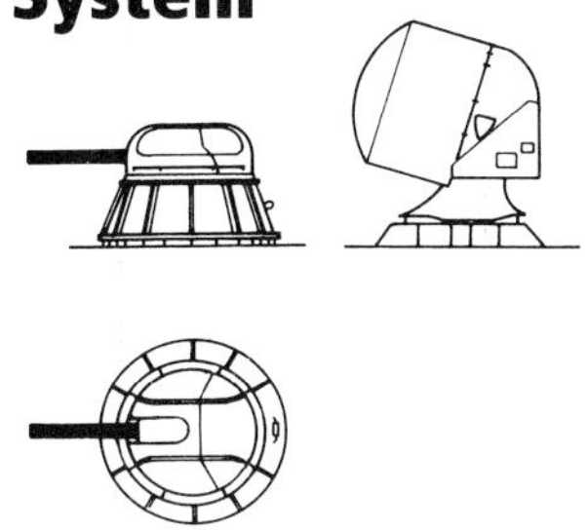

Soviet designation: AK-630
In service: 1970
Angle of elevation: +90/−10 degrees
Rate of fire: 3,000rpm per mounting
Projectile weight: ?
Muzzle velocity: 1,000m/sec(3,280ft/sec)
Range: 2.5km(1.3nm)
Fire control: Bass Tilt

▶ *Admiral Isachenkov* in northern waters. Note the broad funnel structure with its paired boiler uptakes, and the distinctive Side Globe ECM antennae. (MoD, 1980)

Helicopter Hangar Arrangement

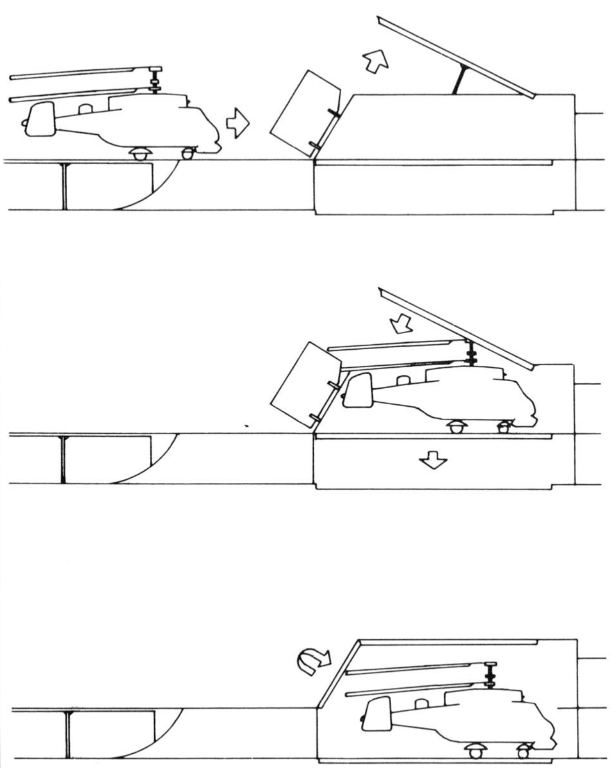

▲The after Head Lights missile guidance radar of a 'Kresta II' BPK. All except the first ship, *Kronshtadt*, are fitted with the Head Lights B variant, which incorporates fire control for the SS-N-14 anti-submarine missile. Beneath the Head Lights installation is the starboard-side AK-257 twin 57mm automatic gun mounting. Fire control for the latter is provided by Muff Cob, seen here on the left of the picture. (US Navy, 1971)

Helicopter Arrangements

Locating the helicopter deck on the low quarterdeck of the 'Kresta I' clearly proved unsatisfactory and arrangements were modified in the 'Kresta II' and in later types, a new raised flight deck being constructed on stanchions above the stern. This not only served to make helicopter operations less dependent on favourable sea conditions, but the flight deck was cleared of capstans and other obstructions associated with line-handling – a solution favoured by west European navies for the new generation of surface ships built in the 1970s.

Unfortunately it proved impossible to raise the hangar floor to the same level because the roof of the hangar would then have blocked the stern arcs of the after SA-N-3 launcher. The designers were therefore compelled to resort to a complex mechanical transfer arrangement involving a split-level hangar. First, the hangar roof, which is hinged at its forward end, has to be raised by hydraulic rams to an angle of some 30 degrees (*see* accompanying diagram). The helicopter is then wheeled into the hangar and secured. The hangar floor serves as a lift between flight deck and quarterdeck level. Once it has descended the hangar roof can be lowered and secured and the hangar doors closed.

This complex arrangement places an additional maintenance workload on the ship's crew and is vulnerable to breakdown, but the Soviet Navy has persisted with it in later types because of the difficulty of accommodating a helicopter with superimposed rotors in ships with 'double-ended' SAM systems.

Service Life

The ten units of the 'Kresta II' class were built exclusively by the Zhdanov Shipyard, construction

▲The 'Kresta II' was the first Soviet class to deploy with the AK-630 anti-missile gun, which was to be standard equipment for all subsequent major surface vessels until the late 1970s, when it was replaced by the CADS-N-1 combined missile/gun system. Seen here is the Pacific-based *Vasily Chapaev*, which also has the two-storey deckhouse between the bridge structure and the tower mast which is a feature of later units of the class. Note the forward Head Lights radar, which is raised in the tracking mode. (US Navy, 1985)

▼The excellent sea-keeping qualities of the 'Kresta II', which is fitted with active fin stabilizers, have made the design eminently suited to operations in the difficult weather conditions which predominate in northern waters. (CinC Fleet, 1975)

ending in 1978. The only significant modification made to the design was the additional deckhouse between the bridge structure and the tower mast incorporated in the last three ships. This suggests that any growth margin there might have been in the original 'Kresta' design was fully taken up by the modifications made to equip the 'Kresta II' for anti-submarine warfare in open waters, and that the final design was somewhat cramped. The new, purpose-built BPKs of the 'Kara' class would have a larger hull and displacement would increase by about 2,000 tons over Kresta'.

The first four units of the 'Kresta II' class deployed to the Soviet Northern Fleet shortly after completion. The next two ships, *Marshal Voroshilov* and *Admiral Oktyabrsky*, joined the Pacific Fleet in 1974 following sea trials and work-up in the Baltic. They were the only major anti-submarine units to serve with the Pacific Fleet until 1978 when a third ship, *Vasily Chapaev*, was transferred to the Far East. The remaining three units all entered service with the Northern Fleet, and 'Kresta II' deployments have remained relatively stable during the 1980s, with the Northern Fleet ships periodically returning to the Baltic for refitting.

The Purpose-Built BPKs

The 'Kara' and 'Krivak' which, unlike the 'Kresta II', were designed as BPKs from the keel up, entered service somewhat later and were similar in a number of respects.

The most striking feature they had in common was that both were powered by gas turbines, whereas the 'Kresta II' had the steam-propulsion machinery of its Rocket Cruiser predecessors. Gas turbines were by no means new to the Soviet Navy, but the adoption of all-gas-turbine propulsion for a major surface unit such as the 'Kara' was clearly an important step. The decision was undoubtedly influenced by the success of the 'Kashin' which was designed and built at the same 61 Kommuna Shipyard.

However, there were other common elements in the new designs that were not features of the 'Kresta II' conversion. Both the 'Kara' and 'Krivak' were equipped with the new SA-N-4 short-range surface-to-air missile system. For the 'Kara' the SA-N-4 was a secondary air defence system which closed the gap between the SA-N-3 area defence system and the AK-630 close-in anti-missile system. For the 'Krivak' it was the primary air defence weapon, these ships being too small to accommodate an area defence missile system.

There was also a return to the twin 76.2mm dual-purpose gun mounting in the 'Kara' and 'Krivak', in preference to the twin 57mm automatic AA mounting installed in the *Moskvas* and 'Krestas'. The larger gun

◀In this stern view of a 'Kresta II', taken during the NATO Exercise 'Ocean Safari' in 1975, the twin doors of the hangar are open and the hinged roof section is raised ready to receive the Ka-25 Hormone A helicopter. In the centre of the hangar is the lift, which is lowered to become the hangar floor. Note the mesh mat covering the landing pad. (HMS *Ark Royal*)

was more effective in the anti-surface role and was therefore a particularly useful weapon for the smaller 'Krivaks' which might be faced by fast patrol boats armed with guns of the same calibre while conducting anti-submarine patrol or escort duties in the Baltic or the Black Sea.

The other common feature of the new ships was that both were fitted with a variable depth sonar. This was derived from the model trialled in the *Moskvas*, and would be particularly valuable in the relatively shallow waters of the Baltic, the Black Sea and the Mediterranean. There was nothing in the 'Kresta II' design (except perhaps weight considerations) to preclude the installation of VDS. However, these ships were designed for service in the North Atlantic and Pacific where the depth of water and the related thermal conditions presented fewer problems for hull-mounted sonar. The 'Kara', on the other hand, appears to have been designed primarily for service in the eastern Mediterranean, and the 'Krivaks' would be distributed between all four Soviet fleets. Variable depth sonar therefore constituted a useful addition to the overall capabilities of these ships.

The 'Krivak'

A fundamental requirement of the 'Krivak' design was that it could be built in shipyards not traditionally associated with the construction of major surface warships. Simplicity was therefore the keynote, and the hull was the minimum platform capable of supporting the required weapons systems.

Even so the 'Krivak' is by no means a small ship, and is comparable in size to ships in the destroyer/frigate class built by west European navies during the late 1970s and early 1980s. Length overall is 125m, beam 14.3m and standard displacement is estimated at 3,300 tons. This represents a significant increase in size over the SKRs of the 'Petya' and 'Mirka' classes (82m and 950 tons) previously built at the Kaliningrad Shipyard, and the 'Poti'-class MPKs (60m and 500 tons) built at Kamysh-Burun.

The appearance of the 'Krivak' is neat and workmanlike, and marks a return to the style of naval architecture exemplified by the 'Kynda'. The hull-form has the full waterplane common to all modern Soviet warships, with a broad transom stern. There is a long forecastle deck with considerable sheer and flare forward, breaking on to a low quarterdeck. A prominent knuckle runs from just forward of the bridge structure to the break in the forecastle. It is a hull better designed for seakeeping than that of the larger 'Kashin' class, with significantly higher freeboard amidships.

In contrast to the rambling superstructures of the 'Kresta' there is a single large superstructure block, of rectangular configuration. The latter incorporates the navigation bridge and is surmounted by a broad, angular lattice mast supporting the air surveillance antenna and fire control radars. Farther aft is a lower single-deck structure incorporating the intakes and uptakes for the gas-turbine propulsion machinery, and additional platforms for fire control radars. The low, squat funnel, which is positioned well aft, contributes to the sleek, racy appearance of the 'Krivak'. As in the 'Kynda', which also favoured the same block superstructures, the weapons are mounted on the centre-line forward and aft, the sole exceptions being the torpedo tubes which are amidships.

'Krivak I' Class: Profile and Plan

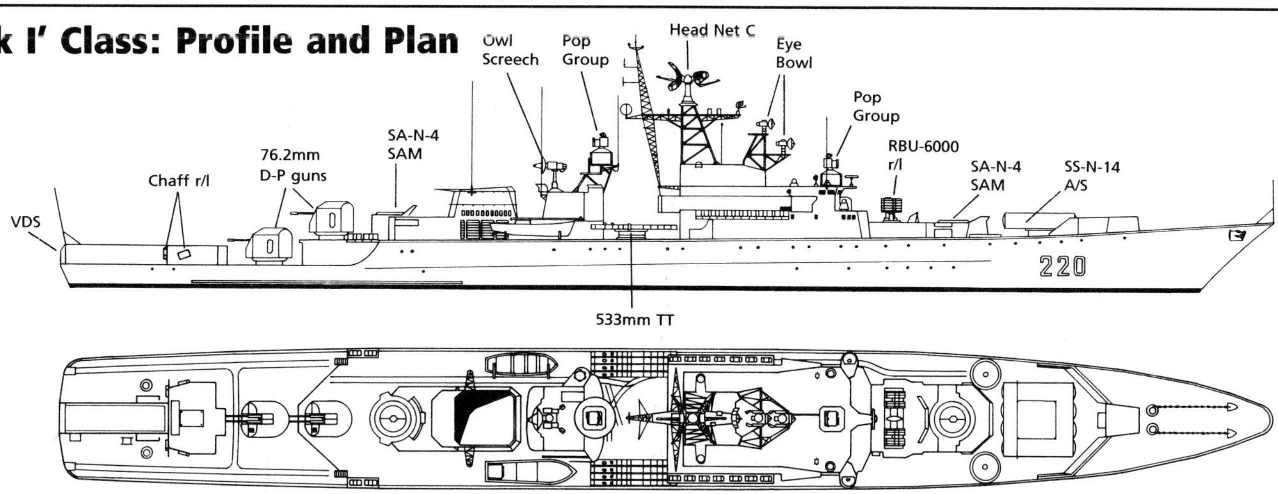

'Krivak I' Class

Soviet Designation: Storozhevoy Korabl' (SKR) – Patrol Ship

Construction

Name	Builder	In service
Bditelny	Kaliningrad Shipyard	1970
Dostoyny	Kamysh-Burun Shipyard	1971
Bodry	Kaliningrad Shipyard	1971
Svirepy	Kaliningrad Shipyard	1971
Silny	Kaliningrad Shipyard	1972
Doblestny	Kamysh-Burun Shipyard	1972
Storozhevoy	Kaliningrad Shipyard	1973
Razumny	Kaliningrad Shipyard	1973
Razyashchy	Kaliningrad Shipyard	1974
Druzhny	Zhdanov Shipyard	1975
Deyatelny	Kamysh-Burun Shipyard	1975
Zharky	Kaliningrad Shipyard	1975
Retivy	Zhdanov Shipyard	1976
Leningradsky Komsomolets	Zhdanov Shipyard	1976
Letuchy	Zhdanov Shipyard	1977
Bezzavetny	Kamysh-Burun Shipyard	1978
Pylky	Zhdanov Shipyard	1979
Zadorny	Zhdanov Shipyard	1979
Bezukoriznenny	Kamysh-Burun Shipyard	1980
Ladny	Kamysh-Burun Shipyard	1980
Poryvisty	Kamysh-Burun Shipyard	1982

Displacement (tons)
Standard: 3,075
Full Load: 3,575

Dimensions
Length: 125m(410ft) overall
Beam: 14.1m(46ft)
Draught: 4.5m(15ft)

Propulsion
2-shaft COGAG; 2 gas turbines each 24,300hp; 48,600hp=30 knots max.

Complement
180

Weapons
ASW: 1 quadruple SS-N-14 launcher (4 Silex missiles); 2 RBU-6000 rocket-launchers; 2 quadruple banks of 533mm(21in) torpedo tubes
AAW: 2 twin-arm SA-N-4 launchers (40 Gecko missiles); 2 twin 76.2mm(3in) D-P guns
ASuW: rails for 20 mines

Sensors
Surveillance radars: Head Net C 3-D; 1 Don-2 or Spin Trough, 1 Don-Kay or Palm Frond
Fire control: 2 Eye Bowl (SS-N-14); 2 Pop Group (SA-N-4); 1 Owl Screech (76.2mm)
Sonars: bow-mounted Bull Nose MF; Mare Tail MF VDS
Electronic warfare: 2 Bell Shroud, 2 Bell Squat; 4 16-tube chaff-launchers

Anti-Submarine Weaponry

The size limitation imposed on the 'Krivak' effectively precluded the installation of an area defence surface-to-air missile system. As a 'low-mix' type, the 'Krivak' was intended to supplement the 'Kresta II' and 'Kara' in the anti-submarine role, and would therefore rely on her larger consorts to provide long-range protection against hostile aircraft. For operations in the enclosed waters of the Baltic or the Black Sea, land-based air cover could be provided. A low-level, quick-reaction system was more appropriate in such an operational context. The 'Krivaks' would therefore have only short-range weapons for self-defence.

Even for the primary anti-submarine mission certain choices had to be made. It was possible to equip the ships either with the SS-N-14 missile, or for ASW helicopter operations, but not both. The ships were too small to accommodate side-mounted quadruple boxed launchers for the SS-N-14 in the manner of the 'Kresta' and 'Kara', so the missile-launchers would have to be located on the centre-line forward of the bridge. The gun mountings would have to be fitted on the centre-line aft, thereby precluding the installation of a hangar and flight deck for a helicopter.

The alternative solution was to install a flight deck and hangar aft, and to suppress the SS-N-14 missile-launchers, thereby enabling the anti-air and anti-surface systems to be fitted forward of the bridge. This was indeed the solution later adopted for the 'Krivak III' variant built for the KGB during the late 1980s (*see* Chapter 7), although it is not clear whether this was seriously considered when the 'Krivak' was designed.

The SS-N-14 missile system was initially preferred for a number of reasons. In the late 1960s the SS-N-14 was clearly regarded as the primary anti-submarine weapon of the BPKs, it being common to all three types. The ship-borne helicopter was as yet largely untried as an anti-submarine system and there were residual doubts about its ability to operate from a small, heaving flight deck in the adverse weather predominating in northern waters. As a small ship, the 'Krivak' was probably considered marginal for helicopter operations in such a climate; there would be many days in the year when the helicopter would be unusable. Anti-submarine missiles, on the other hand, were protected from the elements by their heated launchers and therefore provided a theoretical 100 per cent availability.

Moreover, the 'Krivaks' would not be operating independently, but in conjunction with other larger BPKs of the 'Kresta II' and 'Kara' classes, and possibly with larger helicopter-carrying vessels of the *Moskva* and *Kiev* classes. The Ka-25 Hormone helicopters operated by the latter ships could provide targeting data for all surface units in their area. Collaborative targeting using data from the various sonars fitted in ships of the group was also a well-established Soviet tactical doctrine. The 'Krivak' made a useful contribution in this respect in that it was fitted not only with a medium-frequency bow sonar (the same Bull Nose model fitted in the 'Kresta II'), but also with a variable depth sonar developed from the model initially trialled in *Moskva*. The towing mechanism for the latter is located within a prominent housing on the quarter-deck.

The SS-N-14 anti-submarine missiles are mounted in a 'flat four' arrangement reminiscent of the quadruple SS-N-3 launcher of the 'Kynda'. The launcher can be trained and elevated. It is mounted approximately half-way between the bow and the forward end of the bridge structure and is protected from the worst of the elements by a prominent bulwark, 1.75m high, which is angled out to the sides of the ship.

As in the larger BPKs, there are no reloads. This effectively limits the 'Krivak' to four missile rounds – a figure which has attracted some adverse comment from Western commentators. Even the basic US ASROC system has eight ready-use rounds, while ships fitted with reload magazines have twice or even three times that number available. The French and British vessels fitted with similar systems generally carry 12–13 missile rounds. The comparatively small number of missiles carried by their Soviet counterparts is related to the large size of the SS-N-14, which precludes fitting a reload magazine in all but the largest ships. Presumably the Soviets consider that this constraint is at least partially compensated by the long range of the missile combined with the concept of collaborative anti-submarine operations, in which it is the total number of missiles rounds carried by the group, not by the individual vessel, that is important.

Missile tracking and mid-course guidance for the SS-N-14 missile is provided by a pair of specialized guidance radars designated Eye Bowl. These are mounted on a short lattice structure on top of the bridge structure, and are not dissimilar in configuration to the upper (small) dishes of the Head Lights radar which performs the same SS-N-14 guidance function on ships equipped with the SA-N-3 area defence system.

▶For both the 'Krivak' (seen here) and the 'Kara' there was a return to the twin 76.2mm dual-purpose gun, and both ships were fitted with the Mare Tail variable depth sonar which had been tested in *Moskva*. (MoD, 1983)

◀The appearance of the 'Krivak' is neat and workmanlike, and she appears well-armed for her size. However, most of the weaponry is simply 'bolted on' for ease of installation, and she carries no reloads for her 'flat four' SS-N-14 launcher. Note the empty ECM platforms projecting from the forward end of the bridge structure. (MoD, 1983)

The 'Krivak' has two RBU-6000 anti-submarine rocket-launchers mounted side by side directly beneath the bridge, but there was clearly inadequate space to accommodate the six-barrelled RBU-1000 rocket-launchers which are a feature of both the 'Kresta' and the 'Kara'. Quadruple banks of torpedo tubes, capable of firing either anti-ship or anti-submarine weapons, are mounted to port and starboard in the space between the bridge structure and the air intakes for the gas turbines.

Air Defence

The primary air defence weapon of the 'Krivak' is the SA-N-4 Gecko missile. Developed simultaneously as a land-mobile and ship-borne missile, Gecko has a maximum effective range estimated at 9,000m. The naval system (Soviet designation: Osa-M) was first installed in the Small Anti-Submarine Ships (MPK) of the 'Grisha' class in 1968–9. Besides aircraft, it is probably capable of engaging anti-ship missiles and surface targets.[6]

The SA-N-4 system is unusual in that it employs a twin-arm 'pop-up' launcher which retracts into its own magazine. The magazine comprises a cylindrical 'bin' approximately 5m high and 3.5m in diameter which is countersunk into the deck. The central pillar of the launcher, which is broader at the top than at the base, protrudes above the launcher arms. The hexagonal plate on top of the pillar forms the centre of the bin lid when in the lowered position (*see* drawing). Semicircular plates covering the top of the magazine slide backwards to allow the launcher to be raised, then close around its base to form a weatherproof seal.

The launcher emerges with the missiles ready-loaded on to its twin arms, and has to be lowered into the magazine for reloading. Inside the magazine 'bin' the missiles are stowed in four vertical rings each holding five missiles. The inner positions of the magazine rings line up with the twin arms of the launcher, which reloads while trained to port or starboard (*see* drawing).

Unusually, the missiles are stowed in the nose-down position. This reduces the distance the launcher has to travel to pick up the missiles. It also means that in the event of a fire in the magazine the effects of an explosion could be minimized simply by opening the hinged doors.

The SA-N-4 Gecko Missile System

Soviet designation: Osa-M
In service: 1969
Length: 3.1m(10ft 2in)
Diameter: 0.21m(0ft 8in)
Span: 0.64m(2ft 1in)
Weight: 130kg(286lb)
Warhead: 18kg(40lb) HE fragmentation
Propulsion: dual-thrust, solid-fuel rocket
Speed: Mach 2+
Range: 9km(5nm)
Ceiling: 3,000m(10,000ft)
Min. altitude: 10m(30ft)
Guidance: radar/command
Fire control: Pop Group
Operating frequency: H/J Band

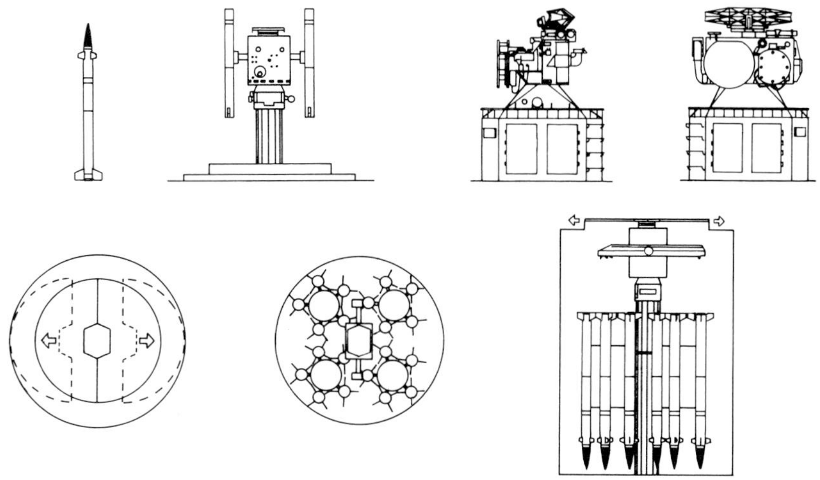

The 'pop-up' launcher concept has a number of advantages and disadvantages. The rationale appears to be that the launcher/magazine complex is weatherproof, thereby improving reliability in what is essentially a quick-reaction system. This is particularly advantageous in cold-weather operations when the missile can be deployed without lengthy warm-up procedures.

The launcher/magazine complex is also completely self-contained and can therefore be installed as a module straight from the factory; it requires only power connections and electronics interfaces with the ship's weapons control centre. This feature must have been particularly attractive given the decision to allocate the 'Krivak' to shipyards that had only limited experience in building surface warships.

▶The Eye Bowl fire control radars for the SS-N-14 missile are carried on a separate lattice structure above the bridge. This unit, *Leningradsky Komsomolets*, completed in 1976, has received her full ECM outfit comprising twin Bell Shroud ESM antennae at the forward end of the bridge structure, and Bell Squat on platforms at the foot of the foremast. (RDN, 1982)

The major disadvantage of the system is the complexity of the reloading arrangements which makes it less robust and more vulnerable to mechanical breakdown than the multiple box launchers adopted for comparable NATO systems such as Sea Sparrow and Sea Wolf. The need for the launcher to retract into the magazine bin in order to reload must also result in a substantial delay before the next pair of missiles can be fired. This is clearly less than satisfactory given that the SA-N-4 is essentially a quick-reaction system which would be expected to engage aircraft attacking in quick succession.

Presumably the 'double-ended' SA-N-4 installation of the 'Krivak' is deemed to compensate the comparatively slow reloading sequence of the launcher because this means that two targets can be each engaged with two missiles simultaneously. However, although the system continued to be fitted in major warships into the late 1980s, the 'pop-up' launcher was abandoned in the SA-N-4 system's lineal successor, the SA-N-9, in favour of vertical launch which is both simpler in mechanical terms and more effective in operation.

The Pop Group guidance radar is reported to use the frequency-hopping monopulse technique. Its adoption signified a great leap forward by the Soviets, placing their radar technology on a par with Western developments of the period. The Pop Group antenna comprises a box housing the transmitter/receiver units with numerous target and missile tracking antennae mounted on it. The whole assembly is stabilized and can be rotated in azimuth, the turning mechanism and power supplies being housed in the larger box structure which serves as its base.

On top of the assembly is a small eliptical reflector of open mesh construction. The reflector, which is about two metres across and can be rotated independently, operates in the H-band and performs a general target search and tracking function with an estimated 30km acquisition range. On the face of the assembly are two broadly circular cassegrain arrays. The larger array reportedly operates in the J-band and tracks the target, effective acquisition range being about 20km; the smaller array tracks the missile. The Osa-M missile has two beacon transponders on the rear fuselage side to facilitate missile tracking, correction commands from the launch ship being transmitted via a small rectangular antenna directly above the missile tracking array. Between the two circular arrays is an electro-optical/LLTV device to provide back-up in hostile ECM conditions.

On the 'Krivak' the forward Pop Group launcher is located on top of the navigation bridge; the second is mounted amidships on a platform above the air intakes for the gas turbines.

Electronic Countermeasures

Without the heavy plated tower mast, which is a feature of both the 'Kresta' and the 'Kara', installation of the bulky Side Globe ECM system was out of the question. A new suite of electronic countermeasures systems specifically designed for smaller surface units was therefore developed for the 'Krivak'.

▼Overhead view of the after section of the Pacific-based *Storozhevoy*, with the square funnel and the circular 'bin' for the after SA-N-4 SAM launcher particularly prominent. Note the four 16-tube chaff rocket-launchers fixed to the sides of the VDS housing. (US Navy, 1980)

Distinctive ECM 'tubs' designed Bell Shroud are located port and starboard at the forward end of the bridge structure, and smaller circular antennae designated Bell Squat are mounted on platforms extending outboard from the deckhouse at the base of the lattice mast. Production problems appear to have been experienced with these systems; a number of 'Krivaks' deployed during the 1970s with empty platforms. The shortage appears to have been exacerbated by the decision to extend installation to those 'Kashins' modified from 1973 to 1976.

There was insufficient space to accommodate an active anti-missile missile such as the AK-630 gatling gun. Neither were these ships fitted with the automatically reloading twin-barrelled chaff-launchers installed on larger contemporary vessels. In their place are small 16-tube 'box' launchers for chaff rockets mounted at a fixed angle. Four of these launchers are fitted on the sides of the low deckhouse above the stern which houses the VDS mechanism. Two launchers are fixed at 90 degrees to the ship's axis to cover the beam, while the forward section of the deckhouse is angled in such a way that the other launchers cover the bow sectors. A similar installation was extended to the modified 'Kashins'.

Anti-Surface Warfare

The 'Krivaks' were designed very much as general-purpose anti-submarine ships, and have a number of traditional 'destroyer' features which distinguish them from the larger, more specialized BPKs of the 'Kresta' and 'Kara' classes. It is perhaps also significant that these ships received traditional adjectival 'destroyer' names, whereas their larger counterparts were named after military heroes or major cities of the Soviet Union, like the Rocket Cruisers which preceded them.

The 'Krivaks' are the only BPKs of their generation to feature mine rails. The relatively heavy torpedo armament, comprising two banks each of four tubes (even the 'Kashin' had only one bank of five), is also indicative of the Soviet Navy's intention to deploy them in the enclosed waters of the Baltic and the Black Sea, where they would be more likely to encounter hostile surface craft at close range, as well as in the more open waters of the Arctic and the north-west Pacific. The reversion to the heavier-calibre 76.2mm gun is also significant, as it would be a useful weapon against the fast pastrol boats of the Baltic NATO allies, most of which were armed with only 40mm weapons when the first 'Krivaks' were laid down.

When the Federal German Navy adopted the

▶ *Druzhny*, completed in 1975, has received her full ECM outfit. The temporary deckhouse abaft the funnel carries additional ESM gear to enable the ship to monitor radar and radio emissions at the NATO Exercise 'Northern Wedding'. (US Navy, 1986)

'Krivak II' Class: Profile View

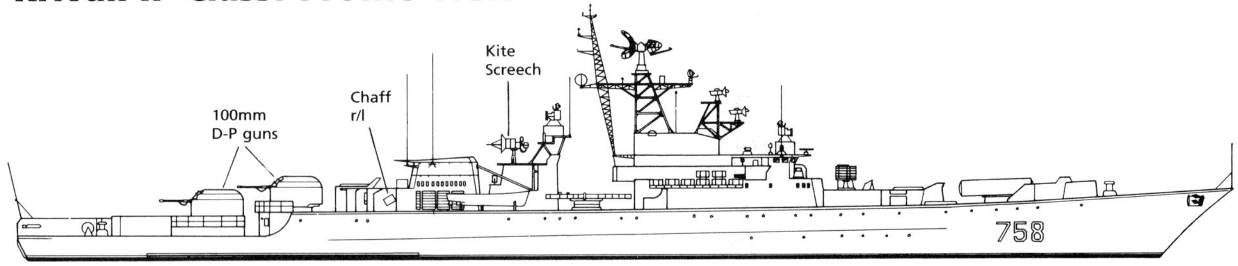

'Krivak II' Class

Soviet Designation: Storozhevoy Korabl' (SKR) – Patrol Ship

Construction

Name	Builder	In service
Rezvy	Kaliningrad Shipyard	1975
Rezky	Kaliningrad Shipyard	1976
Razitelny	Kaliningrad Shipyard	1977
Grozyashchy	Kaliningrad Shipyard	1977
*Neukrotimy**	Kaliningrad Shipyard	1978
Bessmenny	Kaliningrad Shipyard	1978
Gromky	Kaliningrad Shipyard	1979
Gordelivy	Kaliningrad Shipyard	1979
Ryany	Kaliningrad Shipyard	1980
Revnostny	Kaliningrad Shipyard	1980
Pytlivy	Kaliningrad Shipyard	1982

*Renamed *Komsomolets Lityvy* 1988–90

Displacement (tons)
Standard: 3,170
Full Load: 3,670

Dimensions
Length: 125m(410ft) overall
Beam: 14.1m(46ft)
Draught: 4.6m(15ft)

Propulsion
2-shaft COGAG; 2 gas turbines each 24,300hp; 48,600hp=30 knots max.

Complement
180

Weapons
ASW: 1 quadruple SS-N-14 launcher (4 Silex missiles); 2 RBU-6000 rocket-launchers; 2 quadruple banks of 533mm(21in) torpedo tubes
AAW: 2 twin-arm SA-N-4 launchers (40 Gecko missiles)
ASuW: 2 single 100mm(3.9in) D-P guns; rails for 20 mines

Sensors
Surveillance radars: Head Net C 3-D; 1 Don-2 or Spin Trough, 1 Don-Kay or Palm Frond
Fire control: 2 Eye Bowl (SS-N-14); 2 Pop Group (SA-N-4); 1 Kite Screech (100mm)
Sonars: bow-mounted Bull Nose MF; Mare Tail MF VDS
Electronic warfare: 2 Bell Shroud, 2 Bell Squat; 4 16-tube chaff-launchers

Italian 76mm OTO-Melara dual-purpose gun mounting for its new Type 48 missile patrol boats, the first of which was completed in late 1972, the Soviet Navy responded by developing a new single 100mm D-P mounting. This was fitted in place of the two 76.2mm mountings in a new 'Krivak II' variant which began building at the Kaliningrad Shipyard in 1973, the first ship, *Rezvy*, being completed in 1975.

Substitution of the single 100mm mounting for the twin 76.2mm mounting was the only significant change in the 'Krivak II'. The greater weight of the new mounting together with an increase in the height of the VDS housing appear to have resulted in a shift in the centre of longitudinal buoyancy, so that all modified ships appear trimmed by the stern. The revised arrangements aft have also resulted in the chaff-launcher boxes being shifted to the sides of the deckhouse in which the after SA-N-4 launcher 'bin' is located.

Propulsion

Gas turbines appear to have been adopted for these ships in preference to steam-turbine machinery primarily for ease of operation and maintenance, and reduced manning levels. Moreover, the Kaliningrad and Kamysh-Burun Shipyards had acquired considerable experience with the installation of gas-turbine machinery through construction of the 'Petyas', 'Mirkas' and 'Potis', all of which were powered by gas turbines.

The primary unit employed in 'Krivak' is probably a development of the turbine installed in the 'Kashin' which has a similar power output.

There is some disagreement regarding the precise nature of the 'Krivak' machinery installation. Most reference sources now credit these ships with two large gas turbines for high-speed operation and two smaller turbines to provide an economical cruise speed, but it is difficult to reconcile this proposed COGOG arrangement either with established Soviet engineering practice or with the external layout of the ship.

▶ **The 'Krivak II', built exclusively at Kaliningrad, has single 100mm/70 guns in place of the standard twin 76.2mm mounting. This is *Ryany*, photographed in the Mediterranean during the early 1980s. (MoD)**

The 100mm/70 D-P Gun System

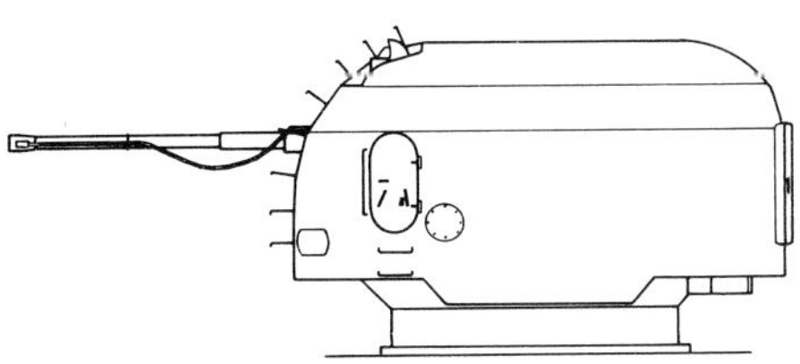

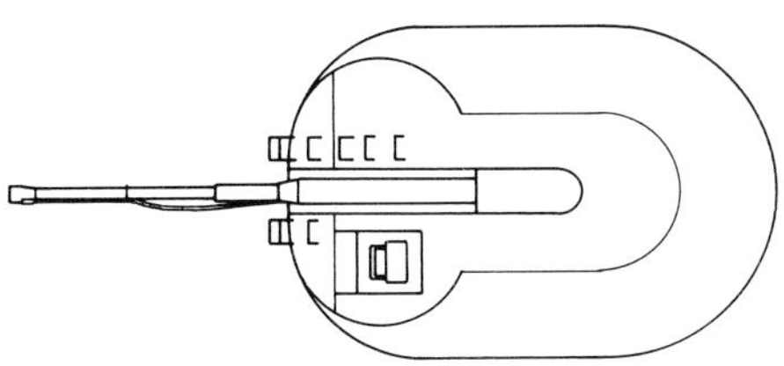

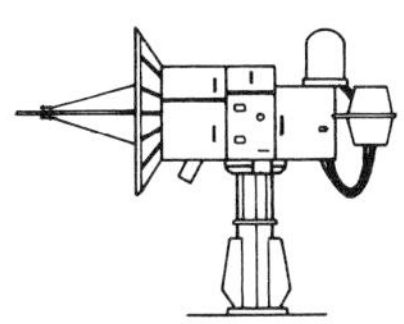

Soviet designation: unknown
In service: 1975
Angle of elevation: +80 degrees
Rate of fire: 80rpm
Projectile weight: 16kg(35lb)
Muzzle velocity: 915m/sec?(3,000ft/sec)
Range: 15km(8nm) max.; 8km(4.5nm) effective AA
Fire control: Kite Screech

West European frigates with the Olympus/Tyne COGOG arrangement mount the turbines in pairs with the gearboxes between them. The main (boost) turbines and the cruise turbines drive in opposite directions so that their exhaust uptakes can be led up together into a single funnel. The direction of rotation of the turbines is unimportant because of the use of controllable-pitch propellers. The intakes for the main turbines are forward of the funnel, and those for the cruise turbines are abaft the funnel.

The 'Krivak' clearly does not have this configuration. The funnel is placed well aft, and the intakes for the gas turbines appear to be located exclusively in the raised deckhouse forward of the funnel (the vents at the base of the funnel are more likely to be intended for cooling the hot exhaust gases, as in the 'Kashin' and other Soviet gas-turbine powered ships). Immediately abaft the funnel is the after SA-N-4 missile 'bin' and the twin 76.2mm gun mountings which have magazines deep in the hull of the ship. It therefore seems probable that all the main machinery spaces are located beneath and immediately forward of the funnel.

The only theoretical possibility is that the four turbines are mounted abreast, but the 'Krivak' does not appear to have sufficient beam to accommodate such an arrangement, particularly as the NK-12 cruise turbines with which they are generally credited are significantly larger than the British Tyne, with a power output in excess of 12,000hp.

Most of the available evidence points to the conclusion that 'Krivak' has a relatively simple COGAG machinery installation based on two large gas turbines. Parallel shafts would enable a mechanical cross-connection to be made, allowing a single turbine to drive both shafts. This would make for economical operation at cruise speed although fuel consumption would still be higher than that of a COGOG ship with low-power cruise turbines. There may also be a separate turbine for reversing, given Soviet persistence with fixed-pitch propellers.

A simple machinery installation such as that described above would result in a ship with limited endurance as compared with Western vessels of similar size, but neither the defence of the Soviet SSBN 'bastions' nor patrol and escort duties in enclosed seas require the level of endurance that might be deemed necessary in a navy with a world-wide 'policing' role. Moreover, the advantages of a simple machinery installation from both the construction and the operational points of view should not be underestimated.

Construction

The overall success of the 'Krivak' design is evidenced by the numbers built. Twenty-one 'Krivak Is' were completed from 1970 to 1982, and a further eleven 'Krivak IIs' from 1975 to 1982 (a third follow-on series, the 'Krivak III', is discussed in a later chapter). This compares with ten 'Kresta IIs' and only seven 'Karas' built during the same period.

The 'Krivak II' was built exclusively at Kaliningrad, parallel construction of the 'Krivak I' continuing at Kamysh-Burun into the early 1980s. At the same time that Kaliningrad began construction of the 'Krivak II', a third production line was opened up at the Zhdanov Shipyard which completed six 'Krivak Is' during the late 1970s as the the 'Kresta II' programme tailed off.

In 1978 the 'Krivak' was reclassified *Storozohevoi Korabl'* (SKR), generally translated in the West as Guard Ship. In the eyes of some Western commentators this constituted a downgrading of the type by the Soviet Navy. However, the reclassification is more properly seen as recognition of the ships' primary mission in support of the SSBN 'bastions', and of their limited capability for blue-water operations. It is

◀*Bessmenny* at speed. The gas-turbine propulsion machinery of the 'Krivaks' is designed for ease of installation, operation and maintenance. Note the revised layout of the 16-tube chaff-launchers on the 'Krivak II', which has a noticeably higher VDS housing. (MoD, 1981)

▲*Nikolayev* photographed in the Mediterranean in 1979. Note the prominent square funnel for the gas-turbine exhausts. The Bass Tilt fire control radars for the AK-630 gatlings had been installed by this time, and the ship has received Square Head IFF interrogators, mounted on platforms projecting from the funnel. (MoD)

perhaps significant that 1978 was the year in which the first of a new larger type of BPK, *Udaloy*, derived from the 'Krivak' but closer in size to the 'Kresta II', was laid down at the Kaliningrad and Zhdanov Shipyards.

The 'Kara' Class

The seven units of the 'Kara' class were built exclusively in the Black Sea at the 61 Kommuna Shipyard, where they replaced the 'Kashin' on the slipways. The first unit was laid down in 1969 and the last was completed in 1980. The period of construction therefore runs parallel to that of the 'Krivak' although each ship took considerably longer to build – almost five years as compared with only half that for the 'Krivak'. The first unit, *Nikolayev*, was completed only in 1973, a full two years after the first 'Krivak' entered service.

Although designed around the weapons suite developed for the 'Kresta II', the 'Kara' was a larger and more complex vessel and incorporated a number of improvements over the somewhat cramped Rocket Cruiser conversion. Differences extended to the nomenclature: whereas the 'Krestas' were named after military heroes, the 'Karas' carried the names of major Soviet cities.

A dominant feature of the new BPKs was their COGAG propulsion plant, based on four large gas turbines. These are rated at 31,250hp by current reference sources, and may be a second-generation turbine derived from the NK-144 aero engine used to power the Tupolev Tu-144 supersonic transport. Maximum sustained speed is probably in excess of 30 knots.

As with the 'Krivak', some reference sources credit the 'Kara' with separate cruise turbines rated at 12,100hp, but although ships of this size could more easily accommodate such an installation the case remains to be proved.

The adoption of all-gas-turbine propulsion machinery was partially responsible for the increase in length over the 'Kresta' which had a particularly compact pressure-fired steam plant. Fitting four large gas turbines into the 'Kara' involved distributing them between two widely spaced machinery rooms. However, the designers clearly wished to dispense with the second set of exhaust uptakes which in the 'Kashin' had effectively precluded the installation of weapons systems on the beam amidships. Instead they opted for a large single funnel, of broadly square configuration, located directly above the after set of turbines. This had the effect of placing the funnel well abaft amidships, thereby keeping the hot exhaust gases clear of the major air surveillance and tracking radars. A prominent lip protruding from the after end of

'Kara' Class

Soviet Designation: Bol'shoy Protivolodochny Korabl' (BPK) – Large Anti-Submarine Ship

Construction

Name	Builder	Laid down	Launched	In service
Nikolayev	61 Kommuna Shipyard	1969	1971	1973
Ochakov	61 Kommuna Shipyard	1970	1972	1975
Kerch	61 Kommuna Shipyard	1971	1973	1976
*Azov**	61 Kommuna Shipyard	1972	1974	1977
Petropavlovsk	61 Kommuna Shipyard	1973	1975	1978
Tashkent	61 Kommuna Shipyard	1975	1976	1979
Tallinn	61 Kommuna Shipyard	1976	1977	1980

*Completed as trials ship (see text)

Displacement (tons)
Standard: 8,000
Full Load: 9,700

Dimensions
Length: 173m(568ft) overall
Beam: 18.6m(61ft)
Draught: 6.7m(22ft)

Propulsion
2-shaft COGAG; 4 gas turbines each 31,250hp; 125,000hp=34 knots max.

Complement
321

Weapons
ASW: 2 quadruple SS-N-14 launchers (8 Silex missiles); 1 Ka-25 Hormone A helicopter; 2 RBU-6000 rocket-launchers; 2 RBU-1000 rocket-launchers (not in *Petropavlovsk*); 2 quintuple banks of 533mm(21in) torpedo tubes
AAW: 2 twin-arm SA-N-3 launchers (72 Goblet missiles); 2 twin-arm SA-N-4 launchers (40 Gecko missiles); 2 twin 76.2mm(3in) D-P guns; 4 AK-630 gatlings

Sensors
Surveillance radars: Top Sail 3-D, Head Net C 3-D; 2 Don-Kay, 1 Don-2 or Palm Frond
Fire control: 2 Head Lights C (SA-N-3/SS-N-14); 2 Pop Group (SA-N-4); 2 Owl Screech (76.2mm); 2 Bass Tilt (AK-630)
Sonars: bow-mounted Bull Nose MF; Mare Tail MF VDS
Electronic warfare: 8 Side Globe; 1 Bell Clout, 2 Bell Slam, 2 Bell Tap (*Nikolayev, Ochakov* only); 4 Rum Tub ESM (*Kerch* onwards); 2 Bell Crown; 2 twin-barrelled chaff-launchers

the funnel serves to protect the after Head Lights missile fire control radar from corrosion.

Reversion to a single funnel was not without cost; substantial hull and superstructure volume is occupied by the exhaust trunking for the forward set of turbines which had to be led some 20 metres aft. However, a beam of 18.6m permitted weapons to be fitted outboard of this trunking, which did not therefore constitute a major problem.

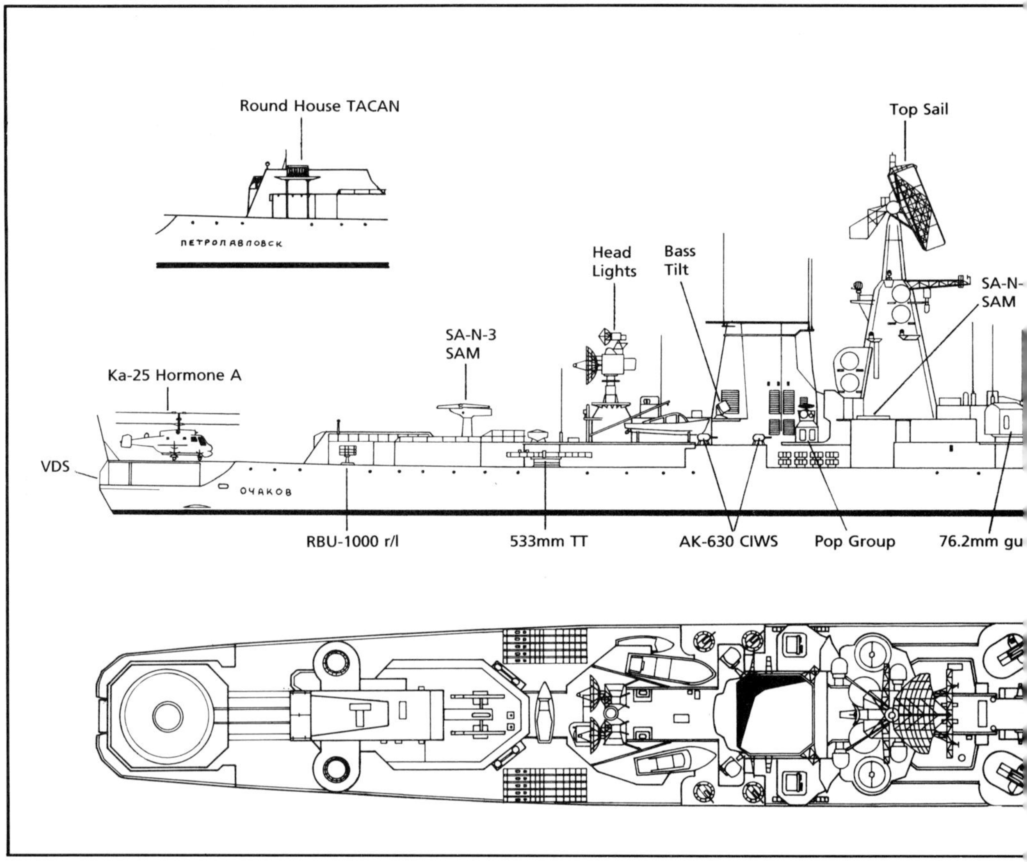

Command and Control

The other feature of the 'Kara' which is responsible for its greater size is the incorporation of elaborate command and control facilities comparable to those in contemporary Western vessels.

The Rocket Cruiser concept did not require extensive command spaces. Long-distance targeting was performed by off-board sensors such as land-based HF/DF networks and maritime bombers. Data were co-ordinated and analysed at shore-based command centres, which subsequently transmitted instructions by radio to surface forces deployed for interception. From its earliest origins, the Rocket Cruiser was essentially a weapons platform operating under shore-based control.

However, blue-water anti-submarine warfare operations required a very different tactical framework. Initial detection of the target would in all probability be made by the sensors of the surface ships operating with the group. Co-ordination both of target data and of the efforts of the various individual surface units involved in the hunt would therefore be most effectively performed on the spot, rather than in some remote shore-

'Kara' Class: Profile and Plan

The profile drawing is of *Ochakov* as first completed. The plan view and inset are of *Petropavlovsk*, with raised helicopter hangar and Round House TACAN replacing RBU-1000 ASW rocket-launchers.

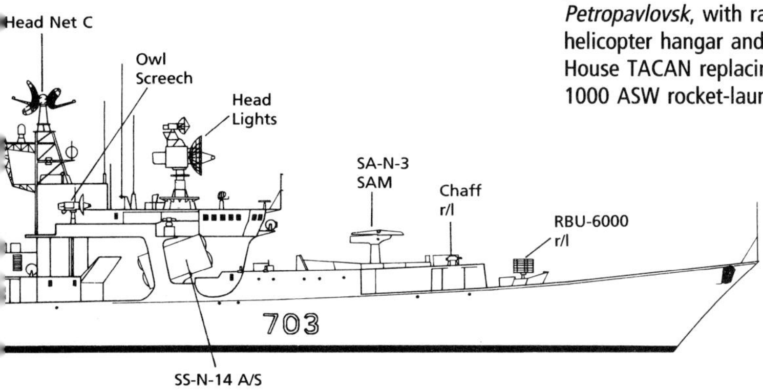

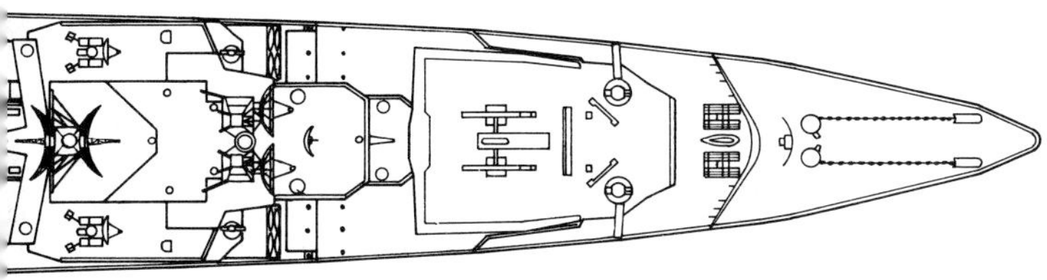

Azov: Profile and Plan (part)

Azov was extensively modified to serve as trials ship for the advanced SA-N-6 missile system, later installed aboard the Rocket Cruisers of the *Kirov* and *Slava* classes. Atop the new deckhouse aft are six square hatches beneath which are rotating eight-missile magazines for the Grumble missile. There is a single Top Dome missile guidance radar in place of Head Lights. Twin torpedo tubes are fitted in place of the quintuple banks of her half-sisters.

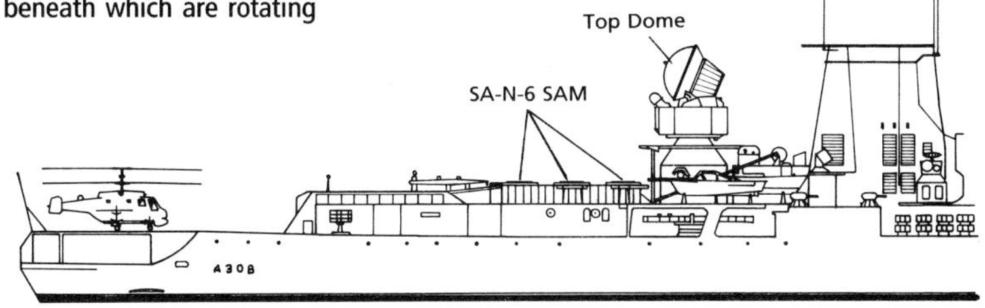

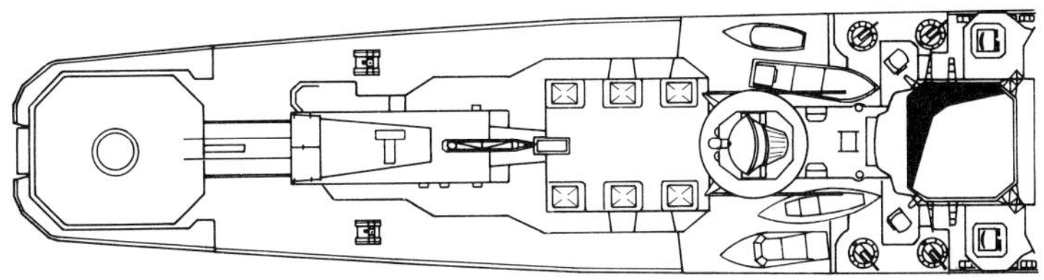

based control post. This created a new requirement for command and control spaces in the larger BPKs.

The 'second-line' anti-submarine ships of the 'Krivak' class would require only a sonar room from which the underwater picture from their own sensors could be monitored, together with a relatively small combat information centre from which their air defence weapons could be controlled. The larger ships, however, would need additional command spaces to enable them to monitor the situation around the task group in three dimensions, and to direct the activities of the other vessels making up the group. More computer-driven consoles were required together with an extensive communications suite.

The 'Kresta II', which was essentially a modification of a Rocket Cruiser design, proved to have insufficient hull volume to meet the new requirements. The double-storey deckhouse added between the bridge structure and the tower mast in the last three units testifies to the 'tightness' of the design. However, the 'Kara', which was designed from the keel up, featured a much-enlarged bridge structure, 22.5m in length (the bridge structure of the 'Kresta II' is only 15m long), with two-tier command spaces of similar dimensions beneath (*see* interior profile).

The result is a ship 15m longer and 2,000 tons heavier than the 'Kresta II'. The difference in external configuration between the two types is almost exclusively related to the insertion of an additional hull section between the bridge and the central tower mast. This has had the effect of shifting the 'balance' of the 'Kara' aft. The forward and after sections of the two ships, in contrast, are virtually identical in terms of both length and layout.

Sonar and Hull Improvements

The anti-submarine weaponry of the 'Kara' is identical with those of the 'Kresta II' in every respect, but the increase in hull size together with a later in-service date made it possible to incorporate both the Bull Nose MF bow sonar and the Mare Tail variable depth sonar of the 'Krivak'.

Large bow-mounted sonars tend to pull down the stem of a ship when operating at higher speeds. The bow section of the 'Kara' has more sheer and flare than the 'Kresta II'. Freeboard amidships, on the other hand,

◀The midships section of *Nikolayev*, with the twin 76.2mm D-P mountings and the SA-N-4 SAM system in the centre of the picture. Beneath the bridge structure are the computerized command and control centres. (US Navy, 1973)

▼The layout of the after section differs significantly from the 'Kresta II'. Both the AK-630 gatlings and the quintuple 533mm torpedo tubes have been shifted farther aft to allow for the installation of the SA-N-4 SAM system and the 76.2mm gun mountings. Note the empty platforms for the Bass Tilt FC radars in this early overhead view of *Nikolayev*. (MoD, 1973)

Layered Air Defence on 'Kara'

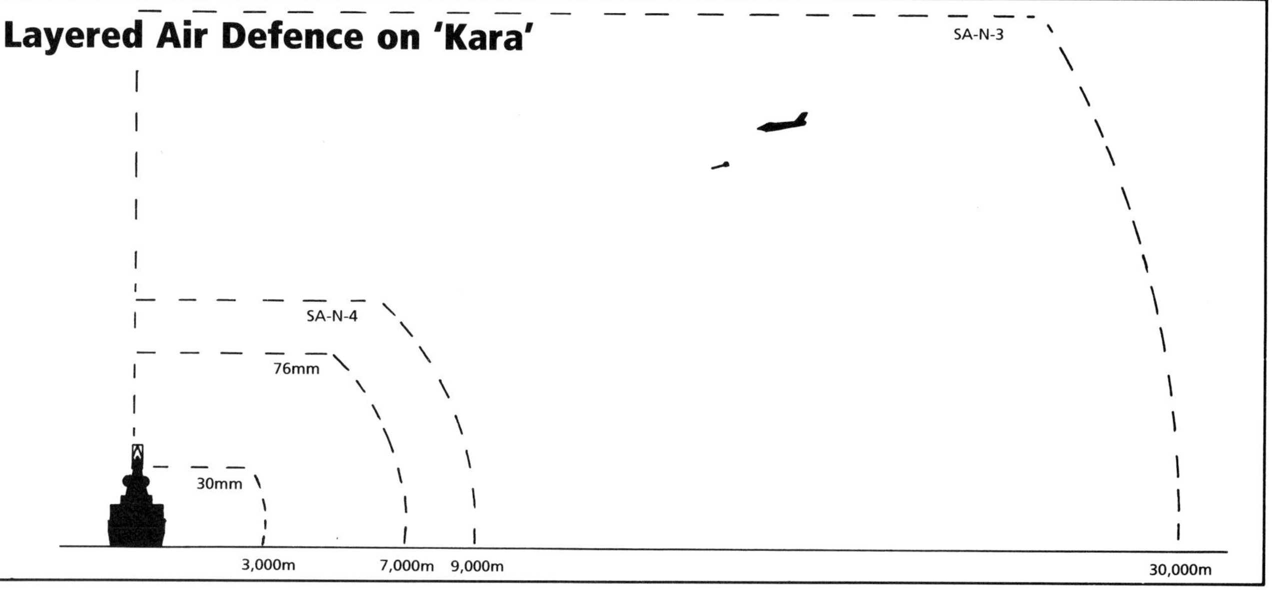

has remained constant at about five metres. In the 'Kara' this means that the height of the bow is double that of the hull amidships. This is an unusual ratio in a flush-decked ship of this size, but is by no means uncommon in Soviet naval construction. The comparatively low freeboard amidships enables more top-weight (in the form of weapons systems) to be carried, while the exaggerated sheer and flare of the bow give these vessels good sea-keeping qualities.

Air Defence

The most obvious differences between the 'Kresta II' and the 'Kara' in terms of air defence capabilities lie in the intermediate-range weapons systems installed on the beam amidships. These constitute an additional element in an elaborate 'layered' air defence system capable of engaging aerial targets ranging from high-flying bombers to low-flying attack aircraft and missiles.

Area defence continues to be provided by the SA-N-3 air defence system fitted in the *Moskva* class and the 'Kresta II', with twin-arm launchers located on deckhouses forward and aft, a Top Sail 3-D tracking radar with Head Net C for back-up, and Head Lights fire control radars above the bridge and atop a short tower structure aft. Close-in defence against low-flying aircraft and missiles is provided by the established combination of AK-630 gatlings, twin-barrelled automatically reloadable chaff-launchers, and Side Globe broad-band electronic jammers.

The third, intermediate layer is provided by two SA-N-4 missile systems and two twin 76.2mm gun mountings. The magazine 'bins' for the SA-N-4 system are located abeam the central tower mast. The Pop Group fire control radars are mounted on pedestals outboard of the funnel and are separated from the bin-mounted twin-arm launchers by prominent curved blast deflectors which are wrapped around the air intakes for the gas turbines. The twin 76.2mm mountings are mounted farther forward, between the tower mast and the bridge structure, with their magazines located deep in the hull of the ship abeam the machinery spaces. Their Owl Screech fire control radars are mounted at the after end of the bridge wings, to port and starboard of the structure on which the short lattice foremast is stepped.

The 76.2mm mountings and their FC radars have effectively displaced the AK-630 gatlings, which occupy a position outboard of the single funnel with their Bass Tilt fire control radars on platforms above them. The first ship completed, *Nikolayev*, entered service without the Bass Tilt radars in place, but these were retro-fitted during the late 1970s.

A feature of these ships which was not immediately apparent is the greater capacity of the SA-N-3 magazines, each of which accommodates 36 missiles. Although a number of commentators had noticed a difference in the number and layout of the reload hatches beneath the twin-arm missile-launchers as compared with earlier vessels armed with the SA-N-3, it was initially thought that this might be related to different types of missile carried. The launchers installed in the *Moskva* class and the 'Kresta II' had four reload hatches and the launcher was reloaded when trained either to

'Kara' Class: Inboard Profile

The drawing is based on photographs of an inboard arrangement drawing displayed on the occasion of a visit to *Kerch* by members of the Western defence press in the spring of 1989.

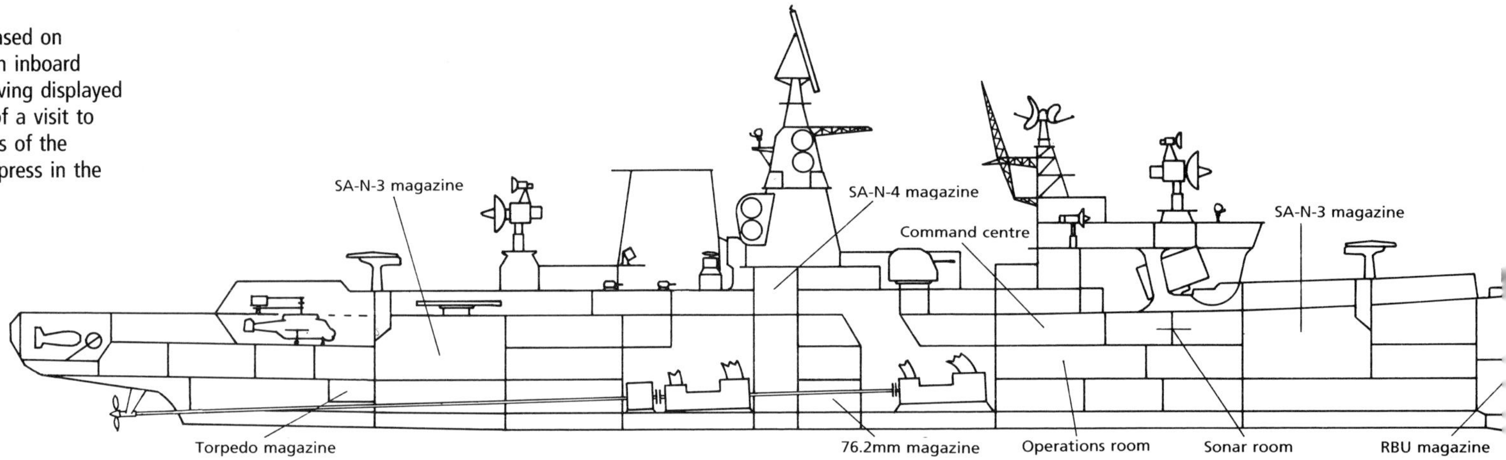

port or starboard. In contrast, the 'Kara' has only two hatches directly beneath the launchers, which reload while trained fore and aft.

It now appears likely that the simplified layout of the reloading hatches on the 'Kara' is related to a revised arrangement of the feed system within the larger-capacity magazine, which is approximately 13m long. The *Kiev* class, which also have the larger-capacity magazine, have identical reloading arrangements. This has led Western Intelligence sources to conclude that later units of the *Kara* class may be armed with the 55km (30nm) SA-N-3B Mod 1 variant of the Goblet missile, with fire control provided by Head Lights C radars (*see* Chapter 5).

Electronics Updates

The first two units of the class, *Nikolayev* and *Ochakov*, completed in 1973 and 1975 respectively, had an electronic countermeasures suite virtually identical with that of the 'Kresta II'. There were eight Side Globe broad-band jammers, a Bell Clout double ESM radome on a platform projecting from the forward face of the tower mast, and pairs of the smaller Bell Slam and Bell Tap antennae mounted on platforms at the four corners of the tower mast below the upper pair of Side Globe radomes.

The third ship, *Kerch*, completed in 1976, emerged with Side Globe but without the smaller radomes of the Bell Series which were replaced by four new ESM antennae designated Rum Tub. Rum Tub resembles the segments of the upper part of a spirits barrel, hence its name. The system was first installed in the Anti-Submarine Cruiser *Kiev*, and appears always to be fitted as a four-piece installation, each antenna covering a 90-degree quadrant around the ship. All subsequent 'Karas' were fitted for Rum Tub, which may be electronically scanned like its US near-contemporary, the SLQ-32. There appear to have been delays in production of the system; the last four ships have all deployed with empty platforms.

The fourth ship, *Azov*, was modified while building to serve as a trials ship for the advanced SA-N-6 area defence system to be installed in a new generation of Rocket Cruisers (*see* Chapter 6). The after SA-N-3 launcher was removed together with its Head Lights fire control radar, and the after deckhouse was remodelled to accommodate six rotating eight-missile

▶ *Ochakov*, the second of the class, photographed during a visit to Istanbul in the spring of 1987. She too had been fitted with Square Head IFF by this time. *Nikolayev* and *Ochakov* have Bell Slam and Bell Tap ESM, whereas later units have provision for the more advanced Rum Tub. (G. Manners)

ECM Arrangements

The first two units of the class, *Nikolayev* and *Ochakov*, have an ECM outfit similar to the 'Kresta II' and *Moskva* classes. Later ships are fitted with (or for) Rum Tub, a more advanced electronically scanned ESM array first installed in *Kiev* in 1975.

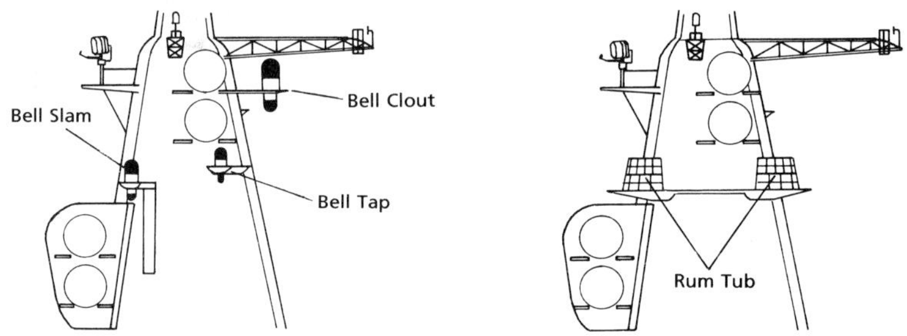

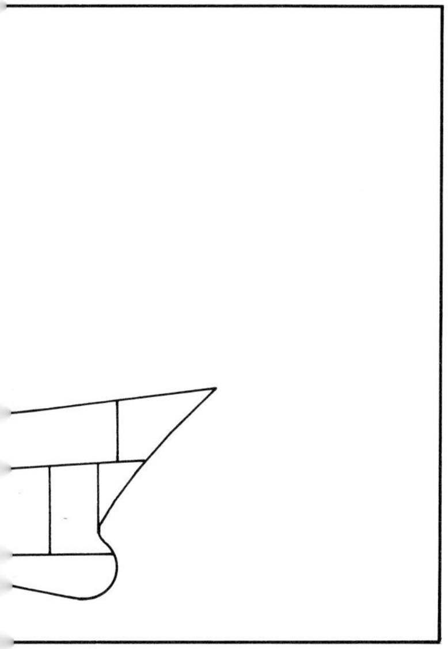

magazine rings, the missile being fired vertically through square hatches. Fire control was provided by a single Top Dome radar, mounted at the top of a short radar tower. The ship remained in the Black Sea until 1986.

Petropavlovsk, the fifth unit of the class, trialled the Round House TACAN system which was subsequently fitted in the Rocket Cruisers of the *Kirov* class and the *Udaloy*-class BPKs. The distinctive cylindrical tubs are mounted on platforms projecting from the upper sides of the helicopter hangar, which is higher than in other ships of the class. The RBU-1000 anti-submarine rocket-launchers have been suppressed to enable Round House to be fitted. This modification was not repeated on the last two ships, which suggests that it was purely for the purpose of trials.

During her first major refit *Nikolayev* received a number of minor modifications to her electronics outfit, emerging in 1979 with two Palm Frond surface surveillance/navigation radars in place of the Don-Kay radars initially fitted, and with Square Head IFF interrogators mounted on platforms projecting from each side of the funnel. *Ochakov* also received Square Head during the mid-1980s, but retained Don-Kay.

Finally, in a 1988–9 refit *Kerch* had Top Sail replaced by a new electronically scanned planar array of broadly square configuration (NATO designation: Flat Screen). This appears to be a more technologically advanced successor to Top Sail and the later Top Pair variant, and is presumably intended for a new generation of missile cruisers. In the same refit *Kerch* received Square Head and Palm Frond.

Service Life

The first three ships of the class, *Nikolayev*, *Ochakov* and *Kerch*, joined the Soviet Black Sea Fleet on completion. All made early deployments to the Mediterranean, serving in rotation with the Fifth *Eskadra*. The

◀ *Kerch*, the third unit of the class, was the first to receive Rum Tub in place of Bell Slam and Bell Tap. Note the twin reload hatches for the forward SA-N-3 launcher in this overhead view; the launchers on the 'Kresta II' have four hatches. (MoD, 1976)

▼ *Tashkent*, one of two 'Karas' to join the Soviet Pacific Fleet in the spring of 1979. The later ships have yet to receive their Rum Tub ESM installations; the platforms between the two groups of Side Globe ECM jammers remain empty. (US Navy, 1988)

fourth ship, *Azov*, remained in the Black Sea to conduct trials of the SA-N-6 area defence missile system and did not transit the Turkish Straits until 1986.

The adoption of 'local' names for these ships suggests that the class was designed primarily for operations within the Black Sea Fleet area. Later deployments, and the nomenclature of the later vessels, show a different pattern, indicating a shift in Soviet anti-submarine priorities away from the Mediterranean and towards the SSBN bastions in the Northern and Pacific Fleet areas.

The Pacific Fleet had suffered considerable neglect since the early 1970s and possessed few major vessels of modern design by comparison with the two major Soviet western fleets, the Northern and the Black Sea Fleets. Construction at the major Komsomolsk Shipyard had long been focused on nuclear-powered submarines, no modern surface units having been laid down since the termination of the 'Kildin'/'Krupny' series.

However, by the mid-1970s the shipyard was completing the first of a series of ballistic missile submarines of the 'Delta I' class, armed with the 4,200nm SS-N-8 Sawfly missile. These submarines would deploy in the Pacific Fleet area, patrolling in the enclosed waters of the Sea of Okhotsk, and would need the same degree of support and protection accorded to their counterparts patrolling in the northern bastion. The surface ships required for this mission would have to be built in shipyards in the western USSR.

Two units of the 'Kresta II' class were duly transferred in 1974, together with a trickle of smaller BPKs of the 'Krivak I' group. It was not until the late 1970s, however, that the trickle became a flood and large numbers of the most modern surface units completed for the Soviet Navy made the lengthy transit around the Cape to join the Pacific Fleet.

A third 'Kresta II', *Vasily Chapaev*, transferred in 1978. She was followed by the largest deployment yet, comprising the new Anti-Submarine Cruiser *Minsk*, accompanied by the latest pair of 'Karas', *Petropavlovsk* and *Taskhkent*, and the first of a new class of landing ships, *Ivan Rogov*. This deployment, which attracted considerable attention from the world's defence press, tended to overshadow equally significant transfers of 'Krivaks' of which several were freshly completed 'Krivak IIs' armed with the new 100mm gun.

The trend continued with the deployment of the last of the 'Karas', *Tallinn*, to the Pacific in 1981. Three years later the Pacific Fleet was further reinforced by the transfer of the first ship, *Nikolayev*, from the Black Sea Fleet in company with the new Anti-Submarine Cruiser *Novorossiysk*, the landing ship *Ivan Rogov* (which had returned to the western USSR for an important series of amphibious exercises), and the new 'Krivak'-class BPK *Poryvisty*.

This final 'Kara' transfer left the Black Sea Fleet with only two ships in active service plus *Azov*, which by this time had completed her missile trials. 'Kara' deployments thus provide a useful indicator of the shifts in Soviet tactical and strategic thinking over this period.

▼*Azov*, the trials ship for the advanced SA-N-6 area defence missile system due to enter service on the *Kirovs*. The after SA-N-3 installation was replaced by a new deckhouse with six vertical launch hatches for the SA-N-6 Grumble missile and a single Top Dome guidance radar. The greater width of the deckhouse is responsible for a reduction to two twin torpedo tube mountings. (US Navy, 1988)

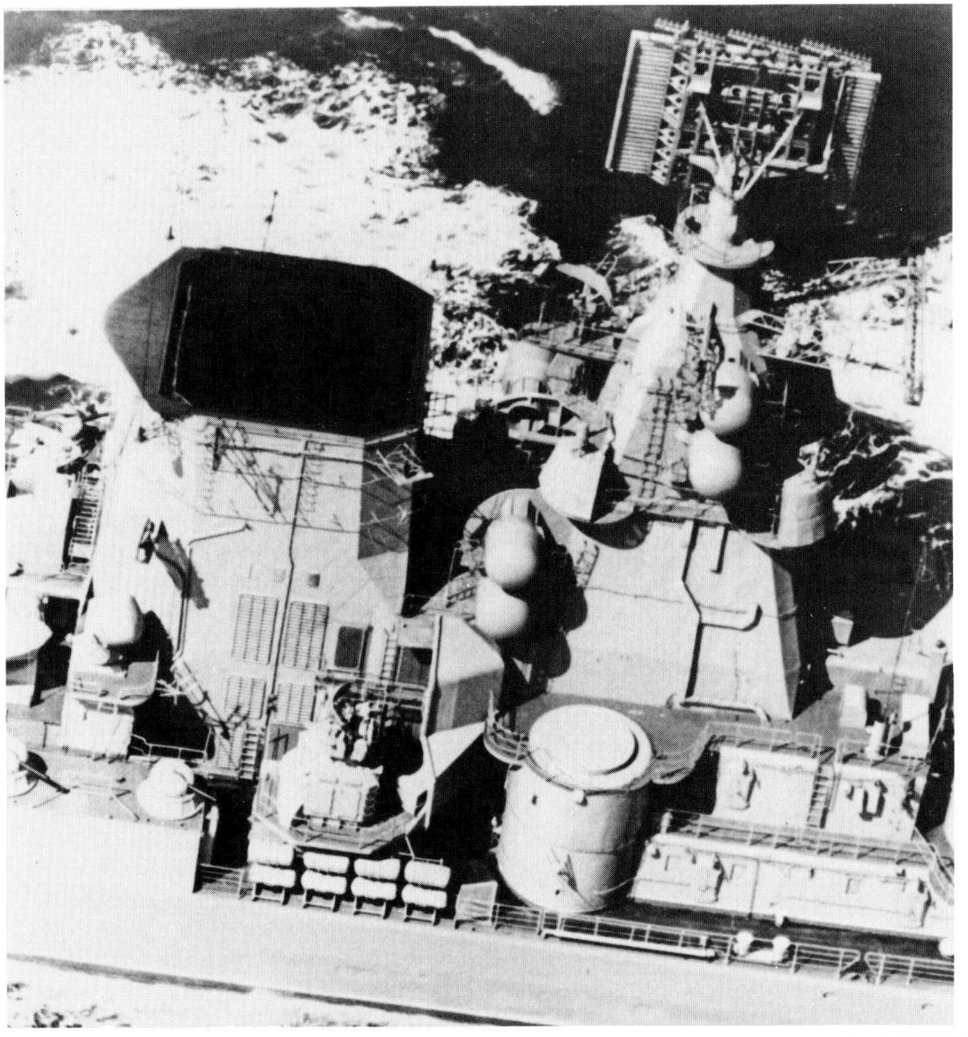

◀ *Kerch*, with the new Flat Screen planar antenna atop the central tower mast. Her electronics outfit is otherwise unmodified. (US Navy)

▼The last unit, *Tallin*, which joined the Pacific Fleet in 1981. She would be followed by *Nikolayev* in 1984. (MoD)

5. The *Kiev* Class

▼ ***Kiev* was the first of a new generation of Anti-Submarine Cruisers. Like her predecessors *Moskva* and *Leningrad* she is a hybrid vessel in which air defence, anti-surface and anti-submarine missile systems have equal status with aviation capabilities. Note the proximity to the waterline of the port-side sponson in this bow view, taken by the helicopter of HMS *Torquay* during *Kiev*'s initial deployment to the Northern Fleet in August 1976. (MoD)**

THE FIRST UNIT of the *Kiev* class was laid down at the Black Sea Shipyard (Nikolayev South) in September 1970. This makes *Kiev* a close contemporary of the 'Kara'-class BPK, the first of which had been laid down little more than one year previously at the near-by 61 Kommuna Shipyard (Nikolayev North). Commentators have remarked on the striking similarities between the respective weapons outfits, but it is also important to note that the two classes were conceived within the same tactical/strategic framework.

In technical terms the *Kiev*s are clearly derived from the earlier Anti-Submarine Cruisers of the *Moskva* class. Equally clearly, the *Moskva*s represent something of a dead-end in tactical thinking, and it was this 'conceptual obsolescence' that was undoubtedly the reason for the early termination of the construction programme. *Moskva* was laid down in 1962 and her sister *Leningrad* in 1964, and it might have been expected that a third and a fourth unit would have been laid down at similar (i.e., two-year) intervals. This suggests that a decision not to proceed further with the programme would have been taken in 1965–6, the year in which there appears to have been a major shift in Soviet thinking about anti-submarine warfare and its place in Soviet maritime strategy, and the year in which the philosophical foundations for the new programme of BPKs were laid.

There were technical aspects of the *Moskva* design that proved to be less than successful. The relatively fine hull sections forward caused the ships to bury their heads in a sea, resulting in a trim of about one metre, and the ships acquired the reputation of being poor seaboats for their size. Moreover, the overall configuration of the ships imposed a number of undesirable constraints on air operations which the Soviets would attempt to rectify in the *Kiev* design. These failings would not, however, have been apparent in 1965–6, as *Moskva* would not be completed until the following year. It therefore appears that the decision to terminate construction was taken on philosophical not technical grounds. The concept of 'offensive' anti-submarine warfare was overtaken by a new, pro-submarine mission, requiring a new, more sophisticated design.

Weapons systems intended for the third and fourth units of the *Moskva* class, and in particular the SA-N-3 area defence missile systems, would be utilized for the BPK conversion of the 'Kresta', 'Kresta II' (*see* Chapter 4), while design work proceeded on a new type of Anti-Submarine Cruiser which would enter service in the mid-1970s.[1]

The Demands of Open-Water ASW

The *Moskva*s were designed to operate in the enclosed waters of the eastern Mediterranean, but they were not expected to contest these waters against hostile surface ships. The latter would be excluded from their operating areas by surface action groups comprising both Rocket Cruisers and submarines armed with long-range cruise missiles, leaving the *Moskva*s to concentrate on their sub-hunting mission. They were well

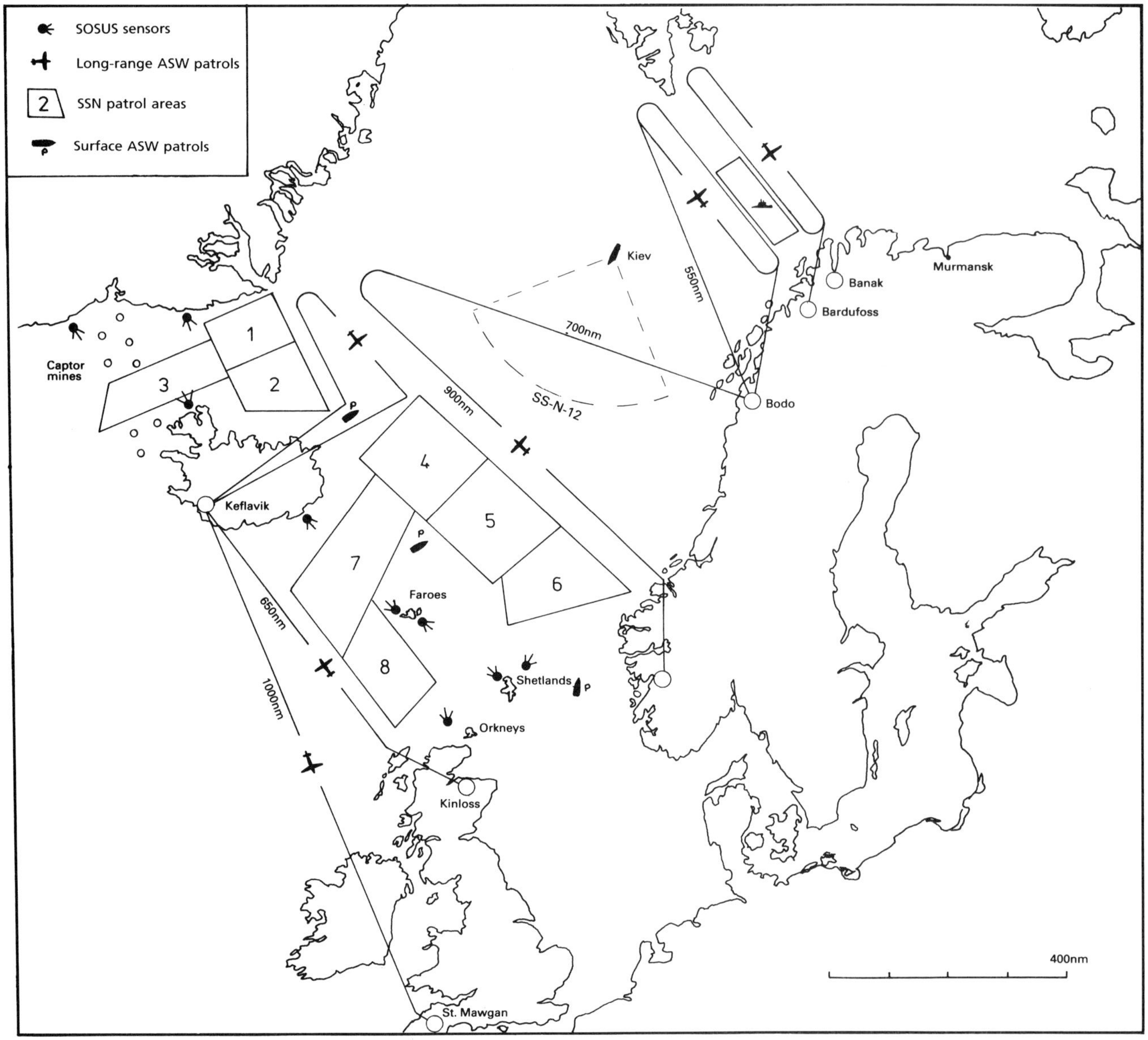

The North Atlantic

A Soviet view of the NATO anti-submarine barrier in the Greenland/Iceland/UK (GIUK) Gap. The Soviet maritime strategy of the 1970s involved a contest for these waters in which powerful surface forces, working in collaboration with land-based aircraft and submarines, would assist the deployment of the Soviet submarine fleet into the North Atlantic, and would simultaneously protect the SSBN bastion in the Arctic against penetration by Western hunter-killers. The drawing shows the potential coverage of the SS-N-12 anti-ship missiles carried by *Kiev*.

equipped to defend themselves against hostile aircraft, but had no dedicated anti-ship weapons.

The new Anti-Submarine Cruisers, on the other hand, were intended to operate in support of the Soviet SSBN bastions in the Arctic and the north-west Pacific, where they would themselves constitute the first line of defence against Western surface forces attempting to penetrate Soviet sea-space. They would therefore be equipped with long-range anti-surface weaponry of their own.

Major surface vessels such as these would also be the natural focus for local control of Soviet surface forces in their area, and would be responsible for co-ordinating anti-submarine warfare, area air defence and anti-surface operations. They would therefore be close in conception to the air-capable Command Cruisers contemplated by the British Royal Navy during the late 1960s, in the wake of abandonment of the CVA-01 carrier programme.

Parallels with British thinking of the time extend to the air complement envisaged for the new PKRs. In the face of the US Navy-led spiral of ever-larger and heavier carrier-borne aircraft, requiring bigger flight decks, longer catapults and more powerful arrestor

▲**Overhead view of *Kiev*, showing the clear division between 'carrier' and 'cruiser' capabilities. The angled deck, which occupies some two-thirds of the ship's length and is angled at 3.5 degrees, is marked out with six numbered helicopter spots plus a seventh for the plane guard helicopter forward. The larger circle aft is for Forger landings. (TASS)**

gear, the Royal Navy had accepted that it could no longer finance a new generation of conventional carriers. Since the early 1960s, however, some had been urging the development of vertical take-off and landing (VTOL) aircraft as a way out of this development spiral. VTOL aircraft did not need a large flight deck from which to operate, requiring neither 90m steam catapults for launch nor arrestor wires to enable them to land. They could therefore be operated from smaller carriers, offering the possibility of an increase in the number of operating platforms for a given financial outlay.

The British were the first to develop VTOL technology based on aero-engines with vectored thrust. The experimental Hawker P.1127 Kestrel began flight trials in 1961 and these were so successful that by 1962 there were plans for an ambitious supersonic follow-on aircraft which would meet the needs of both the RAF and the Royal Navy, the P.1154. Sea trials of the Kestrel took place aboard the carrier *Ark Royal* as early as 1963 and further trials were subsequently conducted from a wide variety of flight decks, including those of naval auxiliaries.

There were certain disadvantages inherent in

vertical take-off operations which would have to be taken into account in assessing the maritime role for which VTOL aircraft would be best suited. The major disadvantage was that of reduced payload in comparison with conventional aircraft using catapult-assisted take-off; the latter could carry more weapons and – of equal importance in a maritime scenario – more fuel. VTOL aircraft would be hampered in the

Kiev Class

Soviet Designation: Bol'shoy Protivolodochny Kreyser (BPKR) – Large Anti-Submarine Cruiser

Construction

Name	Builder	Laid down	Launched	In service
Kiev	Black Sea Shipyard	Sept 1970	Dec 1972	May 1975
Minsk	Black Sea Shipyard	Dec 1972	May 1975	Feb 1978
Novorossiysk	Black Sea Shipyard	Oct 1975	Dec 1978	Sept 1982
Baku	Black Sea Shipyard	1978*		

*Completed to new design (*see* Chapter 8)

Displacement (tons)
Standard: 36,000
Full Load: 43,000

Dimensions
Length: 273m(895ft) overall
249.5m(818ft) waterline
Beam: 53m(174ft) flight deck
32.7m(107ft) waterline
Draught: 9.5m(31ft)

Propulsion
4-shaft geared steam turbines; 8 turbo-pressurized boilers; 200,000shp=32 knots max.

Complement
1,600 (incl. air group)

Aircraft
12 Yak-38 Forger VTOL aircraft; 18 Ka-25 Hormone A anti-submarine helicopters; 1–2 Ka-25 Hormone B missile targeting helicopters; 1–2 Ka-25 Hormone C SAR/utility helicopters

Weapons*
ASW: 1 twin-arm SuW-N-1 launcher (18 FRAS-1 missiles); 2 RBU-6000 rocket-launchers; 2 quintuple banks of 533mm(21in) torpedo tubes
AAW: 2 twin-arm SA-N-3 launchers (72 Goblet missiles); 2 twin-arm SA-N-4 launchers (40 Gecko missiles); 2 twin 76.2mm(3in) D-P guns; 8 AK-630 gatlings
ASuW: 4 paired SS-N-12 launchers (16 Sandbox missiles)

Sensors*
Surveillance radars: Top Sail 3-D, Top Steer 3-D; Don-Kay, 2 Palm Frond
Fire control: 1 Trap Door (SS-N-12); 2 Head Lights (SA-N-3); 2 Pop Group (SA-N-4); 2 Owl Screech (76.2mm); 4 Bass Tilt (AK-630)
Sonars: bow-mounted Moose Jaw LF; Bull Horn MF; Mare Tail MF VDS
Electronic warfare: 8 Side Globe; 2 Bell Clout, 4 Bell Thump, 4 Bell Bash, 2 Bell ?; 4 Rum Tub ESM; 2 twin-barrelled chaff-launchers

*As designed. For modifications to *Novorossiysk* see text.

Kiev Class: Profile and Plan

The profile drawing is of *Kiev* as she appeared in 1976. The plan drawing is of *Minsk* as completed in 1978. On *Minsk* the 14m circle for Forger landings was moved 2.5m to port in order to clear the after deck park.

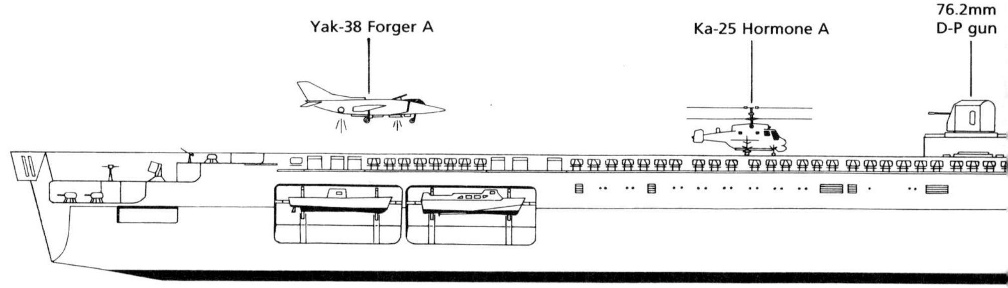

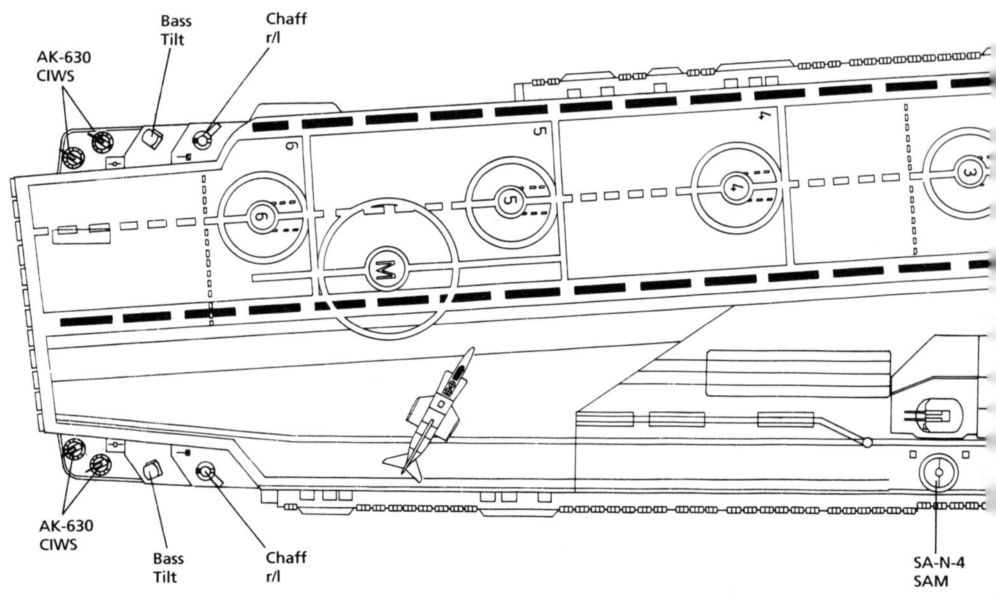

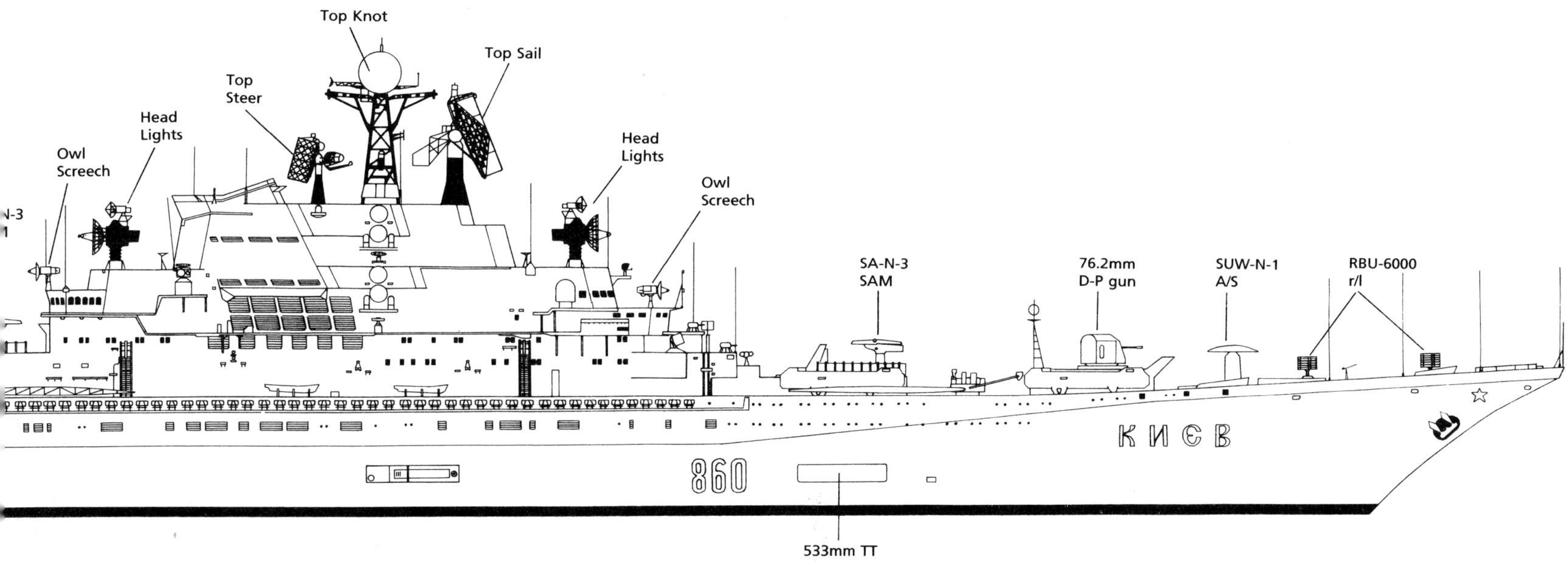

Stern of *Kiev*

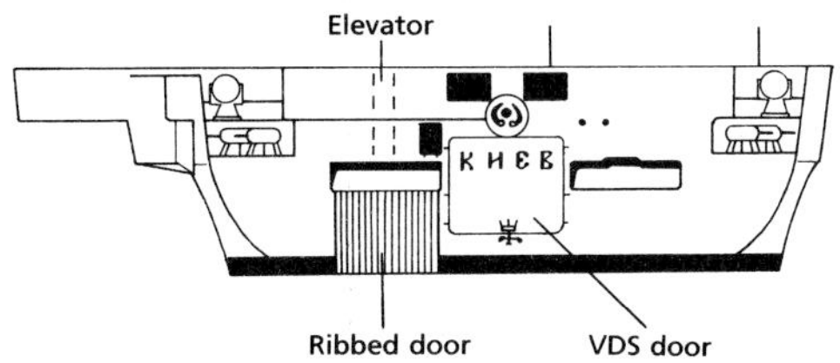

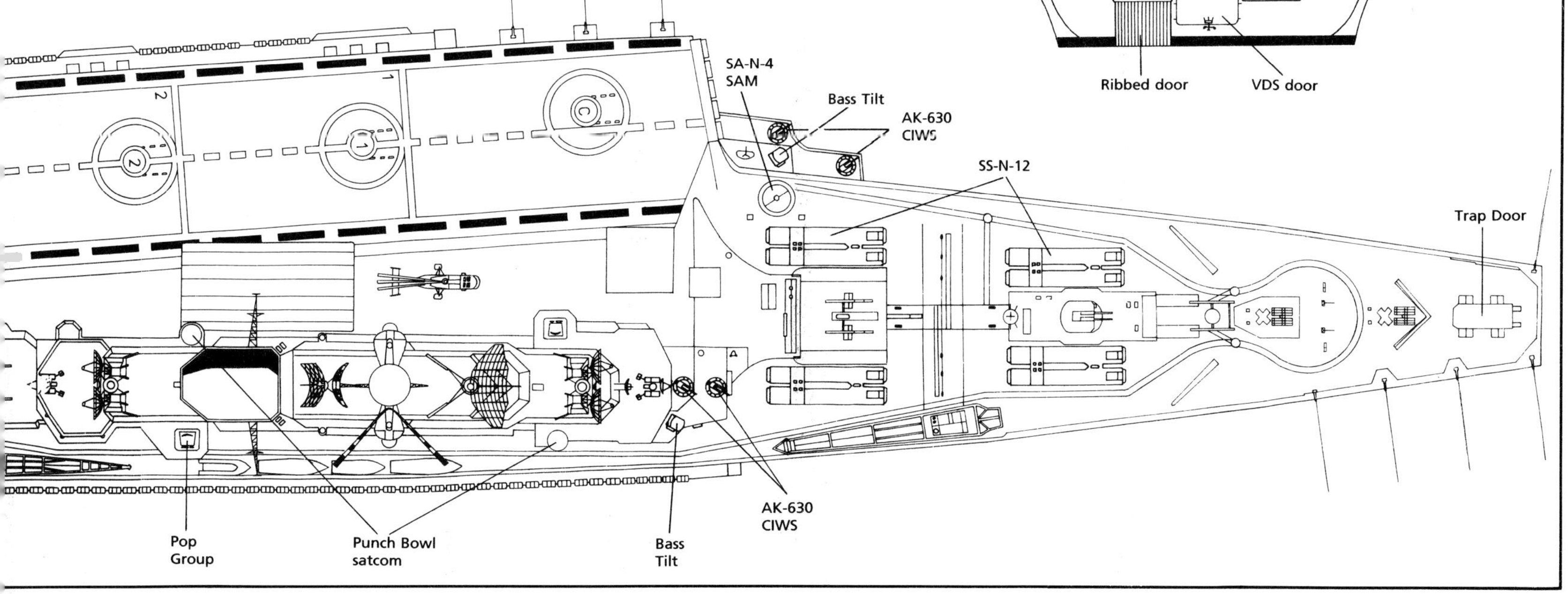

▲*Kiev* **operating in the North Atlantic in the early 1980s. There are four Ka-25 Hormone A anti-submarine helicopters lined up on the angled deck, and two Yak-38 Forger A strike fighters on the deck park aft. (MoD)**

strike role by their small weapons payload and by their lack of operational range, and in the interceptor role by their lack of endurance on station. There was also a significant reduction in high-altitude performance, but this was more acceptable in a maritime environment where hostile aircraft would generally come in low to attack their targets. VTOL aircraft using vectored thrust gained in manœuvrability much of what they lost in high-speed performance.

On balance, it appeared in the early 1960s that the VTOL aircraft had potential as a naval interceptor, provided that some solution could be found to the endurance problem. In opting for an interceptor (first in the form of the P.1154 then, when the latter project was abandoned, in the form of the Sea Harrier) rather than a strike aircraft, the Royal Navy was simply confirming contemporary assessments of the capabilities of VTOL aircraft in a maritime role.

The appeal of VTOL aircraft for the Soviet Navy was of a different order from that which influenced developments in the Royal Navy. The British were, after all, still contemplating the construction of a new generation of fully fledged attack carriers in which the P.1154 would have been but one of the components making up the air group. The Soviet Navy, on the other hand, had rejected attack carriers on the US Navy pattern for both technical and philosophical reasons.

The US Navy continued to regard the manned aircraft as the ultimate tactical weapon system, affording a degree of control and flexibility unequalled by other systems; the anti-surface mission was performed by squadrons of low-level attack aircraft, with fighter squadrons providing protection not only for the carrier but for the long-range attack squadrons themselves. The Soviet Navy had opted instead for a 'technological' solution in which the anti-surface mission was performed by sea- and air-launched stand-off missiles fired under tight central control. By the mid-1960s it was well on the way to building up the ocean surveillance systems which would provide the necessary target data for the ships and aircraft serving as platforms for those missiles.

There was no question of ship-based tactical aircraft taking over the anti-surface mission from the long-range guided missile, although if armed with short-range anti-ship weapons they would constitute a useful complement to these weapons, providing a flexible response against low-value surface targets which might not justify a full-scale missile strike. More useful would be an interceptor capable of fleet air defence. A ship-borne interceptor would enable Soviet anti-submarine groups to operate in hostile, open expanses of ocean where support from land-based fighters was out of the question. This consideration was of particular importance in the all-important Northern Fleet area because Soviet anti-submarine groups operating in the Norwegian Sea would have to contend not only with NATO carriers but with NATO strike aircraft operating from land bases around the periphery.

The first experimental Soviet VTOL aircraft, the Yak-34 Freehand, made its first public appearance at the air display at Domodedovo in July 1967. Develop-

▲**Stern view of *Kiev*, showing the twin recesses in the stern and the prominent black ribbed door which serves as a spray deflector. The central door bearing the ship's name conceals the winch mechanism for the Mare Tail variable depth sonar. (US Navy, 1976)**

ment was therefore well under way by the time the plans for the new Anti-Submarine Cruisers were drawn up. Prototypes of the production model, the Yak-38 Forger, conducted sea trials aboard the PKR *Moskva* during the early 1970s, and *Kiev* would deploy in mid-1976 with a squadron of ten production aircraft together with a single Forger B two-seat trainer.

Configuration

As with the *Moskvas*, 'cruiser' systems were accorded equal importance with aviation-related facilities in the new Anti-Submarine Cruisers, resulting in another hybrid design. However, whereas in the *Moskvas* there was a neat division of the 'carrier' and the 'cruiser' functions at the point where the superstructures ended and the flight deck began, a more complex arrangement was adopted for the *Kievs*. The flight deck occupied some two-thirds of the length of the ship and was angled at approximately 3.5 degrees from the axis, allowing the superstructures to be re-worked as a long but relatively narrow island to starboard.

The revised arrangement had the effect of increasing the number of aircraft operating spots from four in the *Moskvas* to seven in *Kiev* and her sisters – an increase necessary if the ships were to benefit from their more substantial air complement. It also enabled the designers to work-in aircraft and munitions lifts, together with a substantial deck park, to starboard of the angled deck. This constituted a major improvement over the *Moskvas* in which the aviation lifts were adjacent to the helo-spots and their operation therefore interrupted air operations.

A further benefit of the revised flight deck layout was that it more than doubled available hangar space which was essential given the increased size of the air complement. The Soviet Navy was unwilling to adopt the standard US Navy practice of operating with a permanent deck park (in the US 'super-carriers' the deck park generally accounts for at least half the air group) because of the adverse weather prevailing in the Northern Fleet area. Hangar size therefore had to reflect the size of the air group which would include a squadron of VTOL aircraft, utility helicopters for SAR tasks and two or three Ka-25 Hormone B helicopters to provide over-the-horizon targeting for the SS-N-12 anti-ship missiles, in addition to the squadron of Ka-25 Hormone A helicopters carried by the *Moskvas*.

In the *Moskvas* all cruiser weaponry was concentrated in the forward part of the ship. Coverage of the after arcs was therefore poor, with only the 57mm automatic guns having any realistic chance of engaging targets approaching from the stern quarters. The adoption of a large island superstructure for the *Kiev* class enabled the designers to distribute the major air defence systems between the forecastle and the after end of the island, thereby providing all-round coverage. The increased size of the hull also made it possible to locate groups of AK-630 gatlings in all four 'corners' of the ship without obstructing flying operations.

Hull-Form

Kiev has the classic Soviet full waterplane hull-form aft. The forecastle, which carries most of the weaponry, has a marked sheer and the considerable flare of the bow is carried right forward by the squaring-off of the stem at the level of the upper deck. An unusual feature of the hull is the counter stern which sweeps up beyond the waterline before meeting the square transom. In contrast to the *Moskvas*, *Kiev* has a tendency to trim by the stern, particularly at higher speeds. The Moose Jaw hull sonar, which in the *Moskvas* was keel-mounted and fully retractable, is housed in an enormous bow bulge.

Freeboard aft is remarkably low for an air-capable vessel – only 13m as compared with 19m for the US

◀ These aerial views show the major weapons systems of *Kiev*, which are disposed on the forecastle deck and at the after end of the island respectively. The upper photograph is of the foredeck area, with the forward 76.2mm mounting in the centre, flanked by two of the four pairs of elevating launchers for the SS-N-12 missile. The missile elevator, crossed by athwartship tracks for the reloading tray, is visible abaft the 76.2mm deckhouse. Forward of the deckhouse is the twin-arm SUW-N-1 launcher for the FRAS-1 missile. The rectangular reload hatches, which can be seen abaft the launcher, are served by the railway tracks which encircle the forecastle area. Forward of the SUW-N-1 launcher are the two RBU-6000 ASW rocket-launchers. Note the 'branch line' connecting the main track with the SS-N-12 missile elevator.

The lower photograph shows the after end of the island superstructure. The after twin-arm launcher for the SA-N-3 missile is superimposed above the after 76.2mm gun mounting, with the after SA-N-4 launcher 'bin' mounted flush with the flight deck to starboard. Above the SA-N-3 launcher are the Owl Screech and Head Lights fire control radars, while the Pop Group FC radar for the SA-N-4 missile is mounted on a platform projecting from the superstructure at the base of the funnel. (MoD, 1985)

Navy super-carriers. The full waterplane and active fin-stabilization combine to create an exceptionally steady platform. *Kiev* has been observed to have very low overall motions in a seaway, and these are clearly critical for the VTOL operations for which she has been designed.

A disadvantage of the low freeboard is that deck-edge lifts were out of the question. Both aircraft lifts therefore encroach on valuable hangar space. A further problem resulting from the low freeboard is that the gatling sponson immediately forward of the angled deck to port is so close to the water that some loss of speed in heavy weather must result if damage is to be avoided. This sponson has been subject to constant redesign in successive units of the class, finally being suppressed in the fourth unit, *Baku* (*see* Chapter 8).

There are numerous scuttles in the ship's sides forward, and these are complemented by rectangular ventilation grilles, similar to those on the *Moskva* class, alongside the hangar deck. As in the earlier PKRs, quintuple banks of 533mm torpedo tubes are set into the hull behind sliding doors just forward of the island superstructure, and there are open recesses for accommodation ladders amidships. There are deeper recesses aft for the ship's boats, two of which are slung on radial davits on each side of the stern.

In the centre of the broad transom stern there is a large door bearing the ship's name, which conceals the variable depth sonar winch mechanism. Above the VDS door, immediately below the level of the flight deck, are two small recesses which may house a night approach radar or lights.

To port and starboard there are longer, deeper recesses of similar dimensions. The starboard recess contains an outlet for a refuse incineration plant and has also been used for over-the-stern refuelling with the ship at anchor. Directly beneath the recess on the port side is a black, ribbed door which appears to be of rubberized material and extends to the base of the transom. The ribbed door is in line with the centre of the angled deck and is now understood to be a spray deflector which prevents spray from being ingested by the air intakes of the Forger VTOL aircraft as it approaches for the landing-on sequence. The door is also directly beneath the after munitions/stores lift which is set into the angled deck, and probably serves a dual function, giving access to a loading area for the transfer of bulky stores and munitions from fleet tenders and small boats while the ship is alongside.

'Cruiser' Weaponry

The weapons systems of *Kiev* are of the same generation as that of the 'Kara', with the SA-N-3 area air defence system backed up by the close-range SA-N-4 missile system, twin 76.2mm gun mountings and AK-630 gatlings for anti-missile defence. The major differences are the SS-N-12 anti-ship missile system, which occupies a considerable amount of deck-space and hull volume, and the SuW-N-1 anti-submarine missile system which was also a feature of the *Moskva* class, and which is fitted in place of the SS-N-14 system of the 'Kara'.

The air defence systems are distributed between the forecastle and the after end of the island superstructure so as to ensure all-round coverage. The forward SA-N-3 launcher and the forward 76.2mm gun mounting are located in centre-line deckhouses, with a separation of some 13m. The second 76.2mm mounting is located on its own deckhouse at the after end of the island, with the second SA-N-3 launcher superimposed above it. The Head Lights fire control radars for the Goblet missile are mounted above Owl Screech FC radars for the 76.2mm mountings at each end of the island superstructure, which is built up in 'steps' as in the *Moskva* class.

Kiev is credited with an improved variant of the Goblet missile (NATO designation: SA-N-3B), having a maximum range of 55km (30nm) as compared with only 30km (16nm) for the earlier model. The improvement in performance must result purely from advances in Soviet rocket propulsion technology, because the guidance system is essentially unchanged. The Head Lights fire control radar, however, is an improved variant (NATO designation: Head Lights C), capable of handling the Goblet missile at extended ranges.

The two SA-N-4 Gecko short-range missile systems are grouped with the other major air defence weapons. The forward installation is close to the deck edge on the port side of the forecastle immediately forward of the flight deck. The after magazine bin, which like the first is countersunk into the upper deck, is located outboard of the after 76.2mm deckhouse. The Pop Group fire control radars for these weapons are mounted on platforms projecting from the sides of the island. The platforms are diagonally opposite each other, reflecting the position of the launchers themselves.

In each of the four corners of the ship there is a close-in anti-missile system comprising two AK-630 gatlings, a Bass Tilt fire control radar and a Kolonka-II optical sight for local control. Three of the four systems are mounted on sponsons below the level of the upper deck, but the group which covers the starboard bow is located at the forward end of the island superstructure. Twin-barrelled chaff-launchers are co-located with the after groups of gatlings, but there are no comparable systems in the forecastle area, although one might have expected provision to be doubled in a ship of this size.

Anti-submarine weaponry is on a par with that of the *Moskva* class. The twin-arm SuW-N-1 launcher, together with its reload magazine, is located in the centre of the forecastle deck, immediately forward of the 76.2mm deckhouse. Between the launcher and the bow, superimposed on the centre-line, are two automatically reloading RBU-6000 anti-submarine rocket-launchers.

The SS-N-12 Missile System

The broad forecastle of *Kiev* is dominated by the SS-N-12 anti-ship missile system. The eight massive launchers are disposed as pairs outboard of the deckhouses for the forward 76.2mm gun mounting and the forward SA-N-3 missile-launcher respectively. The launchers are elevated to an angle of approximately 25 degrees for firing. Those in the after group are protected from the rocket exhaust of the Goblet missile by angled blast shields, while their own missile exhaust is vented outboard by the curved face of the deckhouse abaft them, which appears to serve as a garage for deck tractors.

Between the two groups of launchers is a capacious below-decks magazine (*see* drawing). The reloads are stowed on each side of a narrow elevator measuring approximately 12.5m by 2.5m. Missile rounds are transferred to the upper deck on a reloading trolley and are then moved sideways on a system of athwartships rails to line up with the launchers. At least eight reloads (sixteen in some sources) are carried, the missiles being stowed on one or two deck levels with four reloads on each side of the elevator.

The SS-N-12 missile is a development of the SS-N-3 Shaddock carried by the Rocket Cruisers of the 'Kynda' and 'Kresta I' classes. It has an overall length of 11.7m and superior propulsion technology gives the missile an increased range estimated at 300nm (550km).

Mid-course tracking is provided by a Trap Door radar similar in configuration to the Front Door/Front Piece antenna fitted in cruise missile submarines of the 'Echo' and 'Juliett' classes. The principal element in the antenna is a 'nodding' heightfinder of parabolic lattice configuration; it is paired with a smaller parabolic antenna which operates in the horizontal plane. The two antennae combine to provide missile tracking in three dimensions. In the *Kiev* class this antenna combination is mounted on the underside of a larged hinged door close to the ship's bows (hence the NATO designation); the door is raised hydraulically and the antenna is deployed only when the missiles are readied for firing. The height at which the antenna is mounted does not appear to be critical, which suggests that the 'cruise' phase of missile flight is conducted at high altitude.

Two or three Ka-25 Hormone B helicopters would normally be carried to facilitate missile targeting and mid-course guidance at intermediate ranges. At longer ranges, however, *Kiev* would continue to depend on the Bear D land-based bomber and satellite reconnaissance for the early location of hostile surface ships and for targeting her SS-N-12 missiles against NATO carrier battle groups. The ship would therefore have to be closely integrated into the Combined Arms framework of combat management, and anti-surface operations would be subject to land-based command and control. Long-range communications provision is markedly superior to that of the *Moskva* class, which had no anti-surface mission, and includes the distinctive Vee Bars high-frequency system and two Punch Bowl radomes that can receive target data directly from ocean surveillance satellites.

The SS-N-12 Reload System

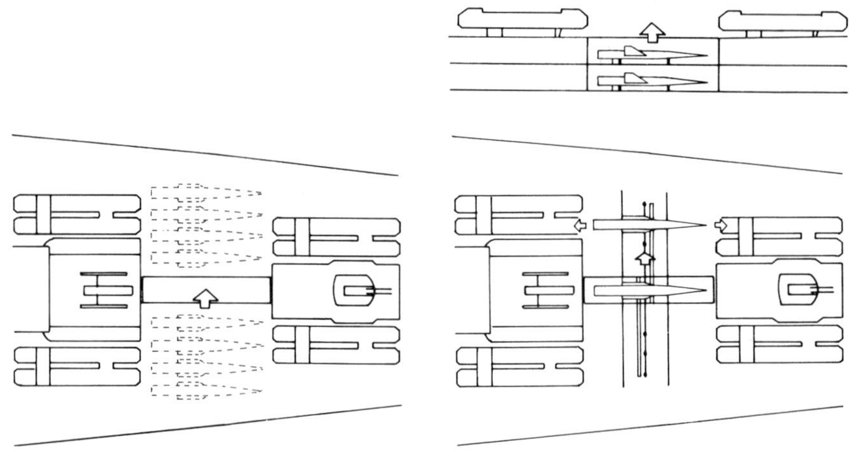

The SS-N-12 Sandbox Missile

Soviet designation: unknown
In service: 1975
Length: 11.7m(38ft 5in)
Diameter: 0.9m(3ft)
Span: 2.6m(8ft 6in)
Weight: 4,800kg(10,560lb)
Warhead: 1,000kg(2,200lb) HE/350kT nuclear
Propulsion: liquid-fuel rocket
Speed: Mach 1.4
Range: 550km(300nm)
Guidance: command/active radar homing
Fire control: Trap Door/Front Door

▶The starboard side of the superstructure of *Minsk*, with Side Globe broad-band ECM jammers flanked by smaller Bell Thump and Bell Bash ESM radomes. At the foot of the central lattice mast is Rum Tub, which made its first appearance on this class. On the far right of the picture is one of two Punch Bowl satcom antennae, which receive targeting data for the SS-N-12 anti-ship missiles. (US Navy, 1979)

▶The principal air surveillance and tracking radars of *Kiev*. On the right is the massive Top Sail antenna; on the left the smaller Top Steer. Between them, housed within a distinctive spherical radome, is the Top Knot precision approach radar, topped by a High Pole B IFF transponder. The prominent V-configuration high-frequency communications antenna has received the NATO designation Vee Bars. (MoD, 1985)

The Island Superstructure

The island superstructure of *Kiev* is approximately 90m long and accommodates the uptakes for the steam propulsion machinery, all surveillance and fire control radars, and a variety of other electronics installations and communications aerials.

The primary air surveillance and tracking radar is Top Sail, which is an essential part of the SA-N-3 area defence system. Similar in configuration but somewhat smaller is Top Steer, which made its first appearance in this class. Top Steer comprises a parabolic lattice antenna with a rectangular frontal aspect, mounted back-to-back with a conventional parabolic antenna apparently developed from Head Net A. Top Steer can track multiple targets in three dimensions and provides a useful back-up to Top Sail in the event of damage or mechanical failure, thereby ensuring that the ship can defend herself against hostile air attack at all times. Top Steer also provides the additional tracking capacity essential in a ship operating large numbers of aircraft.

Between Top Sail and Top Steer is a tall lattice mast surmounted by a large radome designated Top Knot. The latter is thought to house a precision approach radar, enabling the Forgers and Hormones to home in on their mother ship.

The electronic countermeasures outfit is exceptionally complete. The antennae are disposed vertically on each side of the island in line with the central lattice mast. Eight Side Globe broad-band jammers are grouped in pairs to port and to starboard. Beneath each group is a platform on which pairs of Bell Bash and Bell Thump ESM radomes are mounted. At the base of the lattice mast four Rum Tub ESM antennae (*see* Chapter 4) are mounted on platforms and above them are Bell Clout ESM radomes, located on narrow platforms projecting from the sides of the central lattice mast.

Directly beneath the lower ESM platform are Tee Plinth electro-optical sensors, and there are three further Tee Plinth installations along the port side of the island. Between the pairs of Side Globe antennae are Palm Frond surface surveillance/navigation radars which complement the single Don-Kay radar located between the forward Head Lights and Owl Screech fire control radars.

The distinctive Vee Bars HF communications aerials are angled out from the top of the lattice mast directly beneath the Top Knot homing beacon. Punch Bowl radomes, which receive target data for the SS-N-12 anti-ship missile system on a down-link from Soviet ocean surveillance satellites, are disposed *en echelon* opposite the Pop Group fire control radars.

A retractable spherical antenna for a Bob Tail radio sextant is located within a box housing between the after SA-N-3 launcher and the 76.2mm gun mounting. This helps to ensure the accurate position-fixing essential to collaborative targeting under shore-based control.

Replenishment

Throughout the 1960s, Soviet blue-water deployments were sustained largely by the concession of port facilities by friendly countries and by the use of sheltered anchorages in the Mediterranean and the Indian Ocean. Ships were replenished with stores, fuel and munitions while alongside or at anchor, the Soviet merchant marine being widely employed for this task. The few naval tankers capable of under way replenishment used the slow stern refuelling technique abandoned by the Western navies during the 1950s. The transfer of solid stores while under way was unknown.

The Soviet Navy began to build specialized replenishment oilers from the late 1960s, employing the latest abeam transfer techniques to supply their major warships with fuel and solid stores while under way. A number of these were already in service by the time *Kiev* was completed.

Under way replenishment became vitally important with the envisaged extension of Soviet surface operations into the Norwegian Sea and the north-west Pacific. In the *Kiev* design, particular attention was paid to the transfer of stores and munitions to and around the ship. Two constant-tension transfer stations for dry stores and munitions, and three probe-type stations for fuel and fresh water are provided on the outboard side of the island. There are also two large

◀ An overhead view of *Kiev* in 1978, operating off Iceland with a deckload of Ka-25 Hormones. One of the helicopters has just been brought up to the flight deck on the smaller of the two aircraft lifts. Note the area of the flight deck covered by the heat-resistant tiles. (US Navy)

cranes on the starboard deck-edge to assist with replenishment in port or at anchor.

Once aboard, missiles, munitions and dry stores can be moved on wheeled trolleys to any part of the ship via a 'railway track' which encircles the forecastle and runs aft to starboard of the island, directly beneath the transfer stations. The track has branch lines leading off to the SuW-N-1 launcher reload hatches, the SS-N-12 missile elevator and the three smaller elevators aft for aircraft ordnance. The forward crane is capable of transferring SAMs and SSMs to the larger missile elevator, while the after crane handles the ship's boats which are along the outboard side of the island.

Flight Deck

The flight deck is offset to port and angled out at about 3.5 degrees. Helo-circles, 9m across and numbered one to six, are marked out along the centre-line of the angled deck, with a further circle at the forward end offset to port. The latter is marked with the Cyrillic letter 'C' (roman alphabet: 'S'), and appears to be used for plane guard duties during flying operations. There is, in addition, a much larger circle, some 14m in diameter, aft. In *Kiev* this circle was initially centred on the line that marks the inboard edge of the angled deck, but in later units of the class it has been moved 2.5m to port, presumably to clear the deck park aft. The circle has an 'M' at its centre and has generally been used for Forger landings.

Squeezed between the angled deck and the island superstructure is a large lift, 18.5m by 10m, capable of accepting two Forgers or two Hormones. There is a second aircraft lift of similar length, but only half as wide, directly abaft the island. This lift can only just

▶The forward lift, approximately 6m square, can be seen beneath the nose of the helicopters on the deck park. To starboard of the lift is the grid which is thought to provide power to the deck park. Note the steel securing rings around the helicopter spots. The vehicle forward of the parked Hormones is a mobile vacuum cleaner used to clear the flight deck of foreign objects which might otherwise be ingested by the aero engines. (MoD, 1980)

accommodate the Forger with wings folded and therefore suffers the same limitations as the aircraft lifts in *Moskva* with regard to its ability to accept future aircraft. The problem is less acute in *Kiev* because of the comparatively large size of the forward lift, but this is a long way from the after deck park, the area generally used by the fixed-wing aircraft carried by *Kiev*. Moreover, the tightness of the design is illustrated by the less than satisfactory placing of the lifts which obstructs any movement of aircraft along the deck park to starboard of the angled deck.

On *Kiev* and *Minsk* there is a third lift, approximately 6m square, directly abaft the large triangular deckhouse which separates the flight deck from the forecastle area. The deckhouse itself appears to serve as a 'garage' for deck tractors (cf., *Moskva*) and for a large self-propelled vacuum cleaner used to clear the flight deck of foreign objects that might be ingested by the aero engines. The purpose of the lift appears to be to transfer tractors and stores between hangar and flight deck levels. It therefore seems likely that the forward end of the hangar serves as a parking space for tractors, fork-lifts and stores, as it does in the big US Navy carriers.

As completed, *Kiev* and *Minsk* had three munitions elevators located on the outboard side of the after aircraft lift, each about 5m by 1m and connected to the ship's 'railway system'. Their presence suggests that extensive magazines for aircraft ordnance are sited directly beneath them deep in the ship's hull. A fourth elevator of similar size is located at the after end of the angled deck and appears to serve a stores handling bay immediately forward of the ribbed door in the stern.

There are two major areas available for deck parking. The first is alongside the superstructure forward of the main aircraft lift and can accommodate two pairs of Hormones or Forgers. Steel rings disposed in a circular pattern are set into the deck for securing purposes. (Similar rings are provided for each of the helo-circles.) There is also what appears to be a grid running beneath the inner two aircraft spaces, the purpose of which may be to provide an electrical supply for heating or power.

The other deck park is on the starboard side aft. It can accept about five aircraft of either type, angled inboard towards the flight deck, and is frequently used by the Forgers. This is a convenient arming point for the ship's aircraft because of the proximity of the three munitions elevators, which are served by a pair of railway lines for wheeled trolleys.

The after deck park and the entire angled portion of the flight deck were initially covered with brownish heat-resistant tiles about 0.5m square. These were clearly thought necessary to deflect the intense heat generated by the lift engines of the Forger during VTOL operations, although they have since been removed. The extent of the tiling, which covered all seven of the helicopter spots as well as the Forger landing circle, indicates that air operations involving multiple Forger take-offs were envisaged as part of the ship's operational concept from the outset.

Air Group and Hangar

US Navy observations of *Kiev*'s initial deployment established that the air group comprised nine Yak-38 Forger A tactical VTOL fighters, a single Forger B trainer, fourteen Ka-25 Hormone A anti-submarine helicopters and a single Hormone B missile targeting helicopter. This is undoubtedly less than a full complement of aircraft, but is clearly closer to the truth than initial European estimates of up to 50 aircraft.

The contours of the hangar itself can be ascertained from certain visible features such as the position of aircraft lifts, munitions elevators, side recesses and weapons magazines. The hangar is enclosed, on the British pattern, and stretches from the end of the flight deck to the after end of the ship. The width of the forward section of the hangar is constrained by the large island superstructure, the inner edge of which is some 15m inboard. (US Navy carriers have their island placed on the flight deck overhang with access via the supporting sponson, thereby maximizing hangar space.)

The hangar widens abaft the superstructure, but the positioning of the three munitions elevators just outside the starboard edge of the after aircraft lift suggests that the latter marks the outer limit of the hangar to starboard. It would seem logical that the port side of the hangar follows the inner edge of the recess for the ship's boats aft, with aircraft workshops to the side, on the British pattern. This would give convenient hangar widths throughout: 15m for three Hormones or three Forgers alongside forward, and 21m for four of either aircraft aft. The after end of the hangar probably coincides with the narrowing of the flight deck just forward of the gatling sponsons.

If the preceding assumptions hold good, there would be a forward hangar section of about 25m by 15m between the cargo elevator and the larger of the two aircraft lifts, a centre section of the same width but about 55m long extending from the forward aircraft lift to the after end of the second lift, and an after section 45–48m long and 21m wide. These three sections could be isolated from one another by fire curtains for damage control purposes. Such a hangar could accommodate an air group comprising a squadron of twelve

Kiev: Inboard Plan

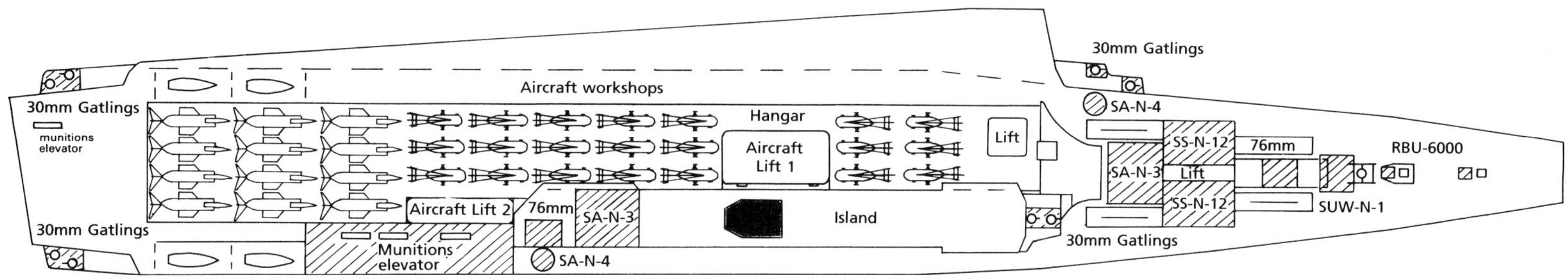

Yak-38 Forger As, plus some 18–21 helicopters, of which perhaps fifteen would be anti-submarine types, the remainder being for missile targeting and SAR duties.

It appears likely that the after lift, like those of the *Moskva* class, can be loaded from the side or from the rear. However, the forward lift-well probably divides the hangar into two separate parts with access only from forward and aft.

Flying Operations

The Yak-34 Freehand developmental VTOL aircraft which made its first public appearance in 1967 was, like the British P.1127 Kestrel, powered purely by vectored thrust. But the Kestrel was not the only Western experiment in vertical lift techniques; other European nations were attempting to develop their own VTOL aircraft using different technology. The French Mirage III prototype of the early 1960s attempted to combine separate vertical lift jets for take-off and landing with a conventional propulsion system for forward flight, while the German VAK-191 tactical reconnaissance fighter attempted to combine both techniques by coupling vertical lift jets with main engines using vectored thrust.

The Soviet Union was well aware of these developments and followed them closely. A Soviet counterpart to the French Mirage III, the Mig-21 PFM Fishbed G, was developed simultaneously with the Yak-34 Freehand, and further prototypes employing lift jet/conventional propulsion followed.

The initial Soviet experiments were not successful. Soviet engine technology lagged badly behind that of the West, especially in the area of advanced, high bypass turbofans such as the Rolls-Royce Pegasus which powered both the Kestrel and the Harrier. Trials with the Yak-34 Freehand, therefore, did not provide the basis for a production aircraft powered exclusively by vectored thrust. On the other hand, the lift jet/conventional propulsion system of the Mig-21 PFM almost certainly suffered similar problems to those experienced by the French with the Mirage III which had difficulty in making the transition from vertical lift to forward flight.

The Soviets opted for the conservative and relatively complex solution of combining separate lift jets housed in the forward fuselage with a separate vectored-thrust engine to provide both lift and forward thrust. This solution allowed the use of the Soviet Union's well-developed turbojet technology for the lift jets, while the vectored-thrust main engine provided stability during the critical transition to forward flight.[2] The model for both the propulsion system and the airframe itself appears to have been the German VFW-191B to which the Forger bears a striking resemblance (and which, ironically, was to be abandoned at the prototype stage).

The Forger appears to be very stable during take-off and landing operations, although the transition from vertical to horizontal flight is carried out very slowly and carefully and appears to be susceptible to turbulence. The multiple-engine configuration brings with it a number of serious disadvantages as compared with a pure vectored-thrust aircraft such as the Harrier. The reliable operation of all three engines is crucial to the aircraft's stability during take-off and landing, yet the likelihood of engine failure in one of the three power units is between six and eight times greater than that of a single-engine aircraft. The problem is exacerbated by the dependence on a successful relight of the two lift engines which are shut down during horizontal flight prior to landing.

The smoothness of the Forger take-off and landing operations suggests an automated handling system to synchronize the lift jets with the vectored-thrust nozzle rotation, but the provision of a level of computerized control sufficient to permit a rolling take-off has clearly presented more of a problem, as this was not achieved until 1985. Even so, the Forger can attain only relatively low speeds by the end of its run, whereas the British Sea Harrier can boost take-off speed to 100–120 knots in the same distance. The latter factor has important consequences for payload because wing lift depends on speed squared. This particular performance gap has been further widened by the British development of the ski-jump which has yielded significant benefits in terms of the take-off weight of the aircraft.

The lengthy transition from vertical to horizontal flight makes for a slower reaction time, increased vulnerability and high fuel consumption. The Forger takes approximately 90 seconds to accomplish its transition – three times as long as the Harrier – and consumes an estimated ten times as much fuel in the process. The small wing area undoubtedly results in a particularly high wing loading. The thrust-to-weight ratio, on the other hand, is exceptionally low and there is only a very limited VIFF (vectoring in forward flight) capability because the Forger has only a single pair of vectored-thrust nozzles for its main propulsion engine as compared with two pairs for the Pegasus which powers the Harrier. This limits combat manœuvrability and, consequently, the effectiveness of the Forger as an interceptor.

US Navy observations of the flying operations conducted aboard *Kiev* during her first deployment established a number of interesting differences from Western practice. There is a distinct lack of ground-crew participation in launching and landing procedures. All instructions for movement, including taxiing, take-off and landing, appear to be directed by radio from the spacious flying control station situated directly beneath the after Head Lights fire control radar. No external electrical power or ground support is used by the Forger during its start-up sequence. Chocks and chains are removed prior to engine start, after which the ground crew retires.

The landing sequence starts far astern with a steady descent until the last 500m. The Forger then decelerates smoothly at a steady 30m above sea-level. Two microwave landing systems are fitted low on the inboard side of the island, directly beneath the flying control station, to help the pilot find the correct approach line. The final approach is made at 10–15m above the flight deck at an aircraft speed of about five knots. The Forger then enters its hover phase and lands vertically. On landing, chocks and chains are re-secured following engine shut-down. Re-spotting is done by a deck tractor.

Arming, servicing and maintenance of the aircraft was initially performed exclusively on the hangar deck, but as the Soviets have gained experience with the *Kiev*s, more and more of these operations have taken place on the flight deck. In addition to the munitions elevators served by 'railway' tracks, fuelling lines and umbilicals for automatic test equipment can be accessed via hinged doors at the edge of the flight deck, and these facilities have been extended as the ships have been refitted.

Missions for the Forger

Even with external fuel tanks, the combat radius of the Yak-38 Forger must be extremely limited – *Jane's All the World's Aircraft* has estimated it at no more than 100nm and 25 minutes on station when equipped with air-to-air missiles and external tanks. The relatively short loiter time together with poor combat manœuvrability make the Forger inherently unsuited to the role of combat air patrol. *Kiev* carries no airborne early warning aircraft (the Ka-25 Hormone B is equipped only for surface surveillance, in keeping with its primary missile targeting mission).

Moreover, the Forger appears to have no long-range sensor systems, either active or passive, with the sole possible exception of a Sirena III Radar Warning Receiver which is fitted in most Soviet aircraft for the early detection of enemy radars that have 'locked' on to them.[3] The primary sensor, housed in the dielectric nose cap, appears to be a range-only radar to assist weapons delivery against surface or ground targets. This feature serves to distinguish the Forger from the Sea Harrier which is fitted with a nose radar for fleet air defence.

It therefore appears that the SA-N-3 area defence system continues to provide first-line defence against high-performance attack aircraft. However, the Forger would be useful if employed as an interceptor against isolated, slow-moving aerial targets such as the E-2C Hawkeye on which US carriers depend to control their own combat air patrols and to direct their attack aircraft against hostile surface targets, and Western carrier-borne or land-based anti-submarine patrol aircraft such as the British Nimrod, the P-3 Orion and the S-3 Viking. The VTOL aircraft could thereby make a useful con-

▶A close-up of three Yak-38 Forger A VTOL aircraft on the after deck park of *Kiev*. A 23mm gun pod is fitted on the outer pylon of the first aircraft. The intake and outlet doors for the vertical lift jets are in the open position.

▶Three Forger pilots in conversation on the flight deck of *Minsk*. They appear to be wearing immersion suits beneath their coveralls and are equipped with backpack parachutes.

▶A Yak-38 Forger takes off on a mission. The vertical lift engines give added stability at the cost of heavy fuel consumption. There are four underwing pylons, and these would generally be occupied by fuel tanks on combat missions. If gun pods are fitted, they too have to be carried on the underwing pylons, the under-fuselage position being unavailable because of the lift jets. (All TASS)

The Yak-38 Forger VTOL Strike Fighter

In service: 1976
Length: 15.5m(51ft)
Span: 7.3m(24ft); 4.9m(16ft) wings folded
Height: 4.4m(14ft 4in)
Weight: 7,485kg(16,500lb) normal; 11,700kg(25,750lb) max. take-off
Engines: one Tumansky R-27V-300 vectored-thrust turbojet, 6,950kg(15,300lb) thrust; two Koliesev/Rybinsk RD-36-35FVR lift jets, each 3,050kg(6,725lb) thrust
Speed: 1,009km/h(545kts=Mach 0.95) at high level; 978km/h(528kts=Mach 0.8) at sea-level
Ceiling: 12,000m(40,000ft)
Combat radius: 185km(100nm) as interceptor (includes 75 min loiter time); 370km(200nm) in high-level attack role; 240km(130nm) in low-level attack role
Armament: as interceptor, two AA-2-2 'Advanced Atoll' or AA-8 Aphid AAMs, two 23mm GSh-23 cannon; as strike aircraft, two AS-7 Kerry ASMs, or 500kg bombs, or rocket pods (16–32 rockets). Two pylons under each wing for 2,600–3,600kg(5,730–7,935lb) max. load
Sensors: ranging radar; Odd Rods IFF; Sirena passive warning

tribution to the defence of friendly surface forces and provide support to Soviet submarines deploying through – or patrolling within – their operational area.

Interceptions would be made under the control of the Anti-Submarine Cruiser itself, which is extensively equipped with air surveillance and tracking radars. While this technique has been abandoned by the post-war US Navy, which has preferred airborne control as a means of extending the defensive ring surrounding its carrier battle groups, it remains the favoured method of control for land-based fighter aircraft, particularly in the Soviet armed forces.

The Forger has four underwing pylons for the carriage of munitions and stores. Each has a load capacity estimated at 250kg (550lb) and a wide variety of air-to-air and air-to-surface weapons can be carried. The primary air-to-air missile is the AA-8 Aphid, a short-range weapon which entered service at about the same time as the Forger and which, like the US Sidewinder, employs infra-red (IR) guidance. GSh-23 23mm gun pods and drop tanks can also be carried. The primary anti-surface weapon is the AS-7 Kerry missile with a maximum range estimated at 6nm (10km), while free-fall iron bombs and 57mm pod-mounted rockets can be used against ground targets.

The Forger therefore has a useful capability against fast patrol craft armed with SSMs or anti-submarine patrol vessels operating without air cover. Such targets might well present themselves if the Soviet surface units were operating close to the coast of Norway,

▲**Yak-38 Forgers on the flight deck of *Minsk*. (US Navy)**

either in the course of routine anti-submarine operations or if acting in support of amphibious landings conducted by the Naval Infantry of the Northern Fleet. It could also be used to provide support for the Naval Infantry or the Soviet ground forces, the most likely theatres being those of the Northern Fleet and the Pacific Fleet, to which all four ships of the class have deployed.

Novorossiysk

Kiev and *Minsk*, completed in 1975 and 1978 respectively, were virtually identical. However, the third ship, *Novorossiysk*, which entered service in 1982, featured a number of important modifications, some of which were clearly intended to correct design defects which had become apparent during operations with the earlier two ships.

When *Kiev* first appeared, Western commentators with experience of fixed-wing carrier operations were quick to point out that the exaggerated sheer of the forecastle and the 'spikiness' of the island superstructure, which carries the first control radars for the 'cruiser' weapons, would result in considerable air turbulence which would adversely affect Forger take-offs and landings. Neither feature of the ship's design could be significantly altered without resort to major reconstruction, but *Novorossiysk* introduced a number of modifications aimed at improving airflow over the flight deck, and these were soon to be retro-fitted to the two earlier ships.

There are three prominent wind deflection plates on the forecastle to port of the after group of SS-N-12 launchers. Farther aft, at the port-side corner of the curved deckhouse which doubles as a tractor garage and a blast deflector for the SSMs, there are triangular shields which appear to have a related function.

A further structural modification is the remodelling of the port-side forward sponson, the 'kink' between

▲ A port quarter view of *Novorossiysk* in the Pacific, with a Yak-38 Forger B taking off vertically from the 14m circle at the after end of the angled deck. Five other Forgers occupy the after deck park. (US Navy, 1986)

the first and the second gatling mountings being eliminated. *Minsk*, in a refit which took place in 1981–2, had her own sponson completely rebuilt, with a gently angled forward section and a streamlined, less angular underside, but the sea-keeping problem was not finally resolved until the sponson was removed altogether in the fourth ship, *Baku*.

Of greater interest than these somewhat makeshift structural modifications to *Novorossiysk* are the radical alterations to the layout of the flight deck. The after deck park, used by both Hormones and Forgers, has been significantly enlarged and now extends over the former deck edge gallery; the cocooned life-rafts have been relocated beneath the overhang of the flight deck. The deck park has also been extended aft and has been squared off above the sponson on which the AK-630 gatlings and the starboard-side chaff-launcher are located.

At the forward end of the deck park more space has been created by the removal of the third munitions elevator. The 'railway track' which previously crossed the deck park has also been removed and now terminates at the second munitions elevator. At the stern, immediately to starboard of the angled deck, a new position has been created to enable the Forgers to warm up their engines prior to taxiing for take-off. This position can be easily distinguished in aerial views of the ship because of the black deposits on the flight deck. There are doors beneath the warm-up spot, and these probably hinge downward to allow the deflected blast from the jet exhausts to be vented through the starboard aperture in the stern.

The enlarged deck park has been given a new set of markings to denote individual parking spaces. One set of lines painted yellow is angled forward, and these lines have tie-down points set into the deck at their outboard ends. There are six positions numbered 1–6 on the main area of the deck park, with a further two

Novorossiysk: After Deck Park

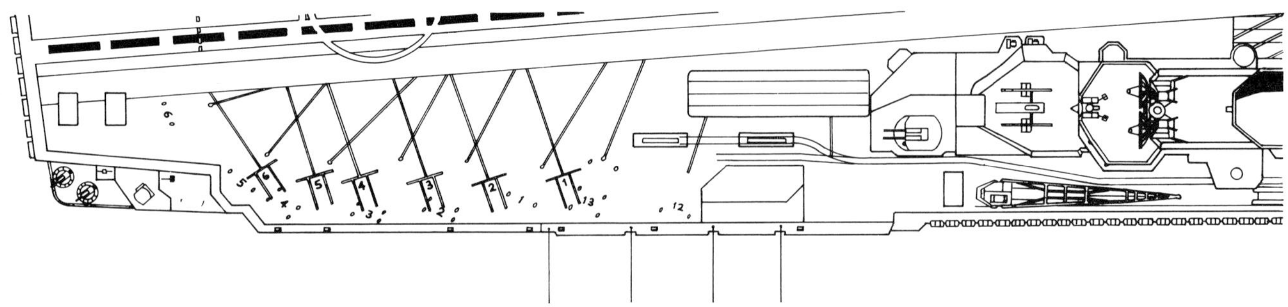

numbered 12 and 13 close to the remaining munitions elevators. Criss-crossing lines 1–6, and angled aft, are six blue lines also numbered 1–6. These lines do not appear to be provided with tie-down points.

A separate control cabin has been constructed at the after angle of the island superstructure close to the aircraft lift, and this may be for control of aircraft movement on the deck park.

Finally, at the forward end of the flight deck the 6m square lifts which on *Kiev* and *Minsk* was located directly abaft the tractor garage, has been suppressed. The rationale for this modification is not entirely clear, but the lift did intrude both on the forward deck park and on the hangar deck, and may possibly have been considered a luxury in view of its proximity to the main aircraft lift.

Weapons Systems Changes

Novorossiysk was the first unit of the *Kiev* class to operate the improved Ka-27 Helix A anti-submarine helicopter, first seen aboard the destroyer *Udaloy* (*see* Chapter 7). Her air group was otherwise unchanged.

A significant difference from the earlier two ships is that the elevator for the SS-N-12 reload magazine has been suppressed. Instead, the wheeled reloading tray which is used to transfer missile rounds to the launchers is kept permanently on deck; presumably it is used only when the ships are alongside. The elmination of the SS-N-12 reload magazine has effectively halved the complement of missiles, but may have served to free hull volume beneath the forecastle deck for additional accommodation in what remains a very 'tight' design. The ready-use missile battery remains unchanged at eight missiles.

A further difference is that the position of the AK-630 gatlings located at the forward end of the island superstructure has been modified to improve the arcs of the upper mounting which can now cover the port side.

The most important modification, however, is the suppression of the SA-N-4 Gecko short-range, surface-to-air missile system in favour of the SA-N-9 missile system developed as the primary air defence weapon of the contemporary destroyers of the *Udaloy* class (*see* Chapter 7). As with the latter ships, problems have been experienced in getting the SA-N-9 into full operational service, and at the time of writing *Novorossiysk* has still to have the system installed.

The positions reserved for the SA-N-9 launcher bins have been covered with steel plates as an interim measure, but it is clear from the configuration of the plates that each installation will comprise a group of six launchers. The forward group of launchers replaces the original SA-N-4 launcher on the port side of the forecastle, but there was clearly inadequate space abeam the island superstructure for the after group which has consequently been moved farther aft; it is located outboard of the after aircraft lift and the munitions elevators which serve the deck park.

The Cross Sword missile fire control radars (yet to be fitted) will occupy the same positions as Pop Group on the earlier two ships, but Cross Sword is significantly larger and heavier than Pop Group, and the platforms have been 'beefed up' accordingly.

Electronics

The major air surveillance and tracking radars show no changes from the earlier two ships, but important modifications have been made to other systems, particularly the electronic countermeasures outfit.

The most significant omission is that of Side Globe, *Novorossiysk* being the first major Soviet warship to be completed without the distinctive ECM radomes since they made their first appearance on *Moskva* and *Vitse-Admiral Drozd* in 1967. Somewhat surprisingly no suc-

▲*Novorossiysk* operating 750 miles west of Midway Island in 1985, at the centre of a nine-ship 'task force'. From the mid-1980s onwards the Pacific-based ships regularly simulated strikes by US carrier task forces against the Siberian coast. (US Navy)

▶Close-up of a Ka-27 Helix A anti-submarine helicopter aboard *Kiev*. Although of broadly similar configuration to the Ka-25, the Ka-27 has greater endurance and more sophisticated avionics than its predecessor. (MoD, 1985)

▶ ***Novorossiysk*, photographed in 1983, shortly after her completion. The angled wind deflection plates on the foredeck are visible, as is the plating over the after SA-N-9 missile installation. The forward edge of the flight deck has been rounded off, and the after deck park extended to starboard. Note the reloading tray for the SS-N-12 missiles, which is permanently on deck on *Novorossiysk* because of the suppression of the missile elevator. (MoD)**

137

cessor is fitted, nor do there appear to be empty platforms reserved for such a system as in the early units of the *Sovremenny* and *Udaloy* classes. In their place are Strut Pair back-to-back air surveillance antennae mounted on prominent platforms directly below the lattice mainmast to port and to starboard.

A third Palm Frond surface surveillance radar has replaced the single Don-Kay antenna previously located beneath the forward Head Lights fire control radar, and the four Tee Plinth electro-optical devices have been superseded by the more advanced Tin Man first seen aboard the Rocket Cruiser *Kirov* in 1980 (*see* Chapter 6).

Service Life

On completion of her trials *Kiev* joined the Soviet Northern Fleet, and although exercises have been conducted in both the Mediterranean and the Baltic, she has remained in that area ever since, returning only once to the Black Sea Shipyard where she was built for a major refit in 1984–5.

The second ship, *Minsk*, deployed to the Far East in the spring of 1979. The Soviets took advantage of this deployment, which was accompanied by the new landing ship *Ivan Rogov*, to assemble a large fleet in the Mediterranean which included both *Kiev* and *Minsk* – the only time these two ships have operated in company with each other. A major amphibious exercise was conducted off the coast of North Africa with *Rogov* as the central element of the assault force. The landings were supported by one of the *Kiev*s, which provided additional helicopters to augment the vertical assault forces and employed Forger VTOL aircraft in ground support operations, while her sister-ship conducted anti-submarine operations around the landing force.

Two years later, in the summer of 1981, *Kiev* entered the Baltic where she joined the PKR *Leningrad* and seventeen other surface warships in supporting ZAPAD 81, a major amphibious exercise involving some 45 amphibious vessels. Opposed landings were made with the assistance of land-based aircraft to suppress coastal defences. Exercises such as the above reveal an interest in testing the full range of capabilities of these ships and, in particular, the added dimension accorded them by their squadron of Forger VTOL strike fighters.

More typical perhaps of their intended deployment was the exercise which took place in the Northern Fleet area in July 1981, in which *Kiev* operated with major surface forces that included four 'Krestas', six 'Krivaks', a *Sverdlov* and a 'Kotlin'. The exercise involved a move outwards into the Norwegian Sea, ending up in the vicinity of Iceland. The focus of this exercise appears to have been to provide support to Soviet submarines deploying into the North Atlantic, and to counter NATO surface forces and land-based anti-submarine aircraft patrolling the GIUK Gap.

It had been expected that the third ship, *Novorossiysk*, would join *Kiev* in the Soviet Northern Fleet, but in early 1984, following completion of her trials in the Black Sea, she sailed for the Far East. The Soviets clearly perceived a major change in US policy during the early 1980s away from a NATO-centred defence strategy and towards a more aggressive maritime

▲The island superstructure of *Novorossiysk* reveals a number of important changes. The electronic countermeasures outfit has been completely revised, and there are Strut Pair air/surface surveillance radars port and starboard. Note the Tin Man electro-optical directors which have superseded Tee Plinth. (MoD)

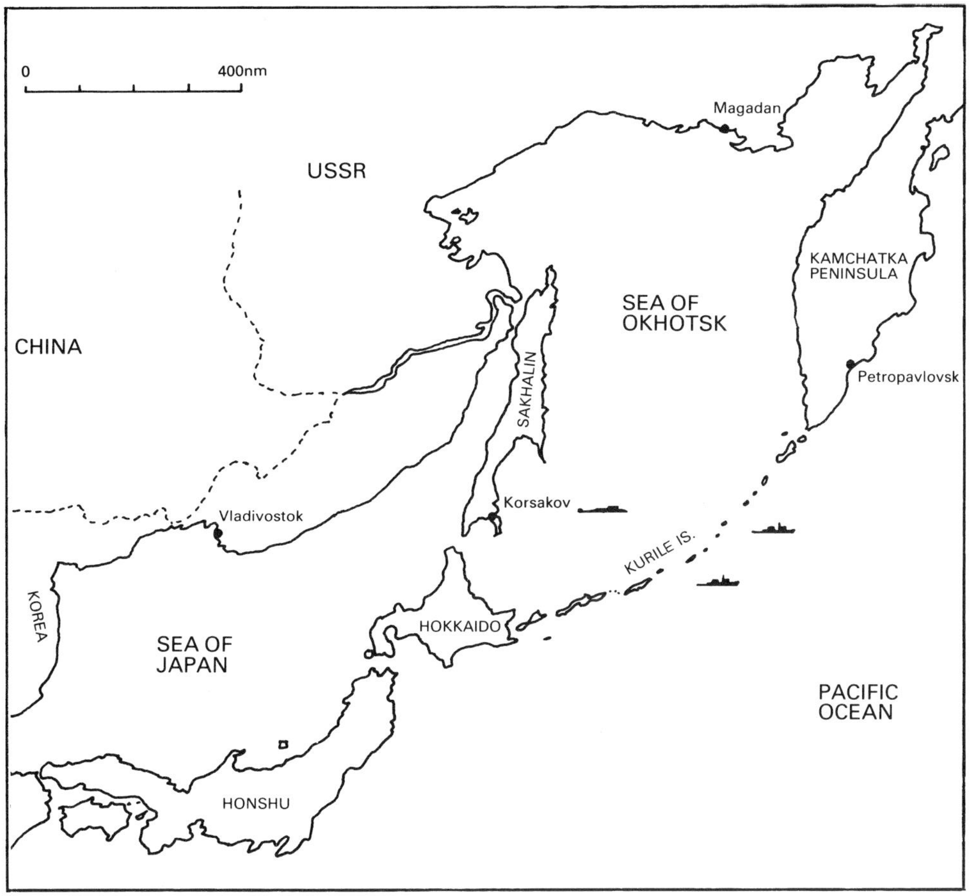

The North-west Pacific

Contrary to expectations, both *Minsk* and *Novorossiysk* have deployed to the North-west Pacific where, in the event of a major conflict, they would operate in defence of the Soviet SSBN bastion in the Sea of Okhotsk.

ECM Outfit

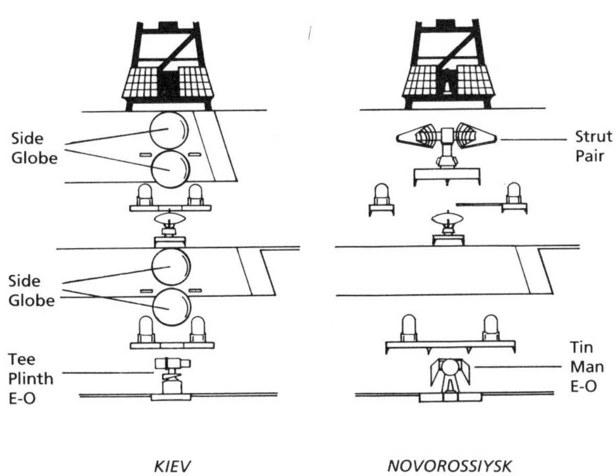

strategy in the Pacific, which aimed to exploit the inherent weaknesses on this relatively open flank of the Soviet Union. *Novorossiysk* was only one of a number of major surface vessels transferred from the 'western' Fleets during this period.

Construction of the *Kiev* class has proceeded more slowly than initially expected – according to early Western Intelligence estimates, all four ships were to have been completed by the early 1980s. However, the fitting out of *Novorossiysk* appears to have been delayed by changes to the original plans submitted by the Soviet Navy, while the fourth ship, *Baku*, was built to a modified design and took a full ten years to complete. Her configuration is now so different from her sisters that she constitutes a sub-type of her own. Redesigned to serve as a trials ship for equipment destined for the new class of through-deck carriers, she will be considered in a later chapter.

Refits for *Kiev* and *Minsk*

In 1981–2, shortly after her arrival in the Far East, *Minsk* underwent a refit in which a number of modifications were made to the forward end of the flight deck. In addition to the rebuilding of the port-side sponson detailed above, she was fitted with angled wind baffles similar to those installed in *Novorossiysk* on completion.

Both *Minsk* and *Kiev* initially had a squared-off leading edge to the angled deck, to which hinged panels serving as wind-breaks were attached. In her refit *Minsk* had the leading edge of her flight deck streamlined by the addition of a prominent curved fairing into which three hinged flaps were incorporated, and this arrangement was repeated in *Novorossiysk*, which emerged in 1983 with an identical configuration.

These modifications were clearly considered the minimum necessary to ensure a smoother airflow to Forger VTOL operations, but in 1984 *Kiev* returned to the Black Sea Shipyard for a more extensive refit which would allow her to operate improved models of aircraft.

She received the wind baffles and the modification to the leading edge of the angled deck, but her forward sponson remained unmodified which suggests that the modifications made to the sponson in her two sister-ships did not result in any significant improvement in sea-keeping.

A line of raised steel studs has been added to the port-side edge of the angled deck, which was originally marked only by a broken line. The purpose of these is unclear, but given the proximity of the helo-circles to the deck-edge, they may have been fitted to prevent aircraft from rolling over the ship's side in heavy weather.

◀ ***Minsk*, the first of the class to deploy to the Pacific Fleet, received a number of modifications in a refit in 1981–2. The leading edge of the flight deck has been rounded, and three wind deflection panels have been constructed on the port side of the forecastle. The port-side forward sponson has been rebuilt to a more streamlined configuration. The heat-resistant tiles have been removed from the angled deck and the after deck park in this photograph, taken in 1985. (US Navy)**

Other structural modifications reflect improvements already made in *Novorossiysk*. A Forger warm-up position, complete with deflector plates, has been installed at the after end of the flight deck. The large-diameter circle for Forger landings has been moved to port to clear the after deck park, although no attempt has been made to enlarge the deck park itself. Rather than suppress the after munitions elevator to create additional space, the Soviets have combined the second and third elevators into a single elevator, some 12.5m long. This is somewhat puzzling given the limited physical dimensions of the ordnance associated with the Yak-38 Forger.

Two prominent metal housings have been installed on the starboard side abeam the munitions elevators. These housings appear to have large-diameter hoses attached which connect them with the flight deck gallery. In view of their proximity to the deck park, it seems likely that they are tanks for aviation fuel and that smaller-diameter hoses are attached to the inboard ends for refuelling aircraft on deck.

It was aboard *Kiev* in 1985, during her post-refit deployment to the Northern Fleet area, that the first rolling take-offs with the Forger were observed. This new development necessitated changes in the flight deck markings which had previously been set up only

▼*Kiev* in the spring of 1985, following her major refit in the Black Sea. She has received similar modifications at the forward end of the flight deck to her two sisters, but the configuration of the port-side sponson remains unchanged. The Ka-27 Helix A anti-submarine helicopter has superseded the Ka-25 Hormone. (MoD)

for vertical take-off. The new element is an inverted 'V', some 150m in length, running from the large-diameter landing circle to the forward edge of the angled deck. The convergent lines appear to be painted red to avoid confusion with the broken white line which marks the centre of the angled deck. The outboard line runs parallel to the flight deck edge so that the 'V' is at approximately 8.5 degrees to the ship's axis.

Besides the new variant of the Forger, *Kiev* has operated a squadron of Ka-27 Helix A anti-submarine helicopters since her refit. This almost certainly entailed modifications to maintenance facilities, because her sister *Minsk* continues to operate the older Ka-25 Hormone A at the time of writing.

Port-Side Sponson: A Comparison

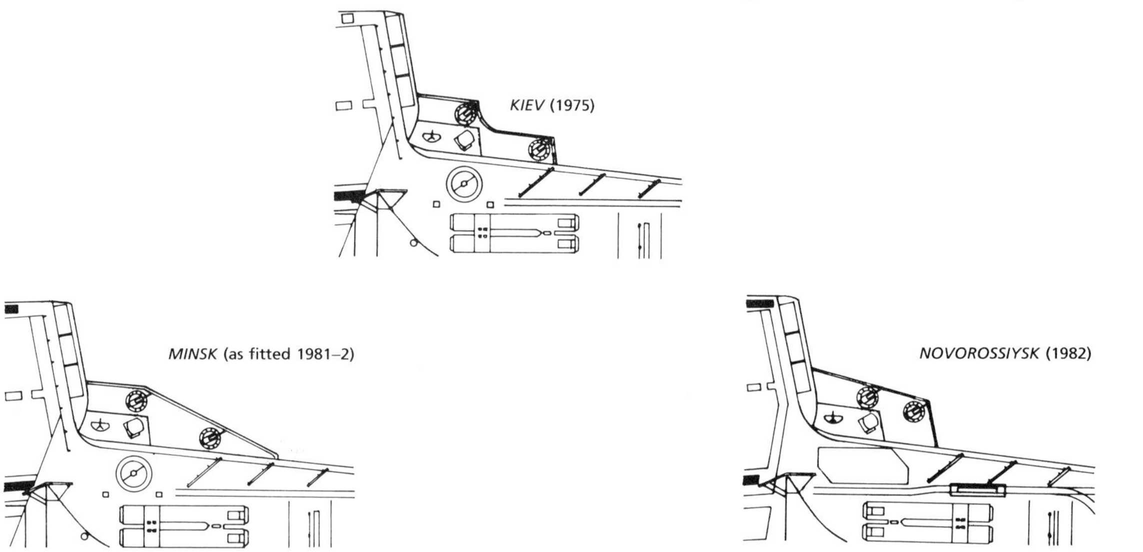

Flight Deck of *Kiev* Post-Refit

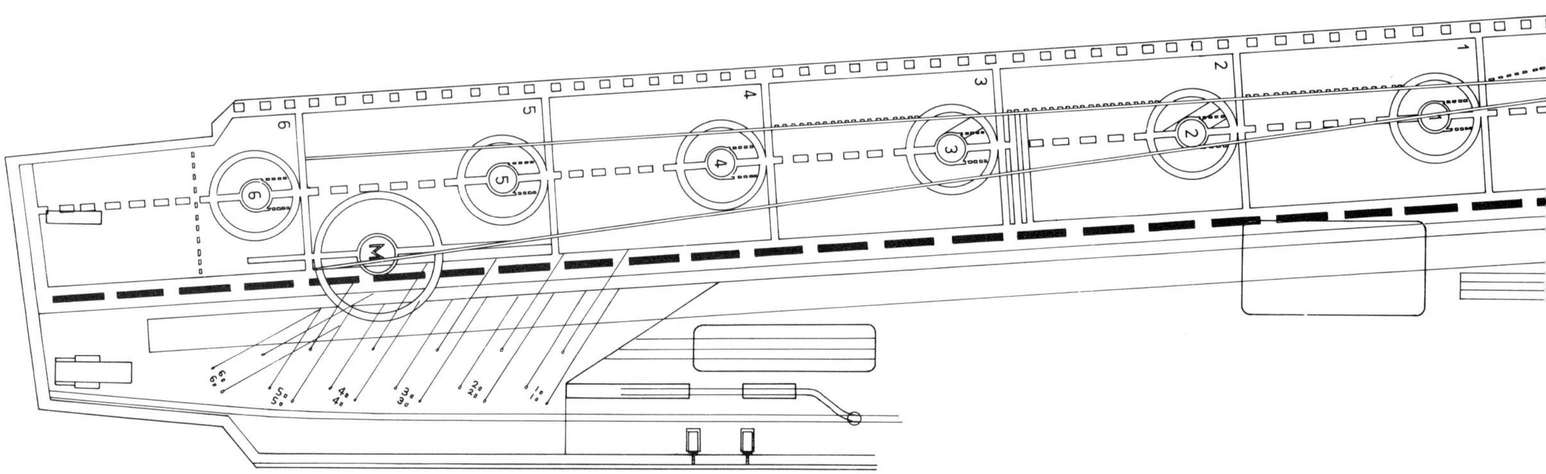

▶An overhead view of *Kiev* as refitted in the mid-1980s. The port side of the angled deck is now lined with raised metal 'studs', and there are new markings in red to enable the Forger to conduct rolling take-offs. (MoD)

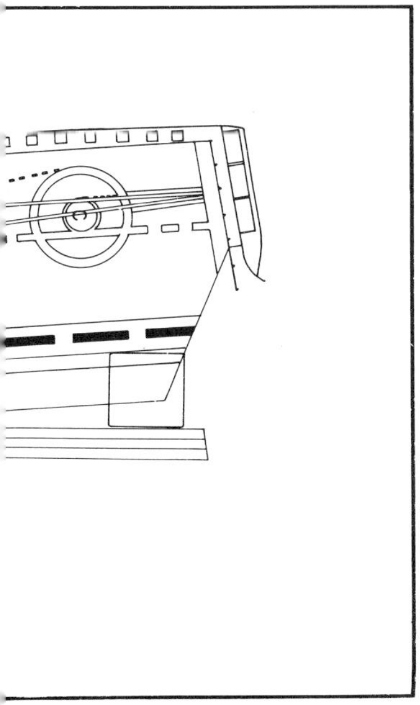

181

6. The New Rocket Cruisers

◀ ***Kirov*** **was the first Rocket Cruiser completed for the Soviet Navy since the 'Kresta I' of the late 1960s. She is the largest conventional surface combatant to be built for any navy since the Second World War. (MoD, 1980)**

BY THE EARLY 1970s the Soviet anti-submarine construction programme was in full swing. The Rocket Cruiser programme which dominated the early 1960s had been abruptly terminated, the last ship of this type having been laid down in 1966. This premature halt in the Rocket Cruiser programme did not signal a loss of faith in the original concept, however. Rather, it indicated a sudden and dramatic change in strategic priorities which demanded that all construction capacity in the shipyards be harnessed to the anti-submarine programme. The conversion of the 'Kresta' from a Rocket Cruiser to a Large Anti-Submarine Ship was symptomatic of this change.

Similar effects were experienced by the Soviet Navy's submarine arm, priority being accorded to the new generation of ballistic missile submarines which the BPKs were designed to support and protect. A new class of submarine with a primary ASW mission (the 'Victor') played a subsidiary, but increasingly important role in the new submarine programme. By way of contrast, the importance of the cruise missile submarine declined. 'Charlie', equipped with the short-range SS-N-7 'pop-up' missile, was better suited to harassment of the US carrier battle groups than to the large-scale combined arms strikes for which its predecessor, 'Echo', had been intended. But the construction rate was only two boats per year, declining further in the mid-1970s to one per year ('Echos' were being completed at an average rate of six boats per year during the mid-1960s, at the peak of the Soviet cruise missile submarine programme).

This did not mean that the Rocket Cruiser programme had been abandoned altogether, merely that it had been temporarily submerged by other more pressing demands. Not only were all the major Soviet naval shipyards fully occupied, but other relatively minor yards had to be upgraded in order to cope with the sheer scale of the anti-submarine programme (*see* Chapter 4). In the interim the Soviet Navy would continue to rely on the surface ships, submarines and long-range land-based bombers completed during the 1960s to defend the Soviet homeland against assault by the carrier task forces, which now posed less of a threat to Soviet territory than the nuclear-powered submarines armed with the Polaris or Poseidon ballistic missiles.

The long-range cruise missile remained the primary offensive weapon of the Soviet Navy against hostile surface forces, however. Its supremacy was in no way threatened by the Yak-38 Forger VTOL aircraft carried by *Kiev*, which reaffirmed established Soviet naval tactics by carrying the heaviest battery of long-range anti-ship missiles yet seen on a Soviet surface combatant. By 1970–1, some five years after the framework of the anti-submarine programme had been decided upon, the Soviet Navy was contemplating the construction of its largest Rocket Cruiser yet, an anti-surface counterpart to the PKRs of the *Kiev* class. Even more ambitious in certain respects than the latter design, the *Kirov* class would have a full-load displacement of 28,000 tons and would be powered by nuclear reactors.

Few of the traditional Soviet naval shipyards would have been capable of building a ship of this size. Of those engaged in surface ship construction during the early 1970s, only the Black Sea Shipyard would have had slipways of sufficient length to accommodate the hull of the new Rocket Cruisers, which would have a length overall of 248m (813ft) and a beam of 28m (92ft) – the dimensions of a Second World War battleship rather than that of a modern cruiser. But this shipyard was already fully occupied with the *Kiev* programme.

The choice therefore fell on the Baltic Shipyard which, as the Ordzhonikidze Shipyard, had laid down battleships of the *Sovietsky Soyuz* class and in the early 1950s had begun work on the battlecruiser *Moskva* as part of Stalin's post-war construction programme. Ordzhonikidze had also been responsible for a substantial part of the *Sverdlov* cruiser programme before it was terminated by Krushchev; the yard was subsequently turned over to mercantile construction. The administration of Leonid Brezhnev was to prove far more responsive to the demands of the military and there seems to have been little resistance to the expansion of the Soviet Navy's construction programme into a shipyard which since 1956 had built only merchant ships.

'Reconversion' to military construction seems not to have been without problems; building a modern naval vessel of the complexity of *Kirov* requires very different skills from those demanded by the construc-

Kirov: Profile and Plan

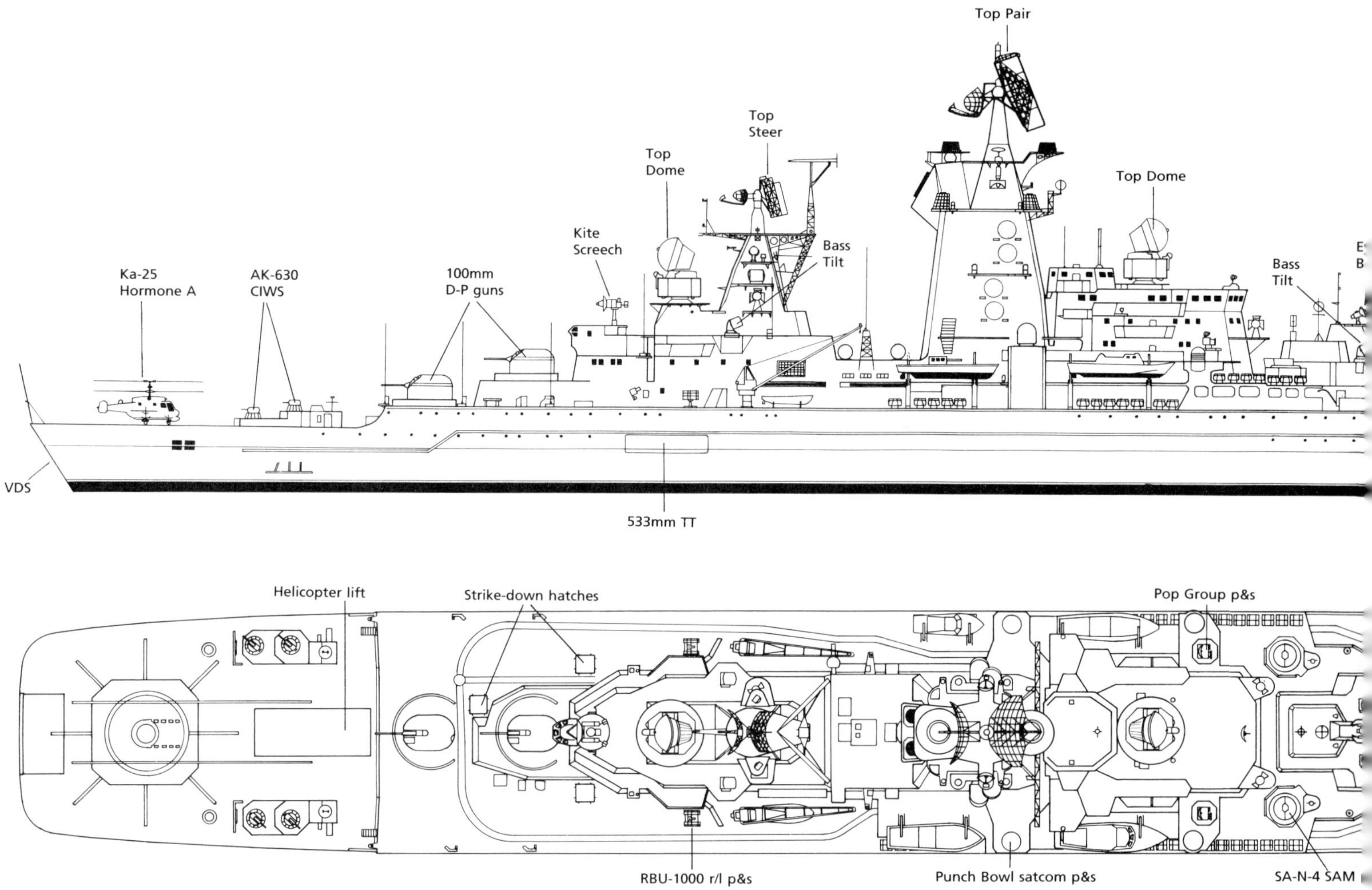

tion of an oil tanker or a bulk carrier. The first unit of the new class took more than seven years to complete (cf., *Kiev*, which was completed in under five years by the Black Sea Shipyard). In later units the construction span has been steadily reduced, construction of the third ship, *Kalinin*, taking approximately five years and six months.

The Baltic Shipyard was not totally without experience of building ships with sophisticated electronics; it had recently completed the civilian space-event support ship *Kosmonavt Yuri Gagarin*, and had earlier performed a similar conversion on *Kosmonavt Vladimir Komarov*, formerly a bulk carrier. More importantly, in 1971 the Baltic Shipyard had laid down the first of a new generation of nuclear-powered icebreakers, *Arktika*, powered by two reactors of new design each with a power output of 37,500shp. The experience gained with the construction of this vessel would prove invaluable when the shipyard laid down *Kirov* in June 1973.

New Missile Systems

The large size of the *Kirov* class was made necessary by the launch mode adopted for the SS-N-19 missile. Earlier Rocket Cruisers such as the 'Kynda' had a ready-use missile battery no greater than that of an 'Echo'-class submarine, but had still experienced

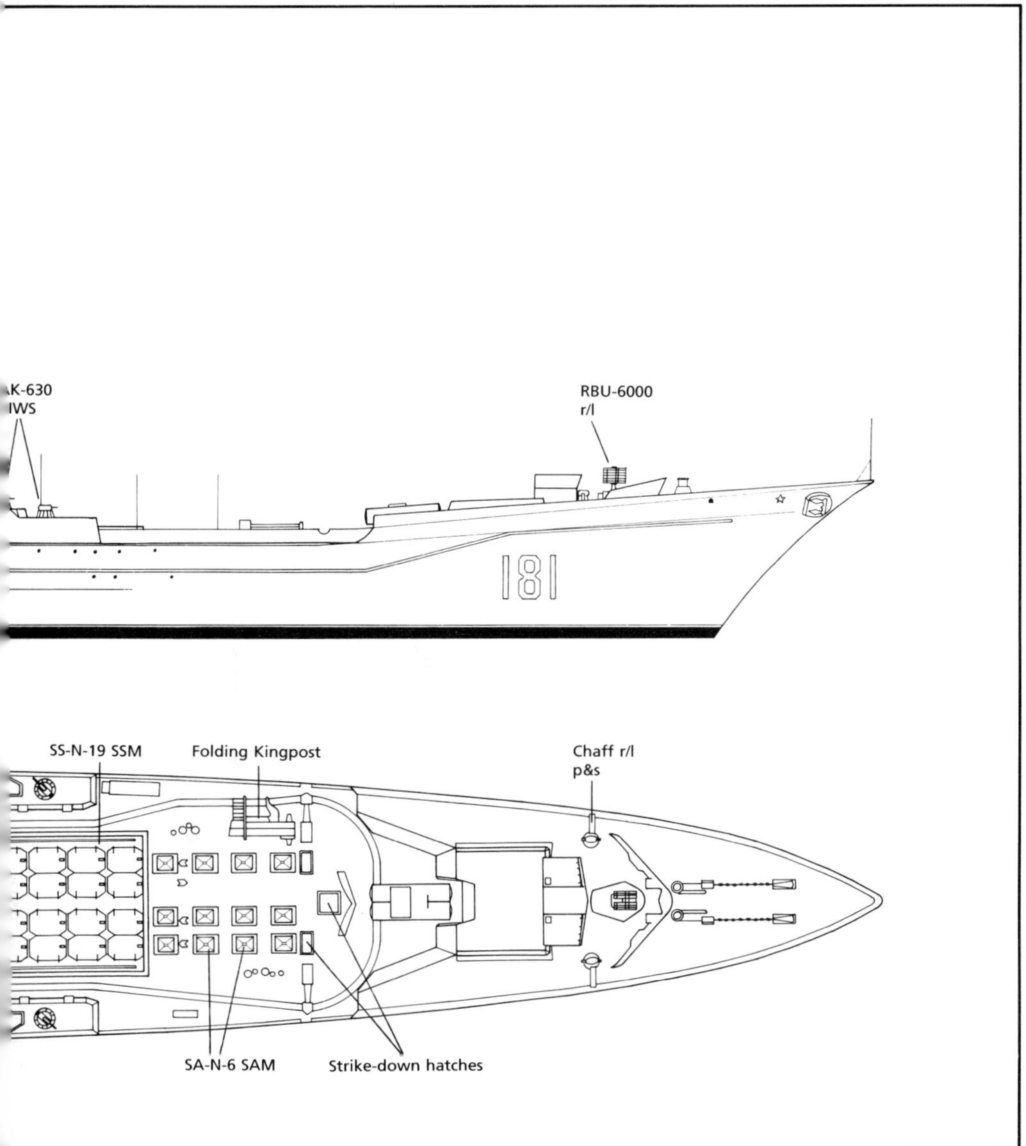

serious topweight problems. In the larger 'Kresta' this problem had been resolved, but only at the expense of halving the ready-use missile battery and dispensing altogether with the bulky reloads. All three types relied heavily on the use of missiles with nuclear warheads for maximum effectiveness against an enemy carrier battle group.

In the interim, however, Soviet war strategy had changed. No longer was it envisaged that nuclear weapons would be used from the outset, in the 'Battle for the First Salvo' outlined by Gorshkov in the books he wrote during the 1970s. Instead it seemed increasingly likely that a conflict between NATO and the Warsaw Pact might begin with a protracted conventional phase. Reliance on nuclear weapons for the 'Battle for the First Salvo' would therefore be unwise, because their use might not be sanctioned by those in political authority.

The logical solution to the problem was to increase the number of missile rounds carried by each weapons platform, and to mount a large-scale missile strike

Kirov Class

Soviet Designation: Raketny Kreyser (RKR) – Rocket Cruiser

Construction

Name	Builder	Laid down	Launched	In service
Kirov	Baltic Shipyard	June 1973	Dec 1977	July 1980
Frunze	Baltic Shipyard	Dec 1977	May 1981	Nov 1983
Kalinin	Baltic Shipyard	May 1983	Apr 1986	Oct 1988
Yuri Andropov	Baltic Shipyard	Apr 1986	Apr 1989	

Displacement (tons)
Standard: 24,000
Full Load: 28,000

Dimensions
Length: 248m(813ft) overall
Beam: 28m(92ft)
Draught: 8.8m(29ft)

Propulsion
2-shaft CONAS; 2 pressurized-water nuclear reactors each 40,000shp; 2 oil-fired boilers each 35,000shp; 150,000shp=32 knots max.

Complement
800

Weapons
ASuW: 20 inclined tubes for SS-N-19 Shipwreck missiles; 1 Ka-25 Hormone B missile targeting helicopter; 2 single 100mm(3.9in) D-P guns in *Kirov*; 1 twin 130mm(5.1in) D-P gun in others
AAW: 12 SA-N-6 vertical launch magazines (96 Grumble missiles); 4 groups each of four SA-N-9 vertical launch magazines (128 missiles) in *Frunze, Kalinin*; 2 twin-arm SA-N-4 launchers (40 Gecko missiles); 8 AK-630 gatlings in *Kirov, Frunze*; 6 CADS-N-1 combined gun/missile CIWS in *Kalinin*
ASW: 1 twin SS-N-14 launcher (16? missiles) in *Kirov* only (see text); 2 Ka-25 Hormone A helicopters; 1 RBU-6000 rocket-launcher (RBU-12000 in *Kalinin*); 2 RBU-1000 rocket-launchers; 2 quintuple banks of 533mm(21in) torpedo tubes (not in *Kalinin*)

Sensors
Surveillance radars: Top Pair 3-D; Top Steer 3-D in *Kirov, Frunze*; Top Plate in *Kalinin*; 3 Palm Frond
Fire control: 2 Top Dome (SA-N-6); 2 Pop Group (SA-N-6); 2 Eye Bowl (SS-N-14) in *Kirov* only; 1 Kite Screech (130mm/100mm); 4 Bass Tilt (AK-630) in *Kirov,* Frunze; 2 unidentified+6 Hot Flash (CADS-N-1) in *Kalinin*
Sonars: bow-mounted Horse Jaw LF; Horse Tail LF VDS
Electronic warfare: 8 Side Globe, 4 Rum Tub ESM, 4 Bell Bash, 4 Bell Thump, 2 Bell ? in *Kirov*; 8 Foot Ball, 4 Wine Flask ESM, 4 Bell Push, 4 Bell Nip, 2 Bell ? in others; 2 twin-barrelled chaff-launchers

intended to saturate enemy tracking radars and defensive systems. However many missiles the enemy might manage to destroy or decoy, in the heat of combat there would always be a handful of missiles that broke through the defences. The greater the scale of the missile strike, the more difficulty the defence systems would have in coping. However, effective shore-based co-ordination of the strike would require that the number of individual missile platforms be minimized. These changes in Soviet tactical thinking would produce both *Kirov*, armed with 20 ready-use SS-N-19 missiles, and its submarine counterpart, 'Oscar', with an even larger complement of 24 SS-N-19 missiles.

Twenty above-decks launchers, each for a missile some ten metres in length, would have occupied a prohibitive area of deck-space. The launchers would have to be mounted at a fixed angle or be capable of elevation. In either case topweight would become a serious problem. The Soviet designers therefore opted for the elegant solution of a battery of angled launch tubes countersunk within the hull of the ship. This enabled them to be mounted close together in a single block, which in *Kirov* is located immediately forward of the superstructures on the broad, flat forecastle deck.

The missile launch tubes are housed within a rectangular 'box', approximately 20m in length and 14.5m wide. Twenty hinged hatches, each 3.5m by 2.5m, are arranged in four rows of five, and the launch tubes beneath them are angled at approximately 45 degrees. It has been suggested that the box containing the launch tubes may have armoured sides.

The arrangement adopted for the SS-N-19 missile tubes has the twin advantages of being remarkably compact while at the same time lowering the centre of gravity of the entire installation. It also facilitates maintenance of the missiles at the correct temperature to ensure reliability even in the coldest conditions. The arrangement is possible only in a large hull, however, and is therefore largely responsible for the jump in size from 'Kresta I' to *Kirov*.

A further factor in determining hull size was the need to provide the ship with a more effective area defence missile than either the SA-N-1 or the SA-N-3. The Yak-38 Forger would give the Soviet Navy a limited air intercept capability against slow-moving aircraft, but there was no question of its providing effective fleet air defence. Soviet surface ships would therefore continue to depend on surface-to-air missiles for defence against hostile air attack.

The SA-N-3 missile system, although far more capable than the SA-N-1 which had equipped the 'Kresta I', was strictly limited in terms of the number of targets it could engage simultaneously. The standard double-

◀This early bow overhead view shows to good effect the layout of the forecastle. The large rectangular hatches forward of the superstructure conceal the angled launch tubes for the SS-N-19 anti-ship missile, and the square hatches forward of them, disposed in three rows of four, are for the SA-N-6 vertical launch area defence missile. Tucked inside the break in the forecastle is the twin-tube reloadable launcher for the SS-N-14 anti-submarine missile. (MoD, 1980)

▼*Kirov* has a hull-form derived from earlier Soviet missile cruisers. Note the broad transom stern and the prominent stiffening strake running the length of the hull. (MoD, 1980)

ended installation could put up two missiles against each of two separate targets, but this was hardly sufficient to defend against a squadron of carrier-based attack aircraft coming in low over the horizon. Moreover, the maximum effective range of the SA-N-3 – 55km in the latest SA-N-3B variant – would probably allow a maximum of two successive engagements before the aircraft were close enough for weapons release.

Western area defence systems of the period were subject to similar limitations, but Western air defence ships were not generally intended for independent operations. The first line of defence for the carrier battle groups was the combat air patrol; area defence missiles provided an insurance policy against 'leakers', and it was not expected that they would be confronted by the entire enemy force. Moreover, carrier battle group operations stressed the importance of integration of air defence countermeasures. US Navy priorities during the 1960s had therefore focused on computer-based data analysis and systems management, combined with effective tactical data links to ensure that hostile air targets were distributed to the appropriate carrier escort for engagement. Thus the limitations of individual units was compensated by the close integration of their activities within a defensive framework which utilized the total capabilities of the battle group.

This was an area neglected by the Soviets in part for technical reasons, but largely because of differences in tactical philosophy. During the mid-1950s Krushchev had concluded that the advent of nuclear weapons signalled the demise of large-scale tactical groupings and had remodelled the Soviet Navy accordingly. Soviet surface ships designed during this period were intended to operate either individually or in groups of two or three ships so as to minimize the effects of a nuclear counter-strike. Tactical co-ordination of these disparate units would be the responsibility of a shore-based command post distributing both target data and instructions using radio communications.

Some erosion of this tactical philosophy had taken place during the late 1960s when Soviet maritime priorities shifted from anti-surface to anti-submarine warfare. Anti-submarine hunting groups and ASW barriers required a higher level of on-the-spot co-ordination, and both the 'Kara'-class BPKs and the *Kievs*, with their extensive command and control spaces, provide evidence of a move towards a much greater degree of local control, if not of autonomy.

However, the fastest nuclear-powered submarine moves considerably more slowly than even a subsonic jet aircraft, and the pace of anti-submarine operations

is necessarily more leisurely than that of air defence where the situation changes by the second rather than by the minute. Collaborative anti-submarine operations by Soviet surface ships of the period almost certainly continued to rely heavily on line-of-sight radio communications between ships rather than on sophisticated data links. It is therefore unlikely that the Soviets would have expended much effort in developing data links for the purpose of integrating air defence, which remained very much a secondary mission for the Soviet Navy. Also, co-operation does not imply integration; for the Soviet Navy, integration was – and continues to be – more important at the operational/tactical level than at the level of tactical groupings. For the purposes of air defence this would effectively mean that individual Soviet surface units would rely largely on their own resources to defend against air attack; and a major surface unit such as *Kirov*, which because of its size and value would automatically become a focus for enemy counter-strikes, would need to defend itself against large numbers of aircraft attacking simultaneously.

Again, the solution adopted centred around a new missile using vertical launch technology, conventional twin-arm launchers being discarded in favour of missile launch from deck hatches served by rotating magazine rings located directly beneath the upper deck. The requirement for multiple-target engagement over an extended period meant that magazine capacity would have to be increased as compared with earlier types; *Kirov* carries one-third more area defence missiles than *Kiev*, and three times the number carried by the 'Kresta I'.

General Configuration

Kirov has a hull-form clearly derived from earlier Soviet cruisers. For three-quarters of the ship's length there is no sheer, but the raised section of the forecastle rises sharply towards the bow. The sides of the hull are flared through the waterline and there is a prominent knuckle running all the way aft from just forward of the superstructures. The after part of the ship is characterized by an unusually broad waterplane which is utilized to provide a broad flight deck for helicopters, with a hangar beneath.

The superstructures are also typical of earlier Soviet construction, rising in gently graduated steps towards a massive central pyramid mast, into which the exhaust vents for the propulsion machinery are incorporated. Forward of the pyramid mast is a massive superstructure block which carries one of two fire control radars for the SA-N-6 air defence system. The navigation bridge is at the forward end of this block. At the after end, sandwiched between the missile fire control radar and the central tower mast, is a hexagonal-shaped admiral's bridge. The navigation bridge is well sited, but the admiral's bridge has poor views forward and aft. Immediately forward of the bridge structure at a lower level is a tall deckhouse on which smaller fire control radars and direction-finding antennae are located. The after superstructure rises in steps occupied by a second series of fire control radars and culminates in a short pyramid mast on which are mounted surveillance radars and aerials for long-range communications.

One of the most striking features of *Kirov* is the long, flat fore-deck area, where both the major missile systems are located. It is in marked contrast to the 'spiky' superstructures which continue to bristle with electronic apparatus in the manner of earlier Soviet designs.

▼The central 'mack' houses the twin uptakes for the oil-fired boilers which make up the conventional half of *Kirov*'s propulsion plant. The large propellers are unusually far forward. *Kirov* is the only ship of the class with the twin reloadable launcher for the SS-N-14 anti-submarine missile and the single 100mm gun mounting. (MoD, 1980)

Propulsion

Kirov features an unusual propulsion system which combines nuclear power with conventional steam-powered geared turbines (COmbined Nuclear And Steam, or CONAS).

The rationale for the adoption of a hybrid propulsion system appears to have been the limited power output of available proven reactors. The model installed in the icebreakers *Arktika* and *Sibir*, which were built at the same shipyard, has a power-rating estimated at 37,500shp. Four reactors of this type would have been required to drive a ship the size of *Kirov* at a speed of more than 30 knots. This would not only have made heavy demands on hull volume, however, but would probably have required four shafts, with the attendant penalties of additional weight and complexity. It would also have made further demands on the production of large nuclear reactors, already required for the *Kirov* and *Arktika* classes and soon to be installed in two classes of giant twin-reactor submarines (the 'Typhoon' SSBN and the 'Oscar' SSGN).

▼The broad transom stern of *Kirov* houses the mechanism for the Horse Tail low-frequency VDS which is concealed behind a large door, 9m square, bearing the ship's crest. (MoD, 1980)

The principal advantage conferred by nuclear propulsion is virtually unlimited endurance. Two reactors each with a power output of about 40,000shp would be sufficient for a constant speed of about 24 knots. It therefore seems likely that *Kirov* uses nuclear power for a high, sustained cruising speed, and steam propulsion in combination with it to boost maximum speed to a figure of 30 knots plus.

Two possibilities have been suggested for the steam element of the propulsion plant. The nuclear reactors and the oil-fired boilers could power separate turbines, operating in parallel and geared to the same drive shafts. An alternative suggestion[1] is that oil-fired superheaters are used to superheat the steam generated by the nuclear reactors, thereby raising the available horsepower by about 50 per cent. This theory is eminently plausible and deserves serious consideration, but appears to be contradicted by the layout of the ship, in which the oil-fired boilers are clearly located forward of the nuclear reactors.

The position of the propeller guards on each side of the broad transom stern not only indicates that the two large-diameter propellers are unusually far forward, but suggests a wide separation of the shafts. The layout of the propulsion plant, in which the nuclear reactors appear to be located on the centre-line immediately abaft the boiler rooms (there are large circular hatches for core removal on top of the deckhouse abaft the tall pyramid mast), suggests that separate turbines for the oil-fired plant may be located outboard of the reactor rooms, while the turbines driven by the nuclear reactors are located directly abaft the reactor rooms, with the CONAS gearing outboard of them in line with the boiler-driven turbines.

Whatever the truth of the matter, the machinery layout adopted for *Kirov* is typically compact. The solution to the damage control problem appears to be that of locating the nuclear reactors inboard of the oil-fuel powered turbines, thereby taking full advantage of the considerable beam of the ship. Longitudinal bulkheads on each side of the centre-line would enable the ship's designers to isolate the operation of the nuclear plant from whatever oil-fired system is employed to provide boost, leaving the ship with more than adequate speed to extricate herself from any potentially dangerous situation in the event of action damage.

The SS-N-19 Missile

The SS-N-19 anti-ship missile (NATO code-name: Shipwreck) has a maximum range estimated at 250nm

◀ **Close-up of the midships section, with the Top Pair and Top Steer 3-D radar antennae particularly prominent. The Vee Tube cones for long-range HF communications have been superseded by Low Ball satcom systems in later units of the class. *Kirov* has the well-established combination of Side Globe broad-band jammers and Bell-series ESM receivers, with Rum Tub ESM antennae covering the four quadrants of the ship; later units have a completely new electronic countermeasures outfit. (MoD, 1980)**

The SS-N-19 Shipwreck Missile

Soviet designation: unknown
In service: 1981
Length: 10m(32ft 10in)
Diameter: 0.9m(3ft)
Span: 1.6m(5ft 3in)
Weight: 4,550kg(10,000lb)
Warhead: 1,000kg(2,200lb) HE
Propulsion: liquid-fuel rocket
Speed: Mach 1.6
Range: 450km(250nm)
Guidance: inertial/active radar homing
Fire control: none

(450km), and can carry either a nuclear or a high-explosive (HE) warhead. Evolved from the SS-N-3/SS-N-12 series of long-range cruise missiles, the SS-N-19 is powered by a liquid-fuel rocket motor which drives it at a cruise speed reported to be about Mach 1.6.

Unlike the SS-N-3 Shaddock and the SS-N-12 Sandbox, the SS-N-19 appears to have no associated missile tracking radar, although clearly a vessel as well endowed as *Kirov* with sophisticated 3-D radar antennae would have little difficulty in tracking the course of the missile. This has led to speculation that missile cruise speed may be so high that mid-course correction is unnecessary. A more cogent reason for the abandonment of dedicated missile tracking radars would appear to be Soviet confidence in the improved quality of target data with the advent of ocean surveillance satellites. Both *Kirov* and her submarine counterpart, 'Oscar', are fitted with Punch Bowl radomes which can receive targeting data direct without the need of radio communications with shore facilities.

Prototype military surveillance satellites had been tested by the Soviets from the late 1960s and an operational system was already in place by 1974. *Kiev* was the first Soviet naval vessel to be fitted with the Punch Bowl receiver, and by the time *Kirov* was completed the satellite surveillance network had been greatly expanded and the capabilities of the individual satellites improved. The Soviet ocean surveillance network is based on complementary pairs of Electronic Intelligence (ELINT) satellites and Radar Ocean Reconnaissance (RORSAT) satellites. The former picks up the radar and radio emissions of surface ships and then keys in the RORSAT to investigate contacts with its active radar. Photographic reconnaissance satellites capable of producing pictures of increasingly sharp definition have also been developed, although their utility in northern ocean areas subject to almost constant cloud cover is clearly limited.

While there can be little doubt that Soviet satellite surveillance will prove increasingly effective in detecting and identifying hostile forces, the continuing use of Bear D reconnaissance aircraft to follow up the initial contact is an indication that land-based aircraft remain an essential component of the Soviet Ocean Surveillance System. Moreover, the *Kirov*s, like *Kiev* and their Rocket Cruiser predecessors of the 'Kresta I' class,

The SA-N-6 Grumble Missile System

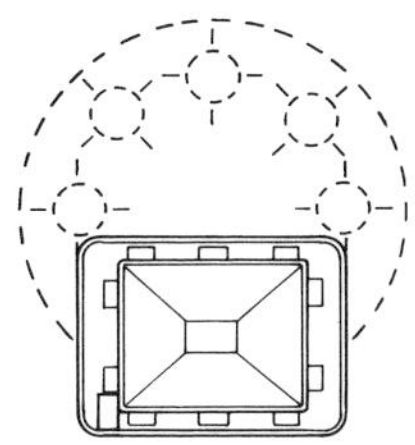

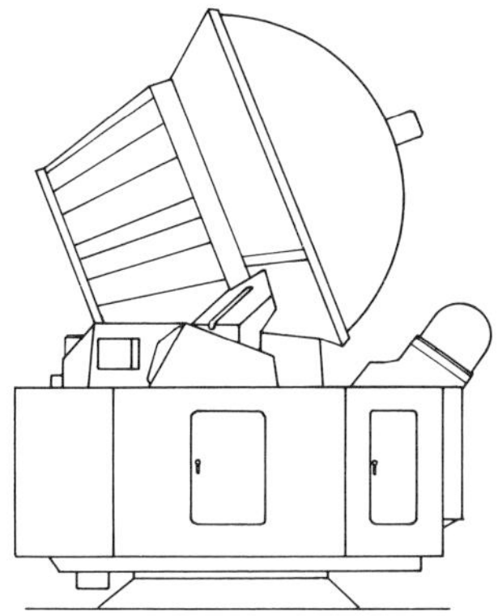

Soviet designation: unknown
In service: 1981
Length: 7m(23ft)
Diameter: 0.45m(1ft 6in)
Span: 0.9m(3ft)
Weight: 1,500kg(3,300lb)

Warhead: 90kg(198lb) HE/nuclear
Propulsion: solid-fuel rocket
Speed: Mach 6
Range: 80km(45nm)
Ceiling: 30,000m(100,000ft)

Min. altitude: 300m(1,000ft)
Guidance: track-via-missile (TVM)
Fire control: Top Dome
Operating frequency: I/J Band

▲The third helicopter embarked on *Kirov* is generally a Ka-25 Hormone B missile targeting variant, photographed here aboard *Frunze* in the mid-1980s. (US Navy)

carry a Ka-25 Hormone B targeting helicopter to provide an on-board targeting facility for their anti-ship missiles.

The SA-N-6 Missile System

The SA-N-6 Grumble missile was first trialled aboard the 'Kara'-class BPK *Azov* from 1977 onwards. Modifications to the fire control radar during development appear to have been necessary. Early photographs of *Azov* taken in the Black Sea, although of poor quality, reveal a radar more closely resembling an enlarged version of Head Lights than the final model.

The naval variant of the missile was developed in parallel with the land-based SA-10 and therefore follows the established pattern of joint missile development for the Soviet Navy and the Soviet Ground Forces – a pattern broken only by the SA-N-3 Goblet. The SA-N-6 is on the large side for a single-stage missile, with a length estimated at about 7m. US artists' impressions of the missile show a slim missile body with fixed long-chord wings indexed in line with small cruciform tail surfaces. The missile is capable of very high speeds, possibly about Mach 6, and has a maximum range estimated at 80–100km (43–54nm) and a ceiling of about 30,000m. It is designed to cope with high-speed manœuvring against small targets at low altitude and can reportedly withstand accelerations of up to 100g.[2] The standard warhead is HE, but there is an optional nuclear warhead for use against large formations at long range.

Both the SA-N-6 and its land-based counterpart, the SA-10, are thought to employ a form of guidance known as Track-Via-Missile (TVM). With TVM a phased-array multi-function radar tracks and illuminates several targets in rapid succession. Each missile launched receives the radar signal reflected from its target and retransmits it down to a computer located in the ship's combat information centre. The computer processes the data and transmits appropriate course corrections to the missile. Active radar terminal homing eliminates the need for target illumination in the final phase. The ability of the tracker/illuminating radars to switch rapidly between targets enables several missiles to be kept in the air simultaneously.

A similar system was proposed for the abortive US Typhon advanced area defence system developed during the late 1950s and early 1960s, and was later to be employed in the US Patriot land-based missile system. The Soviets would undoubtedly have been familiar with American efforts in this direction, and the SA-10 and SA-N-6 represent their attempt to bring the concept to fruition. The SA-10 system is reported to be capable of engaging six targets simultaneously, each with two missiles, and therefore represents a major advance on previous Soviet naval SAMs. It provides evidence both of high-performance missile technology and of increasing sophistication in the computerized tracking of multiple targets and missiles.

On *Kirov* the SA-N-6 vertical launch magazines are located on the forecastle immediately forward of the armoured box which houses the SS-N-19 anti-ship missiles. There are twelve hatches each measuring approximately 2m by 1.5m disposed in three rows of four. Each hatch is served by a rotating magazine ring holding eight SA-N-6 Grumble missiles. The magazine rings are about 4m in diameter. This accounts for the asymmetrical layout of the hatches; the magazine rings for the outer hatches are located on the outboard side of the hatches themselves, while those serving the inner row are located to port of their respective hatches.

Missile guidance is provided by Top Dome, a large fire control radar of distinctive configuration. The main antenna, protected by a hemispherical radome some four metres in diameter, can be mechanically steered in azimuth but is fixed at an angle of 20 degrees in elevation. A smaller radome at its base may house the command link for the missiles. The forward installation is atop the bridge structure and the second Top Dome radar is located directly abaft the after tower mast.

Other Air Defence Systems

The SA-N-6 area defence system is backed up by the customary array of short-range and close-in weapons, although it is in this area that there are most differences between individual units of the class.

Kirov has SA-N-4 launch bins located on each side of the tall deckhouse forward of the bridge, with their Pop Group fire control radars mounted on platforms projecting from the sides of the bridge structure itself. There are two single 100mm gun mountings aft, with their Kite Screech fire control radar immediately above them atop the after superstructure. This follows the pattern established by the 'Krivak II', in which the single 100mm mounting supplanted the twin 76.2mm, and suggests a move on the part of the Soviet Navy towards dual-purpose guns with a superior anti-surface capability for use against small craft such as FPBs.

Paired AK-630 anti-missile gatlings are mounted on their own deckhouses forward and aft. The forward gatlings are located to port and to starboard of the SS-N-19 launch tubes on the forecastle deck; the after gatlings are located on the short quarterdeck, on each side of the helicopter lift. Bass Tilt fire control radars for the forward guns, together with the customary optical sights for back-up, are mounted on each side of the deckhouse forward of the bridge; those for the after guns are located abeam the after radar tower.

Electronic countermeasures systems are on a par with those of *Kiev*, the principal systems being eight Side Globe radomes and four Rum Tub ESM antennae, disposed on each side of the central 'mack'. Twin-barrelled chaff-launchers are located close to the bow on each side of the forecastle.

Electronics and Communications Outfit

The installation of the advanced SA-N-6 missile system has not been accompanied by any radical modification to the outift of air surveillance and tracking radars, which is broadly similar to that of *Kiev*. Atop the central 'mack' is a large 3-D radar designated Top Pair. The latter is basically Top Sail with the addition of a long-range air surveillance antenna of conventional parabolic lattice configuration mounted back-to-back. This is a development typical of Soviet design practice which favours incremental improvements to existing systems rather than radical redesign. The first planar radars would not appear aboard Soviet warships until the mid-1980s, and it would be a further three years before large fixed planars arrays employing electronic scanning would reach the testing stage. This places the Soviets some years behind the United States in radar technology and must limit the effectiveness of an advanced area defence missile such as the SA-N-6.

The second 3-D radar installed is Top Steer which is mounted at the top of the after radar tower. Top Steer provides back-up for Top Pair and is also important to the command and control mission for which these powerful vessels are clearly designed. *Kirov*'s extensive air surveillance and tracking capabilities would enable her to monitor closely all aerial movements in her operational area, and to direct friendly aircraft against hostile ships and aircraft. Tub-mounted Round House TACAN aerials, which were first trialled on the 'Kara'-class cruiser *Petropavlovsk*, are mounted on platforms atop the central 'mack' for this purpose.

Communications systems are exceptionally complete. Two Vee Tube cones for high-frequency communications are angled out from the lattice structure atop the after radar tower, and there are extensive arrays of wire communications aerials and whips for both long-range and for short-range work. To port and starboard of the bridge structure are Bob Tail radio sextants for accurate position-fixing.

Tin Man electro-optical devices are installed in all four corners of the ship. Two are mounted on platforms projecting from the forward end of the bridge structure and a second pair is mounted one each side of the after radar tower.

Like *Kiev*, *Kirov* is designed to serve as a flagship, and command spaces appear to occupy a substantial part of her superstructure volume. Although *Kirov* might find herself working in close company with one of the PKRs, she has in no way been designed as a 'carrier escort'. Rather, one would expect to find *Kirov* at the centre of a surface action group, while *Kiev* would direct the operations of an anti-submarine group.

Anti-Submarine Warfare

Kirov is fitted with a wide range of anti-submarine warfare systems. The philosophy appears to have been to design a ship which, although capable of operating in conjunction with other units, is essentially 'self-escorting'.

The large, low-frequency bow sonar is a new, advanced model code-named Horse Jaw by NATO, while the complementary variable depth sonar (NATO designation: Horse Tail), the mechanism for which is housed behind a large square door set into the stern, is also reported to operate at low frequency. It is therefore possible that for the first time the Soviet Navy has paired these two sonars in the manner of the French DUBV 23/43 so that the data can be co-ordinated and presented on a single display. The variable depth sonar employs a lens-shaped fish to house the transducer; it has a twin boom-mounted empennage with horizontal and vertical control surfaces.

Tucked inside the break in the forecastle is a reloadable twin-tube launcher for SS-N-14 ASW missiles, with its magazine forward of it. It is interesting that the Soviets should have developed a reloadable SS-N-14 system for *Kirov* in preference to installing the SuW-N-1 system which was already in service on the *Moskvas* and the first two units of the *Kiev* class, and would also be fitted in the third ship, *Novorossiysk*. Clearly any modification involving redesign would have further delayed the completion of the latter ship, but the deliberate selection of a non-nuclear missile for *Kirov* in preference to the nuclear-tipped FRAS-1 associated with the SuW-N-1 launcher indicates a further shift in Soviet strategic doctrine towards an acceptance of the possibility of naval operations in situations short of all-out (i.e., nuclear) war.

The reloadable SS-N-14 launcher gives *Kirov* virtually double the missile capacity of the large BPKs built during the 1970s; the number of reloads is generally estimated at 14–16. The missiles are presumably laid out on racks in the below-decks magazine, and are raised horizontally in line with the launch tubes for loading. The abandonment of the SA-N-3 area defence missile system together with its dual-function Head Lights fire control radar, in favour of the SA-N-6, has resulted in the need for separate fire control provision for the SS-N-14 missiles, as on the 'Krivaks'. Two diminutive Eye Bowl radars are superimposed on top of the deckhouse forward of the bridge.

The helicopter hangar, which is located beneath the low quarterdeck, can accommodate three Ka-25 Hormone or Ka-27 Helix helicopters. At least one of the

▼ ***Frunze*, with two Ka-27 Helix A anti-submarine helicopters on deck. The hinged waterproof doors cover the lift-well when the lift is at hangar-level. (US Navy, 1987)**

helicopters is the Ka-25 Hormone B missile targeting variant, but the others are the anti-submarine Hormone A or Helix A models. As with the 'Kresta II' and 'Kara' the lift, which is approximately 13m long and 5m wide and is located at the forward end of the quarterdeck, descends to form part of the hangar floor, the open lift-well being covered by twin hinged doors. This arrangement increases the floor area in the hangar itself. Abaft the lift a single large helo-circle is marked out on the quarterdeck; it is covered by the customary mesh mat.

There is a small flight control cabin for helicopter operations at the after end of the superstructure immediately above the second 100mm gun mounting. Palm Frond navigation radars located on platforms projecting from each side of the central 'mack' are presumably also used for helicopter control, and there is a Fly Screen B microwave landing radar on the starboard side of the after tower mast.

Close to the bow is a single twelve-barrelled RBU-6000 rocket-launcher, and six-barrelled RBU-1000 rocket-launchers are located on each side of the after superstructure at upper-deck level. Just below them, quintuple tubes for 533mm torpedoes are concealed behind sliding doors set into the hull.

Replenishment

Transfer stations for ASW and surface-to-air missiles are located on each side of the bow immediately abaft the break in the forecastle. The SA-N-6 missiles are struck down via rectangular hatches inboard of the transfer stations. There is a large square hatch for strike-down of the SS-N-14 missiles on the centre-line, immediately behind the blast shield for the launcher. On the port side, abaft the transfer stations, is a large kingpost which folds down on to the deck when not in use to avoid interference with the flight of the missiles. It seems unlikely that the bulky SS-N-19 missiles can be taken aboard without specialized portside handling gear, as the silos would have to be loaded at an angle from above, and there are no heavy-lift cranes similar to those aboard *Kiev* and her sisters.

Kirov continues the trend set by *Kiev* towards improved facilities for under way replenishment. There are three stations for liquid replenishment: one at the forward end of the superstructure to starboard, and two further stations on each side of the after superstructure. Amidships to port is a constant-tension transfer station for dry stores, and abeam the after superstructure are cranes which can handle both stores and the ship's boats. Stores and munitions are transferred around the ship via the customary 'railway track' using wheeled trolleys. The track encircles the entire superstructure and foredeck area.

Frunze

Frunze, the second ship, followed *Kirov* on to the slipway at the Baltic Shipyard after an interval of 4½ years. As completed, she displayed a number of important modifications which reflected both technological advances and changes in Soviet priorities.

The most significant of these was the further upgrading of her air defence capabilities by incorporating the SA-N-9 short-range surface-to-air missile system designed to provide the new BPKs of the *Udaloy* class with the ability to defend themselves against hostile air attack (*see* Chapter 7). This modification parallels the changes made in the third unit of the *Kiev* class, *Novorossiysk*. Unlike the latter ship, however, *Frunze* has retained the older SA-N-4 Gecko SAM system which combines with the SA-N-9 system to provide a truly formidable short-range air defence capability.

The SA-N-9 system has been accommodated largely at the cost of the reloadable SS-N-14 launcher and its magazine which means that the installation aboard *Kirov* is, and will in all probability remain, the only one of its kind in the Soviet Navy. There has also been some rearrangement of the after superstructure layout to accommodate the after SA-N-9 launchers and an additional fire control radar.

The deep recess in the forecastle in which the twin SS-N-14 launcher was formerly located has been plated over, and the positions reserved for future installation of the SA-N-9 vertical launch system are covered by two large steel plates, each approximately 5m square. At the time of writing the launchers have yet to

◀ ***Frunze*, the second ship of the class, lowers the 'fish' of her Horse Tail variable depth sonar into the water. Note the twin boom-mounted empennage with its vertical and horizontal control surfaces. (MoD)**

be installed, but the size and configuration of the plates suggests that each of the two launch groups will comprise four v/l magazines each housing eight missiles. The deckhouse forward of the bridge has been remodelled to accommodate the Cross Sword missile fire control radar in place of the two Eye Bowl radars for the SS-N-14.

The after launchers for the SA-N-9 missile effectively replace the AK-630 gatlings formerly located on each side of the quarterdeck, the gatlings being relocated on top of the remodelled after superstructure. The steel plates measure approximately 9.5m by 2.5m and will therefore each house four v/l magazines in line. The after Cross Sword radar will eventually be fitted on top of the after superstructure immediately above the gatlings.

Not only will the SA-N-9 installation provide a magazine capacity of 128 additional surface-to-air missiles, but the missiles will be distributed equally forward and aft. This eliminates a potential weakness in *Kirov*'s design in which both the SA-N-6 and the SA-N-4 missile systems are located in the forward part of the ship.

In order to accommodate the after gatlings and the SA-N-9 fire control radar the after superstructure has been extended by some twelve metres. This has been made possible by the adoption of the new twin 130mm gun mounting first seen on the destroyers of the *Sovremenny* class in place of the single 100mm mountings of *Kirov*. The 130mm has superior anti-surface performance, but is presumably less effective when employed in the anti-aircraft mode. This is of less consequence in *Frunze* because of her new short-range missile system.

The electronics outfit reveals a number of important modifications. *Frunze* has a completely new electronic countermeasures suite in which Side Globe has been discarded in favour of a new bell-shaped radome mounted on a plinth with a back-plate (NATO designation: Foot Ball). This system was first seen aboard the

▼ ***Frunze*, the second ship, differs in a number of important respects from her sister *Kirov*. Air defence capabilities have been enhanced at the expense of ASW capabilities, and the electronics outfit has been upgraded. (MoD, 1984)**

destroyer *Sovremenny* in 1981, but problems with development and/or production appear to have delayed installation in subsequent units of the class, and it did not become standard fit until 1987. *Frunze* has eight Foot Ball radomes which are installed as four pairs on each side of the central 'mack'. Each of the radomes is paired with a smaller ESM radome which shares the same platform. There is also a Bell-type radome of intermediate size disposed centrally between the upper pair with an omnidirectional ESM cone behind it.

Rum Tub has been replaced in *Frunze* by four smaller antennae which have received the NATO designation Wine Flask. Each antenna is in the form of a truncated cone on a broad, circular base. The four antennae are disposed in identical fashion to the Rum Tub antennae on *Kirov* and can therefore be presumed to have a similar function.

The distinctive Vee Tube-C cones for HF radio communications, which on *Kirov* were mounted on a tall lattice framework at the forward end of the after radar tower, have also been discarded. In their place are large-diameter 3m cylindrical radomes for satellite communications abeam the bridge structure. These have effectively displaced the two Pop Group fire control radars for the SA-N-4 missile system which have been moved to the deckhouse forward of the bridge structure. The radomes (NATO designation: Low Ball) appear to be a smaller variant of Big Ball, which was first seen aboard the command cruisers *Admiral Senyavin* and *Zhdanov* (modified *Sverdlov* class) in 1979–81, and which is reported to be associated with the communications satellites of the Molniya and Raduga series. The system provides secure two-way communications between the ship and its command post ashore, and with other units similarly equipped.

▶This view shows the square plates on the forecastle where eight magazine 'bins' for the SA-N-9 air defence missile will eventually be installed. (MoD)

***Frunze*: Profile and Plans**

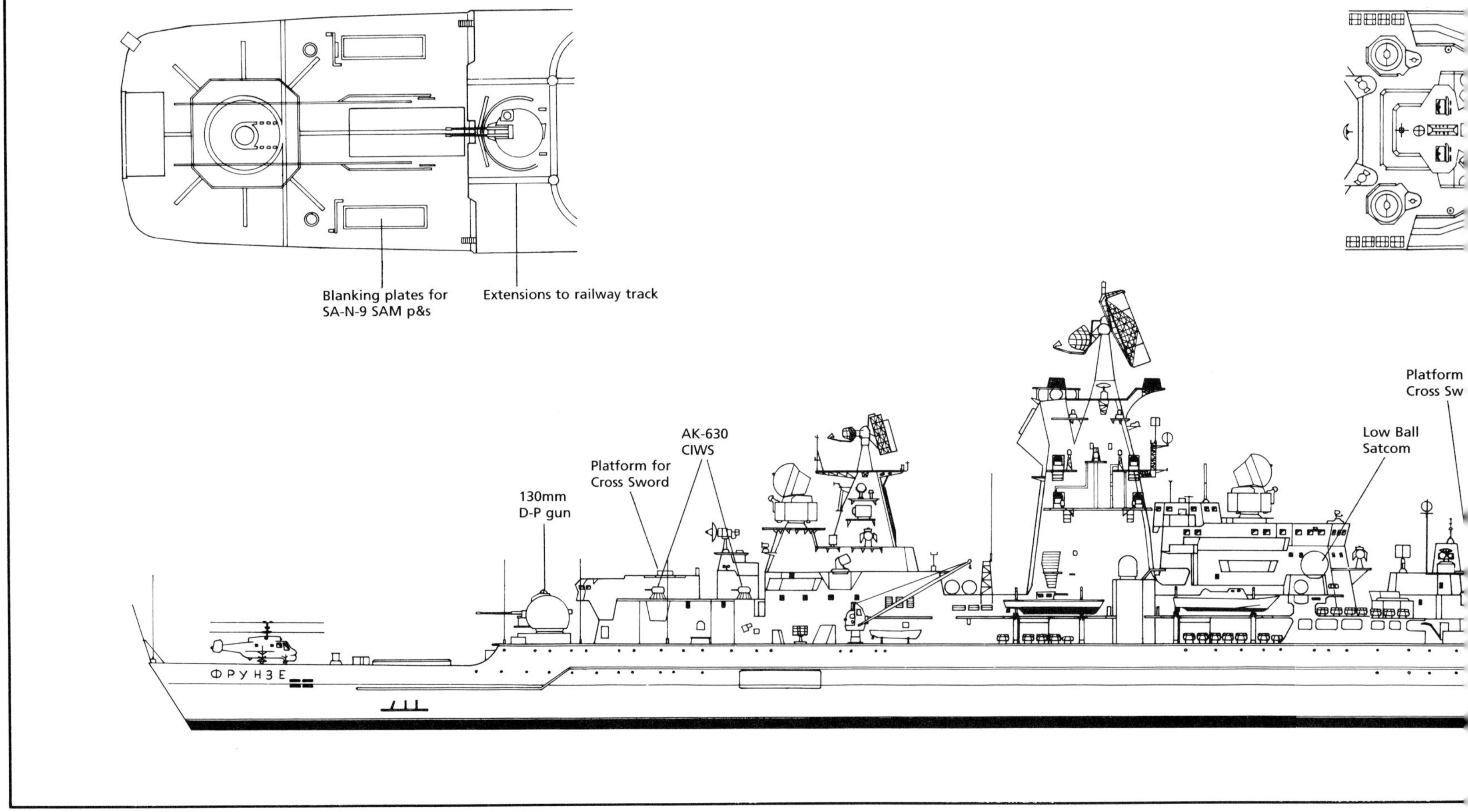

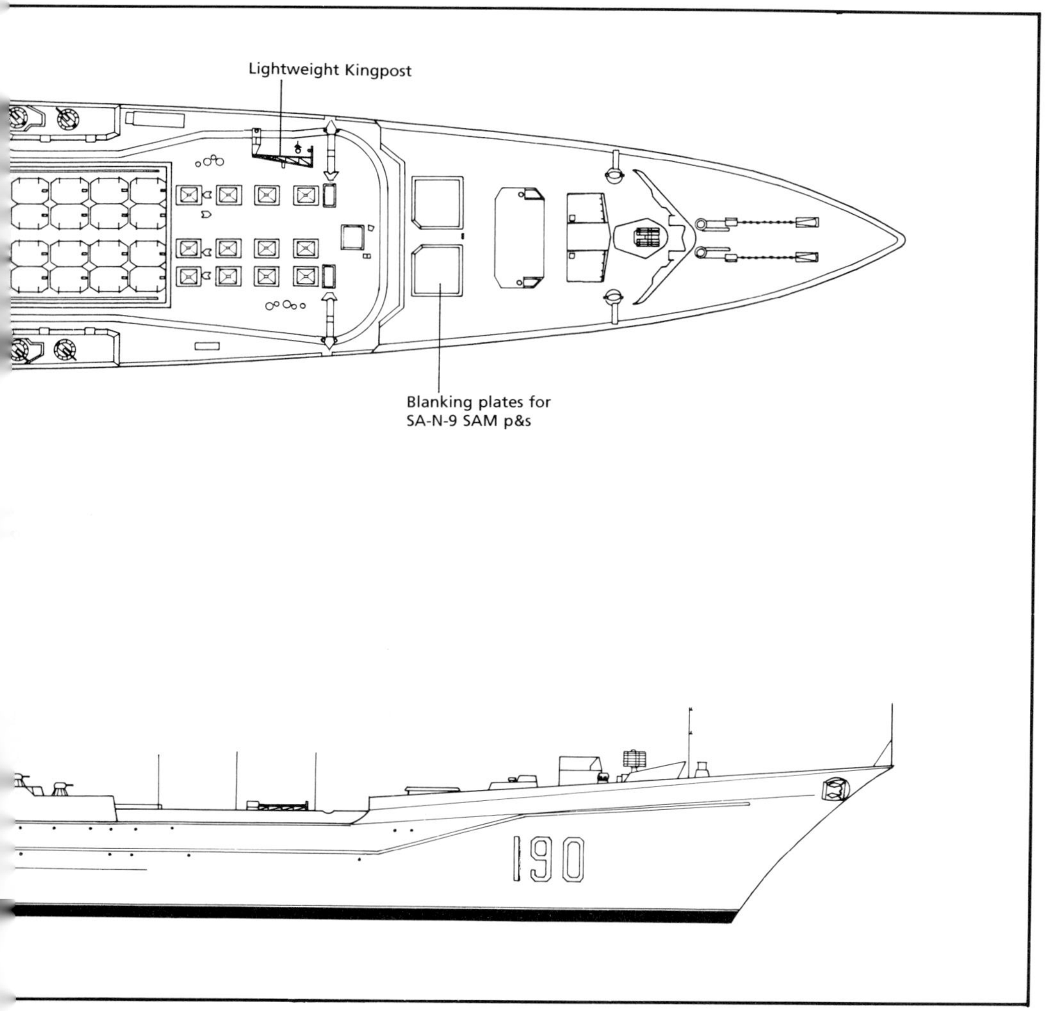

There are a number of minor differences between *Kirov* and *Frunze*. On the forecastle, both the folding kingpost and the missile cradles for the replenishment of the SA-N-6 magazine are of different design. In particular, the folding kingpost of *Frunze* is both lighter and more compact as it no longer needs to handle the bulky SS-N-14 anti-submarine missile.

The 'railway track' for the transfer of stores and munitions around the upper deck has been extended. New branch lines now lead aft from the main track on each side of the 130mm gun mounting, and continue to the break on to the quarterdeck. This presumably allows aviation stores to be taken down directly to the quarterdeck whence they are transferred via the aircraft lift to the hangar.

Directly beneath the after Top Dome missile fire control radar are platforms for two small, apparently rotating antennae. The position of these antennae suggests that they may be associated with helicopter control, although the Fly Screen B microwave radar continues to be fitted on the starboard side of the after radar tower.

An unexplained feature of *Frunze* is the large plated section of broadly rectangular configuration located forward of the SA-N-9 cover plates on the forecastle, in the position formerly occupied by the SS-N-14 magazine.

Kalinin

The latest unit of the class, *Kalinin*, features a number of further modifications in addition to those incorporated in *Frunze*.

The most important change is the replacement of the AK-630 gatlings by a new, more advanced anti-

◀The forward deckhouse has been completely rebuilt to accommodate the forward Cross Sword FC radar for the SA-N-9 missile system, and the Pop Group FC radars for the SA-N-4 system which have been re-sited to make space for the Low Ball satcom system. Below the port-side Low Ball radome can be seen the Bob Tail radio sextant with its hemispherical cover retracted. The ECM array comprises eight Foot Ball combined receiver/jammers paired with complementary ESM radomes of the Bell series. Rum Tub has been replaced by Wine Flask. (MoD)

◀The lower view shows the revised layout aft where two rectangular plates for the SA-N-9 missile system have replaced the original deckhouses for the AK-630 anti-missile guns. The latter have been relocated on top of the lengthened after superstructure. The empty platform for the Cross Sword fire control radar can be seen directly above them. (MoD)

▶A striking overhead view of *Frunze* operating in the Pacific. Note the elimination of the SS-N-14 launcher forward and the twin 130mm gun mounting directly above the quarterdeck, which are the principal distinguishing features of the ship. (RAAF)

▲*Kalinin*, the third unit of the class, has further modifications. Note the large 3m hemispherical radomes to port and starboard of the forward deckhouse in this bow view. (MoD Bonn, 1988)

◀*Kalinin*, photographed by the Norwegian Air Force during her initial deployment from the Baltic to the Soviet Northern Fleet. Like her sister *Frunze*, she has provision for the SA-N-9 missile system, but the system has yet to be installed. Note the Top Plate back-to-back planar antenna atop the after radar tower. (Norwegian Defence Command, 1988)

▲ **Close-up of the CADS-N-1 combined missile/gun mounting seen for the first time on *Kalinin*. Only four of the possible eight SA-N-19 missile canisters are in place. (MoD Bonn, 1988)**

The CADS-N-1 Combined Missile/Gun System

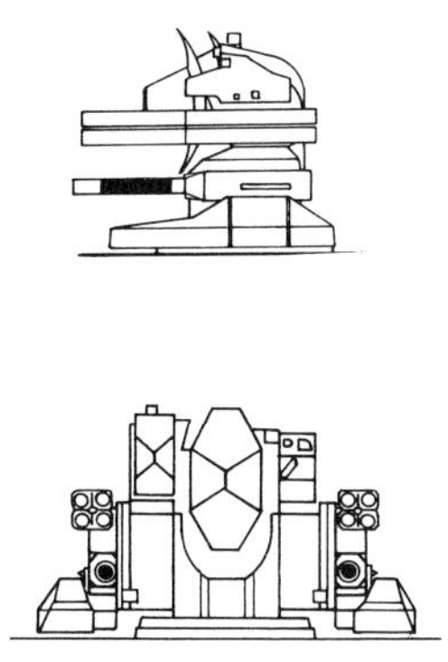

missile system (NATO designation: CADS-N-1) derived from the 2S6 self-propelled air defence system which recently entered service with the Soviet ground forces. Each individual system comprises twin 30mm multi-barrel cannon co-mounted with missile launch tubes; the target acquisition and fire control radars are mounted between the two groups of guns and missiles, thereby eliminating problems of alignment and ensuring improved reaction times. The entire installation is some 4.4m across and 2.8m deep, and is therefore significantly heavier and more space-intensive than the AK-630 mounting although, unlike the latter, it does not require a separate FC radar.

The mounting comprises a central pedestal which rotates on a circular base. The target acquisition/tracking antenna, which is of solid parabolic configuration with cropped corners, is mounted centrally with the smaller missile tracking radar to the left (when viewed from the front). To the right is an electro-optical device reported to be a laser rangefinder/designator. The 30mm cannon and missile tubes are mounted on broad cylindrical arms which enable them to be elevated to 90 degrees. The system appears to be designed to give optimum performance against sea-skimming anti-ship missiles.

The twin cannon are almost certainly the same six-barrelled 30mm model employed in the AK-630. Directly above each cannon are clamps for a maximum of four encapsulated missiles, which are reported to have received the NATO classification SA-N-11. The missile is thought to be a navalized version of the Soviet Army's SA-18, which is the missile used in the land-based 2S6 system. It is thought to have a range of 7–8km and is possibly of the SACLOS (semi-automatic command to line-of-sight) type with an infra-red terminal seeker. In a typical engagement the missiles would engage targets at longer ranges, the twin cannon taking over at about 3,000m.

Six CADS-N-1 mountings are provided in *Kalinin*, of which two are located atop the former gatling deckhouses on each side of the forecastle, and the remaining four are grouped around the after superstructure.

A further modification as compared with the earlier two ships is that the Top Steer 3-D air surveillance and tracking radar has been replaced by Top Plate, a back-to-back planar antenna first observed on the destroyer *Marshal Vasilevsky* of the *Udaloy* class in 1983. Top Plate has been closely associated with the SA-N-9 short-range missile system and is therefore a logical replacement for Top Steer on *Kalinin*.

In place of the former Bass Tilt fire control radars on each side of the forward deckhouse, there are now large hemispherical radomes, each about three metres in diameter. The function of these is as yet unclear, but they may be associated with the CADS-N-1 system (*see* Chapter 7).

The forward anti-submarine rocket-launcher is a new 10-barrelled variant of the RBU-6000, first seen on the carrier *Baku* in 1988. The new rocket-launcher fires a larger rocket than its predecessor and has received the designation RBU-12000, which suggests that rocket range has been increased to match the improved performance of the latest Soviet low-frequency sonars.

Service Life

Kirov entered service in September 1980 following sea trials in the Baltic, and shortly afterwards joined the Soviet Northern Fleet. Since that time the ship has frequently been the focus of large-scale exercises involving the defence of Soviet sea-space against hostile surface forces, although those exercises have decreased in number since the mid-1980s and now take place exclusively in Soviet waters.

The second ship, *Frunze*, entered service in August 1984, deploying to the Pacific Fleet in October 1985. *Kalinin* entered service only in October 1988 and two months later joined *Kirov* in the Soviet Northern Fleet. A fourth ship, reportedly named *Yuri Andropov*, is currently fitting out and is due to enter service in 1992. She will probably deploy to the Pacific Fleet. Apparently no further ships of this class will be built. In 1989–90 the Baltic Shipyard laid down the first of eight Kronshtadt Ro-Ro ships, indicating that the yard has reverted to mercantile construction.

Slava

The first unit of the *Slava* class was laid down at the 61 Kommuna Shipyard, Nikolayev, in 1976, probably on the slipway formerly occupied by the 'Kara'-class BPK *Tashkent*, which was launched in the same year. With a main armament comprising a battery of 16 SS-N-12 Sandbox long-range anti-ship missiles plus the advanced SA-N-6 area defence missile system, these ships are clearly intended as a 'low-mix' counterpart to the 'high-value' Rocket Cruisers of the *Kirov* class with which they have a number of features in common.

The decision to build *Slava* was probably taken in about 1974, following a wide-ranging debate within the Soviet Defence Council regarding the future strategy, shape and structure of the Navy. Throughout the previous decade Soviet naval construction programmes had been dominated by the BPK. No fewer than 49 major surface warships with this classification

◀ ***Marshal Ustinov*, the second ship of the class. The *Slava*s, with their powerful battery of SS-N-12 anti-ship missiles, provide a useful 'low-mix' complement to the larger, more sophisticated nuclear-powered rocket cruisers of the *Kirov* class. (MoD, 1987)**

would be completed in the twelve years from 1970 to 1982, together with three PKRs of the *Kiev* class. The only Rocket Cruiser completed during this period was *Kirov*, whose construction had to be undertaken outside the mainstream of Soviet naval shipyards.

The new programme would have a very different balance. There would be two distinct strands: an anti-surface strand, based on a new generation of Rocket Cruisers, backed up by a new class of large destroyers armed with medium-range anti-ship weapons (the *Sovremenny* class, *see* Chapter 7); and an anti-submarine strand, based on the PKRs of the *Kiev* class, and backed up by a second new class of large destroyers with a primary armament of anti-submarine weapons (the *Udaloy* class).

Shipbuilding capacity, which throughout the previous decade had been fully occupied with the anti-submarine programme, would have to be re-allocated accordingly and some adjustments made to the particular specializations of individual yards. The termination of the 'Kresta II' and 'Kara' programmes would provide just such an opportunity. Construction of the new destroyers would be concentrated in the Baltic, at the Zhdanov and Kaliningrad Shipyards.

The Zhdanov Shipyard had been responsible for many earlier large destroyer types including the Rocket Cruisers built in the 1960s. However, the new classes of Rocket Cruiser planned by the Soviet Navy, with their multi-missile anti-ship batteries and their sophisticated air defence systems, would be considerably larger than their predecessors and it is questionable whether the slipways of the Zhdanov Shipyard could have coped with vessels of this size. The shipyard at Kaliningrad had been steadily upgraded to enable it to build ships of destroyer size, but would have to be even further upgraded to build *Udaloy*.

The first of the *Kirovs* was already under construction at the Baltic Shipyard which specialized in large nuclear-powered hulls. Construction of these complex vessels was slow, however, and the parallel production of a new class of nuclear-powered icebreakers would preclude any expansion beyond a single-stream programme.

The 61 Kommuna Shipyard at Nikolayev, which had traditionally built ships of cruiser size, therefore became the logical choice for a second, smaller type of Rocket Cruiser. *Slava*, the name-ship, would be built as a follow-on to the 'Kara' programme and have a number of important features in common with the latter ships.

General Configuration

The hull-form of *Slava* is characteristic of all recent Soviet surface warships, with a length to beam ratio of about 9:1 and a broad waterplane aft. It is, however, unusual in that the ship's sides run parallel for a considerable part of her length. This reflects the layout of the SS-N-12 missile-launchers which are disposed in pairs on each side of the forward superstructure and are supported by fixed-angle ramps which extend to the deck edge. It is also dictated by the need to accommodate two rows of SA-N-6 missile magazines side by side in the after part of the ship.

There is considerable flare in the hull beneath the first pair of missile-launchers; this serves to restore some degree of fineness forward, thereby improving powering characteristics. The concentration of weight in the forward part of the ship is quite marked, however, and it is doubtful whether the pronounced flare and sheer of the bow section provides adequate compensation in extreme weather conditions.

Like the 'Kara', *Slava* has a short quarterdeck topped by a helicopter flight deck. The larger size of the *Slava* hull together with the adoption of a vertical launch air defence system has made a split-level hangar unnecessary, and the latter is significantly taller than those of the BPKs. Even so, the floor of the hangar is beneath the level of the flight deck, which is raised some two metres above the quarterdeck in order to provide clearance for the variable depth sonar mechanism, and to eliminate mutual interference between flight operations and line-handling. An inclined ramp connects the flight deck with the hangar floor, and the helicopter is moved using a chain-operated traverse system.

The superstructure is arranged as three blocks. The forward superstructure, which culminates in a large plated pyramid mast, is long and narrow, its shape being determined by the deployment of the anti-ship missiles along the ship's sides. It is linked to the central superstructure block, comprising a tower mainmast with large twin funnels directly behind, by a single superstructure deck which terminates abaft the funnels. Stowed between the twin funnels is a large crane which handles the ship's boats and which is also used to reload the SA-N-6 missile magazines when the ship is alongside. There follows a flat expanse of upper deck where the SA-N-6 missile magazines are located, which separates the central mast/funnel structure from the tall helicopter hangar. The latter is flanked by deckhouses and topped by the single Top Dome missile fire control radar.

The overall effect is not as pleasing to the eye as earlier Soviet designs and *Slava* is certainly a less handsome ship than the 'Kara'. The ship's profile is broken up by the flat expanse of deck for the SA-N-6 missile system, while no attempt has been made to

SA-N-6 Launcher on *Slava*

Slava Class:

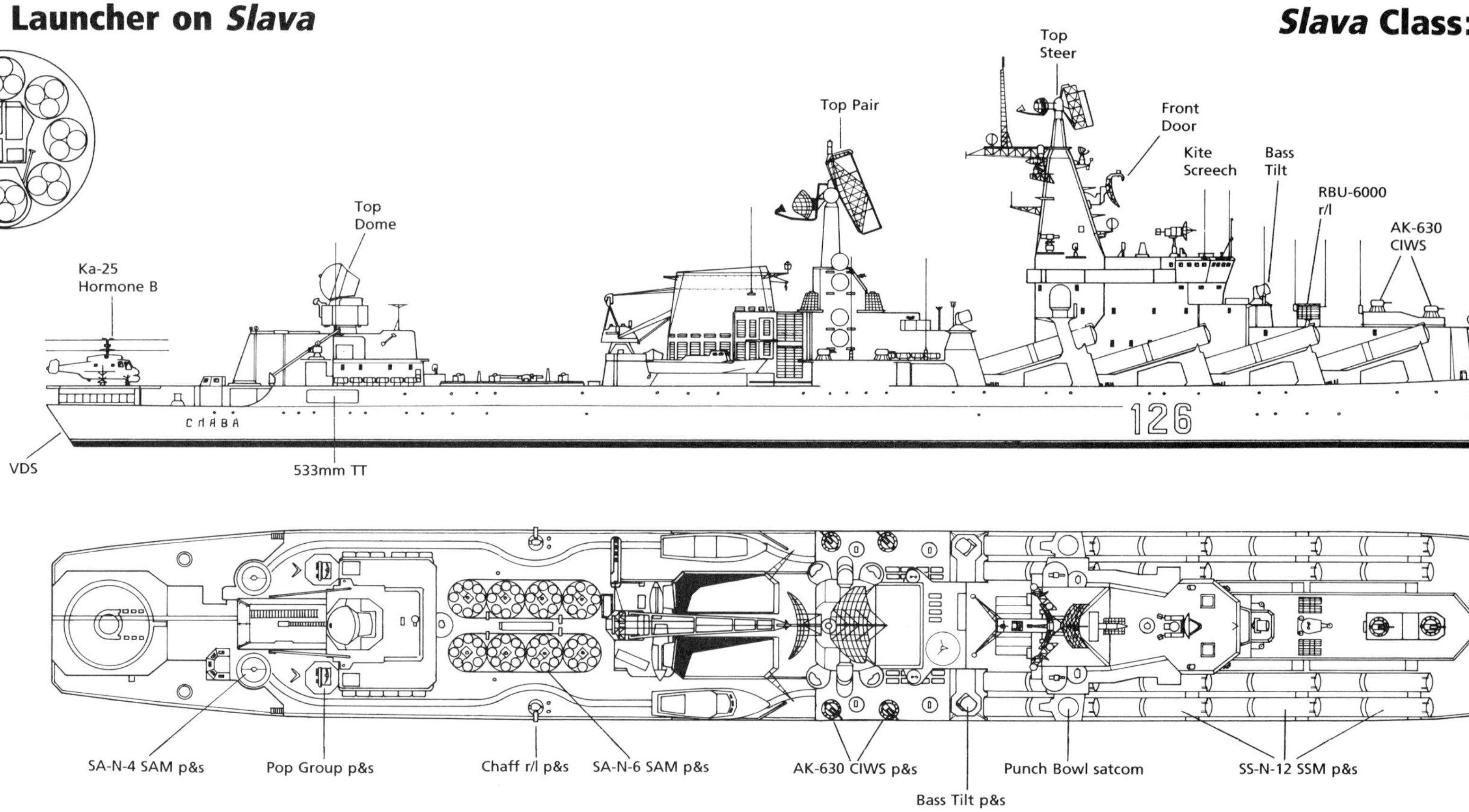

◀ ▶The layout of the missile launch tubes forward means that the hull sides run parallel for most of the ship's length. Note the layout of the SA-N-6 vertical launch magazines sunk deep into the hull abaft the funnel. (US Navy, 1984)

integrate the angled SS-N-12 launch tubes into the overall design, so that they give the impression of having been simply bolted on to the hull.

The SS-N-12 Missile System

The adoption of the SS-N-12 Sandbox missile in preference to the faster, more advanced SS-N-19 Shipwreck missile fitted in *Kirov* initially occasioned some surprise. However, this is by no means the only example of an older, less capable weapons or electronics system being adopted for *Slava* in preference to more recent models.

Among the factors which appear to have been responsible for these choices one must include the characteristic inertia of the industrial establishment which serves the Soviet armed forces, with its predilection for long production runs of tried and tested systems. But some of these choices appear to have been dictated by size constraints imposed on the design. Even at 12,500 tons full load displacement, *Slava* is a relatively small ship to accommodate such a powerful battery of weapons.

Kirov was able to accommodate her own anti-ship missile in angled launch tubes countersunk within the hull, but *Kirov* has nearly 50 per cent more beam and significantly greater hull depth than *Slava*, and the armoured box for her SS-N-19 launch tubes occupies an estimated 3,000m^3 of internal volume. This would

'rofile and Plan

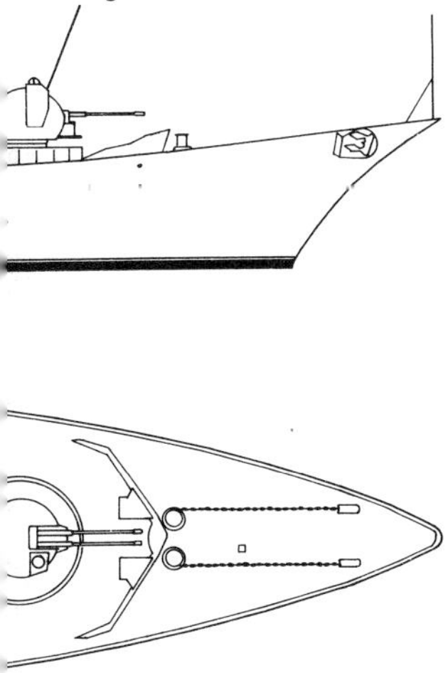

Slava Class

Soviet Designation: Raketny Kreyser (RKR) – Rocket Cruiser

Construction

Name	Builder	Laid down	Launched	In service
Slava	61 Kommuna Shipyard	1976	1979	Aug 1982
Marshal Ustinov	61 Kommuna Shipyard	1978	1981	Apr 1986
Chervona Ukraina	61 Kommuna Shipyard	1979	1983	Jan 1990
Admiral Lobov	61 Kommuna Shipyard	1984	Aug 1990	

Displacement (tons)
Standard: 10,000
Full Load: 12,500

Dimensions
Length: 187m(614ft) overall
Beam: 20.8m(68ft)
Draught: 7.5m(25ft)

Propulsion
2-shaft COGAG/COGOG?; 4 gas turbines each 31,250hp; 125,000hp=32 knots max.; 2 NK-12 cruise turbines each 12,100hp(?)

Complement
515

Weapons
ASuW: 8 paired SS-N-12 launchers (16 Sandbox missiles); 1 Ka-25 Hormone B missile targeting helicopter; 1 twin 130mm(5.1in) D-P gun
AAW: 8 SA-N-6 vertical launch magazines (64 Grumble missiles); 2 twin-arm SA-N-4 launchers (40 Gecko missiles); 6 AK-630 gatlings
ASW: 2 RBU-6000 rocket-launchers; 2 quintuple banks of 533mm(21in) torpedo tubes

Sensors
Surveillance radars: Top Pair 3-D, Top Steer 3-D (Top Plate in *Lobov*); 3 Palm Frond
Fire control: 1 Front Door C (SS-N-12); 1 Top Dome (SA-N-6); 2 Pop Group (SA-N-4); 1 Kite Screech (130mm); 3 Bass Tilt (AK-630)
Sonars: bow-mounted Bull Nose MF; Mare Tail MF VDS
Electronic warfare: 8 Side Globe; 4 Rum Tub ESM; 2 Bell Crown; 2 twin-barrelled chaff-launchers

have been out of the question for *Slava*, hence the lightweight above-decks launchers (the launch tubes are reported to be constructed of fibreglass, with steel liners).[3] Fixed-angle launchers eliminate the requirement for bulky, power-consuming mechanisms for elevating and training, and the missile exhaust is vented beneath the launch tube immediately behind it. This is an ingenious way of accommodating a ready-use missile battery equivalent to the total missile capacity of a 'Kynda' (including reloads).

For mid-course missile tracking and guidance there is a Front Door C parabolic lattice antenna mounted on the face of the forward tower mast. Front Door C is virtually identical in configuration with the Trap Door antenna (also designated Front Door B) which accompanies the SS-N-12 missile system in *Kiev*. Data for targets beyond the horizon are provided either by the land-based Bear D reconnaissance bomber, or by the ship's own helicopter, which is the Ka-25 Hormone B variant. As with *Kiev* and *Kirov*, *Slava* is fitted with a Punch Bowl satellite communications antennae which can receive target data on a direct down-link. The twin

▼Close-up of the midships section of *Slava*, showing the massive SS-N-12 launcher tubes mounted in pairs on their fixed-angle ramps. The missile exhaust is vented beneath the following ramp. Directly above the after pair of tubes is the starboard-side Punch Bowl satcom antenna, which can receive target data for the missiles on a direct

down-link. Above Punch Bowl is an elderly Tee Plinth electro-optical device; other contemporary large combatants have received the more advanced Tin Man. (US Navy, 1983)

▶ Midships section of *Slava*. The large square air intakes immediately forward of the funnel serve two engine rooms, each housing two powerful 31,250hp gas turbines. The ducting for both the intakes and the exhaust uptakes for the forward engine room are presumably led up through the deckhouse which projects forward from the tower mainmast. (US Navy, 1983)

Punch Bowl radomes are located on cantilevered platforms projecting from the after end of the bridge structure.

The SA-N-6 Missile Installation

Accommodating the SA-N-6 missile system in *Slava* presented the designers with a number of problems. At 7m the Grumble missile is more than 50 per cent longer than the US Standard SM-2 MR missile – the only current Western area defence missile to employ vertical launch – and therefore requires considerable hull depth for its rotating magazine. The rotating 8-round magazine also consumes more deck-space than the fixed 8-cell module employed in recent US Navy vessels. It has been estimated by Kehoe and Brower[4] that 151 Standard SM-2 missiles could be fitted into the same area occupied by the 64 missiles carried by *Slava*.

The arrangement adopted by the *Slava* designers has the eight SA-N-6 magazines disposed in two rows of four between the machinery spaces and the helicopter hangar, with the twin shafts running outboard of them. This places them close to the centre-line, leaving adequate space outboard for the railway track which encircles the upper deck and for passage-ways between decks.

The magazines differ substantially in configuration from those in *Azov* and *Kirov*. Whereas the rotating magazine rings of the latter ships were located beneath a continuous upper deck broken only by a single rectangular hatch for each magazine, those of *Slava* and her sisters are located inside self-contained cylindrical 'bins' similar in conception to those associated with the SA-N-9 missile. This serves to minimize the depth of each module, and must also simplify installation at the shipyard.

The circular bin-lid shows clearly the position of the eight missile rounds, there being a circular cap above each missile incorporating three smaller circular blow-out panels (*see* drawing). One of these caps is set into a hinged hatch which can be raised for missile firing, the magazine rotating to place the next missile beneath the hatch. The magazine is reloaded via the same hatch when the ship is alongside, with the aid of the large crane stowed between the twin funnels.

The air surveillance, tracking and missile fire control radars associated with the SA-N-6 system on *Slava* are identical with those of *Kirov*. A Top Pair antenna for 3-D air surveillance and tracking is mounted at the top of the tall plated foremast, while the larger Top Pair antenna is at the top of a smaller radar tower amidships, reminiscent of the 'Kara'. The single Top Dome missile tracking and fire control radar is located on the helicopter hangar and enjoys excellent all-round arcs against high-flying aircraft, but the reduction to a single FC radar means that *Slava* can engage simultaneously only half as many aerial targets as *Kirov*.

Propulsion Machinery

The central section of the ship is occupied almost exclusively by the gas-turbine propulsion machinery.

A number of theories regarding the precise layout of the machinery installation have been advanced, American sources favouring a COGOG arrangement combining four large gas turbines for high-speed operation with two smaller turbines for cruising,[5] while other sources credit the ships with an installation identical with that of the 'Kara'.

A comparison of profiles reveals that the midships sections of the 'Kara' and *Slava* are strikingly similar,

▲Stern view of *Slava*, showing the tall helicopter hangar flanked by launcher bins for the SA-N-4 missile system. (MoD, Bonn)

and that the distance between the combined mast/funnel structure and the after end of the bridge structure is virtually identical. This suggests that there are two turbine rooms, separated as in the 'Kara' by three closely spaced bulkheads.

Slava has two large square air intakes immediately forward of the funnel on each side of the ship. These clearly serve the four main turbines, and an analogy with the arrangements aboard the 'Kara' suggests that the trunking for the lower intakes is led down to the after machinery room directly below the funnels, while the trunking for the upper intakes is led horizontally through the prominent deckhouse forward of the central radar tower, and down into the forward machinery space.

The after generators are probably located in the after machinery space, while the forward generators may be housed in a separate machinery space directly beneath the upper deck as in the 'Kara'. This arrangement would appear to leave little space available for cruise turbines, especially if the Soviet ships also incorporate separate reversing turbines for their fixed-pitch propellers. *Slava* has more beam than the 'Kara', but the increase is not significant – approximately 1.7m. Moreover, the configuration of the *Slava* hull and the layout of the SA-N-6 missile magazines suggests

▲Close-up of the stern of *Slava*. Note the angled ramp leading from the flight deck to the hangar, with its chain-operated traverse system. To starboard of the ramp is the helicopter control cabin. (MoD, Bonn)

that the propeller shafts are parallel and that a relatively simple mechanical cross-connection would enable a single large turbine to power the ship at an economical cruising speed.

It is difficult to reach a firm conclusion because there are features which support the existence of separate cruise turbines. There appear to be smaller air intakes directly abaft those for the main turbines, although these may be for gas-turbine generators. Moreover, US Navy photographic analysis suggests that there are two small gas-turbine exhausts in addition to the exhausts for the four main turbines and the four gas-turbine generators. If there are indeed separate cruise turbines, it would seem logical that they would be located alongside the main turbines in the after machinery space.

Minor Air Defence Systems

Besides the SA-N-6 area defence system, *Slava* is equipped with the customary array of active and passive systems for short-range and close-in air defence.

Two SA-N-4 Gecko surface-to-air missile systems are fitted abeam the hangar, the magazine 'bins' being countersunk into the hull with the Pop Group fire control radars located immediately forward of them. The SA-N-4 system is also a feature of the 'Kara' and *Kirov*, but by the time of *Slava*'s completion had already been superseded by the more advanced SA-N-9 system, which had been initially developed for the BPKs of the *Udaloy*-class BPK but which had been incorporated in both *Novorossiysk* and *Frunze*. However, the SA-N-4 was a tried and trusted system which was still in production; its launch system was more compact than the multiple-magazine configuration of the SA-N-9, and its missile guidance radar was significantly lighter and less bulky than Cross Sword.

Although ships the size of *Novorossiysk* and *Frunze* could easily accommodate the SA-N-9 system, it is difficult to see how it could have been fitted into *Slava*, which is already a very 'tight' design with little space or weight available for modifications. It is worth noting that the SA-N-4 system, which is now more than 20 years old, has continued to be fitted in later units of the *Kirov* class in addition to the new SA-N-9 system.

In contrast, *Slava* has the new twin 130mm automatic gun mounting first fitted in the destroyers of the *Sovremenny* class. The mounting is located close to the bow with its Kite Screech fire control radar on top of the bridge structure. It is primarily an anti-surface weapon, but has a useful capability against aircraft.

For anti-missile defence there are the customary pairs of AK-630 gatlings controlled by Bass Tilt radars.

There are two groups abeam the midships tower mast, and a third on top of the long, narrow deckhouse forward. ECM provision is on a par with later units of the 'Kara' class, with four Side Globe broad-band jammers on each side of the central radar tower, and Rum Tub ESM antennae mounted on platforms projecting from the four corners. Two smaller Bell-type radomes are mounted at the top of the plated foremast and there are twin-barrelled chaff-launchers outboard of the SA-N-6 missile magazines. This outfit has remained unmodified in the later units, even though no other Soviet warship has received Side Globe since 1980.

Anti-Submarine Warfare

It is in the area of anti-submarine warfare where the greatest sacrifice in capabilities has been made. Whereas the *Kirovs* are essentially 'self-escorting' in this respect, the *Slavas* would certainly require the support of specialized ASW vessels in order to perform their duties in hostile waters.

In particular they lack the long-range anti-submarine weaponry of the *Kirovs* and the BPKs. There are no anti-submarine missiles and the single helicopter carried is reported to be the Ka-25 Hormone B missile variant. The sonar outfit, which comprises the Bull Nose medium-frequency bow sonar plus Mare Tail VDS, is on a par with the 'Kresta II' and 'Kara' classes, but cannot compare with that of *Kirov* and the latest BPKs of the *Udaloy* class which have matching low-frequency models. This is yet another instance of the installation of relatively elderly, second-line equipment in this class.

Anti-submarine weapons comprise only rockets and homing torpedoes. The two RBU-6000 automatically reloading 12-barrelled rocket-launchers are located on the forward deckhouse, and the torpedo tubes are behind sliding hull doors abeam the hangar. This arrangement parallels that of the *Kirovs* and the PKRs, and the tubes can presumably be reloaded from internal magazines without exposing the crew or the torpedoes to the elements. The hull doors of *Slava* are abeam the helicopter hangar. The torpedo reload magazine could therefore serve both the tubes and the hangar itself, enabling *Slava* to embark a Ka-25 Hormone A or Ka-27 Helix A anti-submarine helicopter instead of the Hormone B missile targeting variant should the situation require it.

Service Life

The first unit, *Slava*, entered service with the Black Sea Fleet in 1982 following trials, and since that time has conducted regular deployments to the Mediterranean culminating in her projected employment as the host ship for the Bush/Gorbachev summit of December 1989 – a move which was ultimately frustrated by severe weather, compelling the superpower leaders to conduct their talks aboard a Soviet cruise liner being used as an 'hotel' by the Soviet delegation.

The second unit, *Marshal Ustinov*, was completed in 1986 and was subsequently transferred to the Soviet Northern Fleet. The third ship, *Chervona Ukraina*, was completed in January 1990 and transferred to the Soviet Pacific Fleet in October 1990 in company with the destroyer *Bystry* of the *Sovremenny* class. A fourth ship is currently fitting out at Nikolayev. Until recently

▼*Marshal Ustinov*, photographed on her initial deployment to the Soviet Northern Fleet in 1987. (US Navy, 1987)

she was reported to have been named *Admiral Kuznetsov*, but this name has now been transferred to the first of the new through-deck carriers (formerly *Tbilisi, see* Chapter 8), and it is reported that the fourth *Slava* has been renamed *Admiral Lobov*.

Construction of the *Slava* class has been exceptionally slow, shipyard labour having been transferred to other work with a greater priority – the new carrier programme at the neighbouring Black Sea Shipyard presumably taking precedence over the construction of other surface vessels. *Slava* took six years to complete, *Ustinov* eight and the third unit, almost ten.

Although these powerful vessels serve as flagships, they are by no means as well equipped as either the *Kirovs* or the *Kievs* in terms of long-range communications. The large forward superstructure block undoubtedly houses extensive command spaces, and there appears to be a separate admiral's bridge directly beneath the navigation bridge. However, long-range communications antennae such as Vee Tube and the more recent Big Ball/Low Ball satellite communications radomes are notably absent, and these omissions inevitably make them less capable in the command role than their larger counterparts.

670

480

7. The New Destroyers

◀ ***Sovremenny*, the first of the new destroyers, with electronics systems in place but with her major weapons still to be fitted. The boiler uptakes for her pressure-fired steam turbines are combined into a single broad funnel, located abaft of amidships, as in her immediate predecessor, the 'Kresta'. (MoD, Bonn)**

◀ ***Udaloy*, first of a new generation of BPKs. The hull-form is derived from the 'Krivak', and she is powered exclusively by gas turbines. Like *Sovremenny*, *Udaloy* was completed without some of her intended weapons systems. The circular 'blanking plates' on the forecastle and forward of the hangar are for the SA-N-9 short-range air defence system, and there are empty platforms forward and aft for the associated Cross Sword fire control radar. She also has a temporary surveillance radar outfit comprising two Strut Pair back-to-back antennae. (MoD, Bonn)**

THE LEAD-SHIPS of two new classes of large destroyer, *Sovremenny* and *Udaloy*, appeared within a few months of each other during 1980–1. Both had an estimated full load displacement approaching 8,000 tons and a length overall of about 160m. Despite superficial similarities, however, two more different ships would be difficult to imagine.

Sovremenny is fitted primarily for anti-surface warfare; she is armed with anti-ship missiles, an area defence missile system and twin 130mm dual-purpose gun mountings, and is powered by steam turbines. *Udaloy* is fitted almost exclusively for anti-submarine warfare; she is armed with anti-submarine missiles, carries two ASW helicopters and has only short-range air defence systems. Anti-surface capability is limited to two 100mm dual-purpose gun mountings, and she is powered by gas turbines.

The new destroyers exemplify the fundamental changes in Soviet naval policy which took place in about 1974–5, when the twin strands of anti-submarine warfare and anti-surface warfare were established as having equal priority in future construction programmes (*see* Chapter 6). They also illustrate an aspect of Soviet military hardware that is rarely given the attention it deserves; the close, often inextricable links between design and production.

The very different configuration of *Sovremenny* and *Udaloy* can be explained only partially by reference to their respective missions. Even with their different weapons systems, one might have expected at least a common hull and a common propulsion system, as with other contemporary European types such as the Dutch 'Standard' frigate and the French C70. However, in keeping with traditional Soviet practice of allocating military projects to a particular bureau, the designs of *Sovremenny* and *Udaloy* were drawn up in two different shipyards.

The lead yard for *Sovremenny* was the Zhdanov Shipyard, Leningrad, which had been the major Soviet shipyard responsible for large destroyer-size ships since the Defence Review of 1959. Beginning with the 'Kynda', Zhdanov had built the initial series of Rocket Cruisers and had then converted to the modified 'Kresta II' BPK, which was still in production during the mid-1970s. All the Zhdanov ships had a similar hull-form and featured large, plated tower masts. All were powered by a turbo-pressurized steam propulsion plant producing 100,000shp which had probably been little modified in fifteen years.

The *Sovremenny* hull-form is clearly derived from the 'Kresta II', the only significant difference being the full waterplane aft characteristic of all recent Soviet surface construction. The deckhouse, which on the 'Kresta II' carried the forward SA-N-3 missile-launcher, has been extended to the bow on *Sovremenny*, so that the forecastle is one deck higher; and the cut-down quarterdeck has also been abandoned. These modifications have served both to improve sea-keeping and to increase internal volume, which proved to be inadequate in the 'Kresta II'. The adoption of a broad transom stern without the cut-down quarterdeck has also made it possible to reinstate mine-rails – a traditional feature of Soviet destroyers.

The single broad, angular funnel of *Sovremenny* is located slightly abaft amidships, the distance between the funnel and the stern being identical with that of the 'Kresta II'. Not only is *Sovremenny*'s propulsion machinery of the same type and power output as that of the 'Kresta', but the layout below decks is probably essentially the same.

The adoption of a common hull and an identical machinery installation had clear benefits in terms of continuity of production. The last of the 'Kresta IIs', *Admiral Yumashev*, was launched in 1976 and *Sovremenny* followed her on to the slipway without a break. She was launched in November 1978 and ran her sea trials only two years later. Indeed, the hull appears to have been completed in advance of expectations because *Sovremenny* is the only major Soviet surface type to have run sea trials without any of her major weapons systems installed.

The lead yard for *Udaloy* was Kaliningrad which had been responsible for the 'Krivak'. Kaliningrad had never before built ships of this size and production may have been delayed by the need to expand and upgrade facilities at the shipyard. The first hull was not laid down until 1978 – some eighteen months after the laying down of the first of her anti-surface counterpart. Moreover, Kaliningrad did not have sufficient capacity to be able to cope with the envisaged construction programme, so a second production line had to be opened up at Zhdanov alongside the *Sovremenny* class.

The derivation of *Udaloy* from the 'Krivak' is unmistakable. The ship is powered exclusively by gas turbines, and the hull is essentially an enlarged version of the 'Krivak' hull, with a broad square transom stern and a quarterdeck which breaks immediately abaft the machinery spaces. In marked contrast to *Sovremenny*, *Udaloy* has 'clean', squared-off superstructure blocks and her major radar antennae are carried on tall lattice masts.

The 8,000-ton Destroyer

Although similar in size and displacement to the 'Kresta', the new destroyers are very different in conception. This difference is partially disguised by the Soviet Navy's system of ship classification under which both 'Kresta II' and 'Udaloy' are designated Large Anti-Submarine Ships (BPK). However, it is recognized by the Soviets in the nomenclature, the lead units of both the *Sovremenny* and *Udaloy* classes being given adjectival names traditionally used for destroyer types (later units of the *Udaloy* class have reverted to the pattern established by the 'Kresta II', and have been named after military heroes of the Soviet Union).

The *Sovremennys*, despite their medium-range anti-surface missiles, are not designated as Large Rocket Ships (BRK), as might have been expected, but as *Eskadrenny Minonosets* (EM), the traditional Soviet destroyer classification. They should therefore be regarded as latter-day successors to the ageing 'Kotlins'.

The conceptual difference between the new destroyers and the BPKs of the 'Kresta II' and 'Kara' classes is that whereas the latter were designed to be relatively self-contained in terms of their military capabilities, the new ships are more specialized in their respective missions and are clearly designed to operate as part of a larger group of surface warships. Whereas the 'Kresta II' and 'Kara' were equipped with major area defence systems, requiring bulky surveillance and tracking radars, the *Udaloy* class has only a short-range system for self-defence, and *Sovremenny* has a medium-range system employing lightweight fire control radars which is essentially the linear successor to the SA-N-1 system of the 'Kashin' rather than the SA-N-3 system of the large BPKs of the 1970s.

Dispensing with a long-range area defence system has released both weight and volume for other systems, *Udaloy* being heavily biased towards ASW and *Sovremenny* towards anti-surface warfare.

The extent to which the large size of these ships was influenced by trends in the US Navy is difficult to gauge. Certainly *Udaloy* bears a striking resemblance both in size and overall conception to the US battle-group escorts of the *Spruance* class. The anti-submarine frigates and air defence destroyers in service with the major west European navies are markedly smaller; even the latest variant of the British Type 22 frigate, generally regarded as an extremely capable ocean-going ASW vessel, displaces a mere 5,250 tons at full load.

US Navy battle-group escorts tend to be larger than their European counterparts because large size can be equated with superior endurance, and the geo-strategic position of the United States requires that forward-deployed ships be able to conduct lengthy transits and operational patrols with the minimum of underway replenishment. It has therefore been assumed by some US commentators that the construction of these new destroyers marks the culmination of Soviet Navy ambitions to create a fleet capable of world-wide operations which might eventually challenge the US Navy on the high seas.

Changes in the pattern of Soviet maritime operations since the accession of Admiral Chernavin to the post of naval C in C make it difficult to assess the validity of these theories. If a world-wide challenge to US naval supremacy was ever seriously considered – and under Gorshkov, it is possible that it was – it is clear that this is no longer the case. Moreover, the construction of large destroyers would have been insufficient to support such a policy. One would have expected a major programme of underway replenishment vessels and this has not materialized. Indeed, not a single major vessel of the latter type has been completed for the Soviet Navy since *Berezina*, completed in 1978.

Clearly there were other factors that drove up the size of these ships. One important consideration was undoubtedly the comparatively large size of Soviet weapons systems. The SS-N-14 missile, although more capable in terms of range and payload than its Western counterparts, is both heavier and bulkier and is fired from individual launch canisters mounted above decks. The SS-N-22 anti-ship missile system which equips *Sovremenny* occupies a similar amount of space, in contrast to the diminutive US Harpoon missile which performs a similar function aboard many NATO surface vessels, and which can simply be bolted on to decks or superstructures. The helicopters of the Kamov Ka-25/27 series, while not as large as the US Navy's LAMPS III Seahawk, are significantly larger than the Westland/Aérospatiale Lynx in service with the west European navies, and have an unfavourable height to length ratio which makes them difficult to accom-

▲*Otlichny*, the third unit of the *Sovremenny* class, carries her full complement of weapons in this photograph taken in the Baltic shortly after completion in 1983. These ships have the adjectival names traditional to Soviet destroyers; *Otlichny* means 'Perfect'. (MoD, Bonn)

▶ *Vitse-Admiral Kulakov*, the second unit of the *Udaloy* class. Designated BPK by the Soviets, these ships lack the area defence system of their linear predecessors, and in conception are strikingly similar to the US anti-submarine battle group escorts of the *Spruance* class. Note the distinctive twin helicopter hangars and the large stern door for the Horse Tail variable depth sonar – the same model installed in the *Kirovs*. (MoD, Bonn)

Sovremenny Class: Profile and Plan

The drawing shows *Sovremenny* as fully fitted out, with Foot Ball ECM radomes in place.

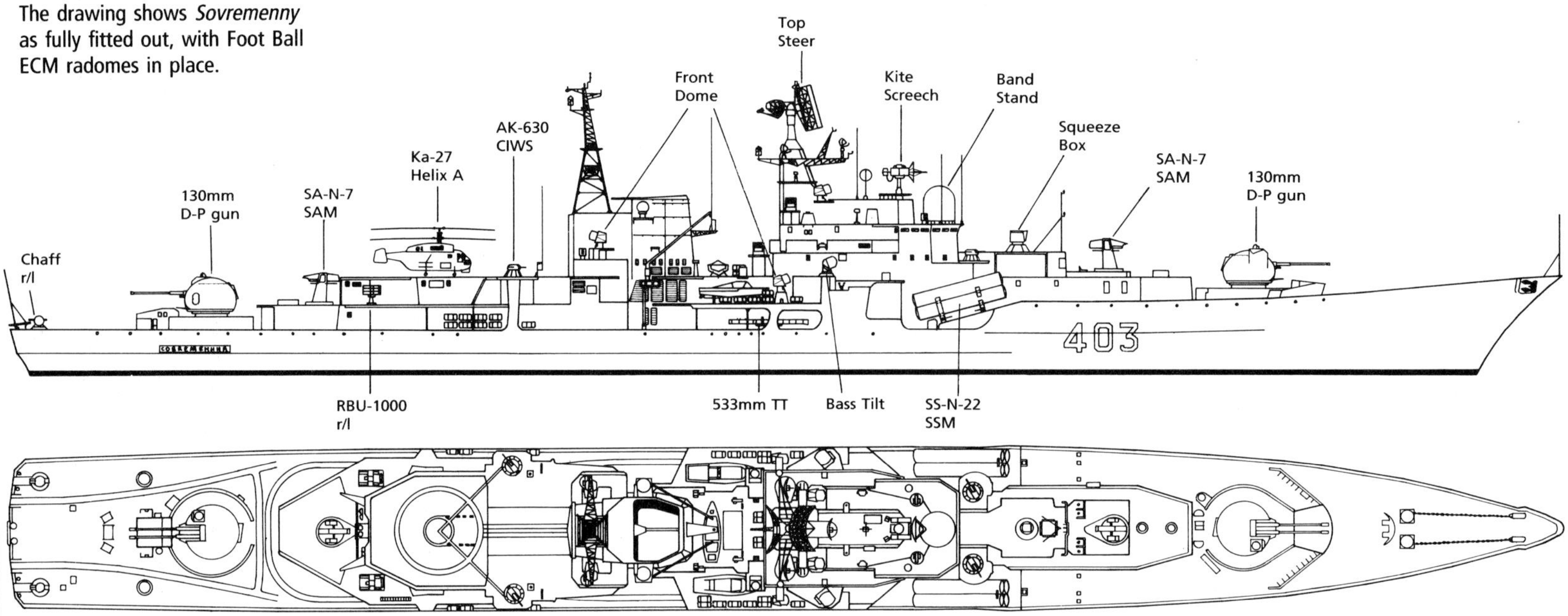

modate in a small surface vessel. The large bow sonar dome of *Udaloy*, and the bulky hoist mechanism of its towed counterpart, would have been equally difficult to accommodate in a smaller hull.

A further factor was undoubtedly the difficult weather patterns prevailing in the Northern Fleet area and in the north-west Pacific. Large hull size can generally be equated with superior sea-keeping, and since the new destroyers were intended for lengthy patrols rather than high-speed interception runs of limited duration, a large hull would bring with it significant benefits in terms of crew comfort, and would provide a stable weapons platform which could be maintained at a high state of readiness over a long period. It is perhaps significant that all units of these two classes so far completed have deployed to one or other of these two fleet areas, where they will increasingly provide the backbone of the surface forces intended for defence of the SSBN bastions.

The latest Soviet cruisers, as we have seen in Chapters 5 and 6, have outgrown traditional surface ship construction yards such as the Zhdanov Shipyard. This has effectively freed the slipways for larger destroyers. Moreover, the construction of these ships has undoubtedly served to reinforce the Soviet Union's image as a naval 'superpower'.

Sovremenny

Sovremenny has the classic 'double-ended' layout of the 'Kresta', with a twin 130mm gun mounting paired with a superimposed single-arm SAM launcher forward and aft. The quadruple launch tubes for the SS-N-22 anti-ship missile are mounted at a fixed angle abeam the bridge structure. The latter is significantly larger than in earlier types, but the plated foremast is correspondingly small because the Top Steer air surveillance and targeting radar initially fitted is lighter and less bulky than Top Sail.

The major innovation is the location of the helicopter deck and hangar on 02 level immediately abaft the single funnel. This is an arrangement favoured by the US Navy for its large battle-group escorts. It has the advantage of placing the flight deck close to amidships, where ship motion is much reduced, and higher above the waterline, thereby improving flight safety. The midships location also frees the quarterdeck for guns and missile-launchers, whereas ships with the hangar and flight deck above the stern have only limited options for the placing of their aft-mounted weapons systems (hence the 'sunken hangar' of the 'Kresta' and 'Kara').

However, even a ship the size of *Sovremenny* is limited in the number of weapons systems that can be

▲In the *Sovremenny* class the layout of the major weapons systems is essentially symmetrical, with twin 130mm dual-purpose gun mountings being paired with superimposed single launchers for the SA-N-7 area defence missile. The quadruple launchers for the SS-N-22 anti-ship missiles are abeam the bridge structure. Unusually, the helicopter deck and hangar are between the funnel and the after weapons systems; the telescoping hangar is fully extended in this view of *Osmotritelny*, taken during her deployment to the Pacific Fleet in 1985–6. (RAAF)

***Sovremenny* Class**

Soviet Designation: Eskadrenny Minonosets (EM) – Destroyer

Construction

Name	Builder	Laid down	Launched	In service
Sovremenny	Zhdanov Shipyard	1976	Nov 1978	Aug 1980
Otchayanny	Zhdanov Shipyard	1977	Apr 1980	May 1982
Otlichny	Zhdanov Shipyard	1978	Apr 1981	May 1983
Osmotritelny	Zhdanov Shipyard	1979	Apr 1982	June 1984
Bezuprechny	Zhdanov Shipyard	1980	Aug 1983	June 1985
Boyevoy	Zhdanov Shipyard	1981	Aug 1984	June 1986
Stoyky	Zhdanov Shipyard	1982	Aug 1985	Sept 1986
Okrylenny	Zhdanov Shipyard	1983	June 1986	Sept 1987
Burny	Zhdanov Shipyard	1984	Feb 1987	Aug 1988
Gremyashchy	Zhdanov Shipyard	1984	June 1987	Nov 1988
Bystry	Zhdanov Shipyard	1985	Dec 1987	Feb 1989
Rastoropny	Zhdanov Shipyard	1986	June 1988	Dec 1989
Bezboyaznenny	Zhdanov Shipyard	1986	Mar 1989	Sept 1990
Bezuderzhny	Zhdanov Shipyard	1987	June 1990	
...	Zhdanov Shipyard	1987	Oct 1990	
...	Zhdanov Shipyard	1987		
...	Zhdanov Shipyard	1988		
...	Zhdanov Shipyard	1988		

Propulsion

2-shaft geared steam turbines; 4 turbo-pressurized boilers; 100,000shp=32 knots max.

Displacement (tons)

Standard: 6,300
Full Load: 7,850

Dimensions

Length: 156m(512ft) overall
Beam: 17.5m(57ft)
Draught: 6.2m(20ft)

Complement

364

Weapons

ASuW: 2 quadruple SS-N-22 launchers (8 Sunburn missiles); 1 Ka-25 Hormone B missile targeting helicopter; 2 twin 130mm(5.1in) D-P guns; rails for 40 mines

AAW: 2 single-arm SA-N-7 launchers (40 Gadfly missiles); 4 AK-630 gatlings

ASW: 2 RBU-1000 rocket-launchers; 2 twin banks of 533mm(21in) torpedo tubes

Sensors

Surveillance radars: Top Steer 3-D in first 3 units; Plate Steer in *Osmotritelny, Bezuprechny*; Top Plate in others; 3 Palm Frond

Fire control: 1 Band Stand (SS-N-22); 1 Kite Screech (130mm); 6 Front Dome (SA-N-7); 2 Bass Tilt (AK-630)

Sonars: bow-mounted Bull Horn MF

Electronic warfare: 4 Foot Ball; 2 Bell Shroud, 2 Bell Squat; 2 twin-barrelled chaff-launchers accommodated on the centre-line. The desire to mount both a gun and a missile-launcher aft has compelled the designers to adopt a telescoping hangar for the single helicopter, the first of its type in the Soviet Navy. The telescoping hangar ensures protection from the elements, but is limited in the level of maintenance and stowage it can provide.

The SS-N-22 Missile

The SS-N-22 anti-ship missile (NATO designation: Sunburn) is derived from the SS-N-9 missile carried by the 'Nanuchkas' and other corvette-sized ships intended for the defence of the Soviet maritime zones. The two missiles are of approximately the same length, but the SS-N-22 is thought to fly faster and lower than its predecessor. Although almost certainly not a true 'sea-skimmer' in the manner of Exocet, the missile is reported to incorporate an altimeter which enables it to fly at low altitude out to a range estimated at 60nm (110km).

The Band Stand tracking radar, housed within a distinctive GRP radome 3m in diameter and 3.75m high, is also a standard feature of the SS-N-9 missile system. On *Sovremenny* it is mounted directly above the navigation bridge. The continued pairing of Band Stand with the SS-N-22 missile suggests that the latter does not have as low a flight profile as some commentators have suggested, because the missile would otherwise disappear over the horizon too quickly to justify a dedicated onboard tracking radar. Smaller spherical radomes which have received the NATO designation Light Bulb and which are reported to house a missile-associated data link are mounted on each side of the funnel.

The helicopter, when embarked, is reported to be the Ka-25 Hormone B missile targeting variant. Clearly some method of off-board OTH targeting would be required if the missile were to be effective out to its maximum range. *Sovremenny* is not equipped with the Punch Bowl satellite receivers fitted in the latest Soviet cruisers and in the SSGNs of the 'Oscar' class. The most likely targets for these missiles are therefore hostile

The SS-N-22 Sunburn Missile

Soviet designation: unknown
In service: 1981
Length: 9.15m(30ft)
Diameter: 0.85m(2ft 9in)
Span: 1.6m(5ft 3in)
Weight: 3,500kg(7,700lb)
Warhead: 500kg(1,100lb)
Propulsion: liquid-fuel rocket
Speed: Mach 2.5
Range: 110km(60nm)
Guidance: command/active radar or IR homing
Fire control: Band Stand

▼An overhead view of *Otchayanny* running speed trials in the Baltic. The single launchers for the SA-N-7 area defence system have yet to be installed. Note the broad transom stern with its network of mine rails. On the outer corners are the twin chaff-launchers.

▶Bow view of *Bezuprechny*, the fifth ship of the class, showing the powerful battery of weapons mounted in the forward part of the ship. On the centre-line, between the two quadruple launch tubes for the SS-N-22 missile, can be seen the Squeeze Box electro-optical device which is reported to combine LLTV, a laser range-finder and infra-red sensors. (MoD)

surface units operating independently, not the carrier battle groups which would have to be engaged at longer ranges.

The SA-N-7 Missile System

The SA-N-7 medium-range air defence system (NATO code-name: Gadfly) was developed in conjunction with the land-based SA-11, to provide a high-performance missile with a maximum range reported to be about 30,000m (16nm).

The radar/command fire control mode employed in the SA-N-1/SA-N-3 missile systems and in the land-based predecessors of the SA-11 had proved vulnerable to electronic countermeasures during the Middle East War of 1973 when entire missile batteries were effectively disabled. This problem had been recognized by Western navies many years previously, and even by the early 1960s radar/command guidance was being phased out in favour of semi-active guidance, in which the missile homed on the reflected radiation of small illuminators which could switch rapidly between targets. Semi-active guidance using

Provorny as SA-N-7 Trials Ship

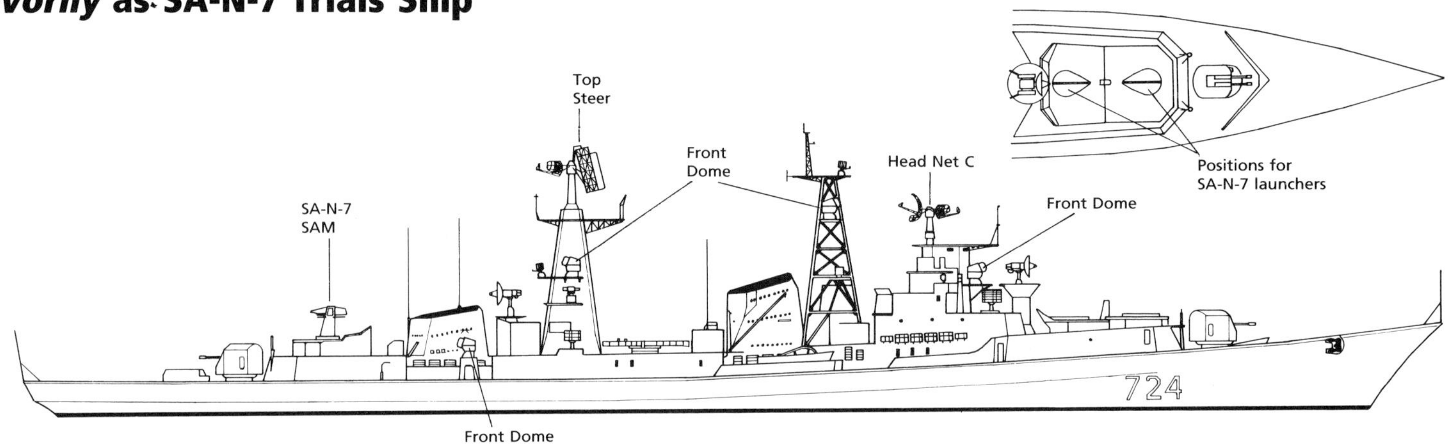

The SA-N-7 Gadfly Missile System

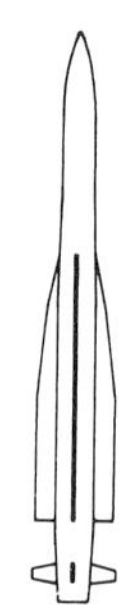
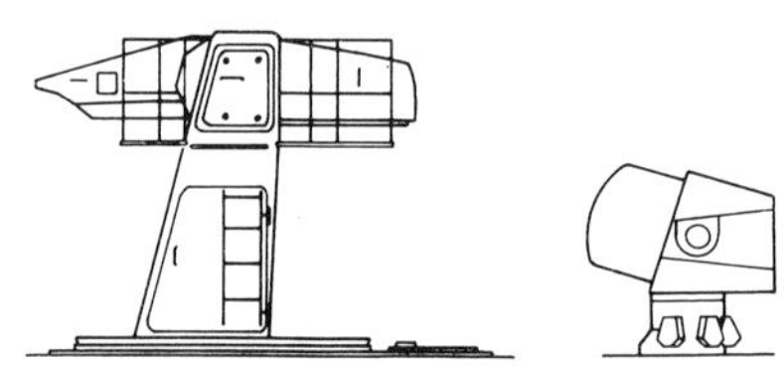

Soviet designation: unknown
In service: 1981
Length: 5.8m(19ft)
Diameter: 0.4m(1ft 4in)
Span: 0.9m(3ft)
Weight: 650kg(1,430lb)
Warhead: 54kg(120lb) HE
Propulsion: solid-fuel rocket
Speed: Mach 3
Range: 30km(16nm)
Ceiling: 15,000m(50,000ft)
Min. altitude: 30m(100ft)
Guidance: semi-active radar
Fire control: Front Dome
Operating frequency: H/I Band

monopulse techniques effectively reduces the size and complexity of fire control radars which in consequence can be provided in greater numbers.

The SA-N-7 missile system was first installed in a single unit of the 'Kashin' class, *Provorny*. The ship was taken in hand in the mid-1970s, and conducted missile trials in the Black Sea probably from about 1980, emerging into the Mediterranean only in the autumn of 1981.

Reconstruction was extensive; both the original SA-N-1 Goa missile systems were removed, and *Provorny* was given a remodelled bridge structure and a new plated mainmast. Only a single SA-N-7 launcher/magazine module was fitted, although there are base plates for two further systems forward. The tall plated mainmast carries a Top Steer air surveillance and tracking radar, and four pairs of Front Dome tracker/illuminating radars are fitted. These are distributed forward and aft on the sides of the superstructures and masts. A Head Net C 'V'-beam radar has replaced the original Head Net A, and has been relocated on the remodelled bridge structure.

The Soviets were clearly anxious to retain the ship's operational capabilities; all four anti-submarine launchers and the torpedo tubes have been retained, and although the original electronic countermeasures outfit was removed, there is provision for a new ECM outfit on a par with other modified ships of the 'Kashin' class (*see* Chapter 2).

The single-arm missile-launcher – the first of its type in Soviet Navy service – is strikingly similar in configuration to the US Navy's Mk 13 launcher, which has been widely fitted in ships armed with the Tartar and Standard missile systems. It was initially thought that this implied a missile of similar size and configuration to Standard, but a recently released photograph of the SA-N-7 missile shows that this is not the case, and that not only does Gadfly closely resemble the SA-N-6 Grumble missile in its general configuration, but that it has a length of about 5.8m as compared with only 4.5m for Standard. Only the lower section of the missile body is secured to the launcher arm, and the magazine is significantly deeper than that of its US counterpart – a feature which explains why the launchers are located at 01 deck level in *Sovremenny*.

Magazine capacity is currently estimated at 20 missiles per launcher. This is a comparatively low figure by Western standards, but for the Soviet Navy it is clearly deemed adequate given that *Sovremenny* has a 'double-ended' installation and a mission that is not primarily concerned with area air defence.

▶ *Provorny*, on her first sortie into the Mediterranean in late 1981. Note the new plated mainmast for the Top Steer 3-D tracking radar, and the paired Front Dome illuminators. The single SA-N-7 launcher installation has a canvas protective cover. (MoD)

The Front Dome fire control radar has a configuration reminiscent of the smaller Soviet gun fire control radars such as Drum Tilt and Bass Tilt, and is similar in size. Six are fitted in *Sovremenny*, two pairs being located on the sides of the forward superstructure and the third on platforms alongside the funnel. This arrangement gives all-round coverage, but as the tracker/illuminators are mounted hard against the superstructures the ship is limited to the use of three on either beam, while the bow and stern arcs are covered by only two. Again the problem is centre-line space, the only available positions being occupied by the Band Stand and Kite Screech fire control radars above the bridge structure.

Early units of the *Sovremenny* class were fitted with Top Steer which provides target data to the Front Dome illuminators. The fourth and fifth units, *Osmotritelny* and *Bezuprechny*, were fitted with a new hybrid antenna designated Plate Steer, in which the large square lattice reflector of Top Steer was replaced by a small solid planar array. Later ships, beginning with *Boyevoy* (completed 1987), have received the Top Plate back-to-back 3-D planar array first trialled in *Marshal Vasilevsky* of the *Udaloy* class (q.v.). In what constitutes a break with previous Soviet practice there is no secondary air surveillance antenna for back-up.

Anti-Missile Defence

For anti-missile defence *Sovremenny* has the standard four AK-630 multi-barrel guns controlled by two Bass Tilt radars. Unusually, the gun mountings are not paired, the forward mountings being located on 03 level forward of the bridge, while the after mountings are at 02 level on platforms projecting from the helicopter deck. The Bass Tilt fire control radars are between them abeam the bridge structure. The adoption of this arrangement appears to have been influenced by a number of factors related to the layout of the midships section, not least of which was the need to locate the boat deck between the funnel and the forward superstructure. It has effectively improved coverage of the forward and after arcs, although the wide spacing between the individual mountings and the fire control radar may prove problematical, given the need for rapid response against a fast-manœuvring target.

Sovremenny ran her trials without much of her electronic countermeasures equipment. As completed, she was fitted only with the diminutive Bell Squat and Bell Shroud antennae of the 'Krivaks'. By late 1981, however, she had received four Foot Ball antennae which were subsequently to replace Side Globe on the second unit of the *Kirov* class, *Frunze*. Foot Ball, which comprises a large hemispherical radome on a plinth

678

◀◀ Close-up of the Top Steer 3-D air-surveillance/tracking radar fitted in *Sovremenny* and early units of the class. Directly beneath are two of the six Front Dome illuminators for the semi-active SA-N-7 Gadfly missile. (MoD)

◀ A close-up of the Plate Steer radar on *Bezuprechny*, showing the distinctive honeycomb pattern on the planar array, which is tilted at approximately 20 degrees to the vertical. In the foreground is the Band Stand radome which houses electronics associated with the SS-N-22 missile. (Author)

▶ *Sovremenny* in late 1981, fully fitted out. *Sovremenny* ran the most extensive sea trials of the new Soviet surface ships because the design incorporated a whole new generation of equipment. The prototype Foot Ball ECM antenna, designed as the successor to Side Globe, was fitted at this time; there were to be no further installations aboard ships of this and the *Udaloy* class until late 1986. (US Navy)

◀ *Boyevoy*, the first of the class to receive the Top Plate radar antenna, comprising two planar arrays mounted back-to-back at a relative angle of about 30 degrees. (MoD Bonn, 1986)

with a prominent backing-plate, is thought to be a combined receiver-jammer and is significantly less bulky than Side Globe. The antennae fitted in *Sovremenny* are located at the foot of the plated foremast and on each side of the deckhouse between the funnel and the bridge structure.

Problems associated either with development or with production appear to have delayed the availability of this system. The next four units were completed with empty platforms, and only with the sixth unit, *Boyevoy* (completed 1987), was installation resumed.

Sovremenny has the standard pair of twin-barrelled chaff-launchers which are located at the corners of the transom stern, the mine-rails running inboard of them.

The 130mm Gun Mounting

The continued value of large- and medium-calibre guns was demonstrated during the Vietnam War when US Navy fire support operations against enemy positions ashore were most effectively performed by elderly war-built cruisers armed with 8-inch and 6-inch guns. The success of these operations led the US Navy to develop a Major Calibre Light Weight Gun (MCLWG) for installation aboard ships of destroyer size and upwards during the mid-1970s. The 8-inch gun, in a lightweight glass-reinforced plastic (GRP) mounting, was to have been capable of firing a variety of munitions, including rocket-assisted laser-guided projectiles capable of accurate long-range fire. The high cost of development together with a general unease about

The 130mm/70 D-P Gun

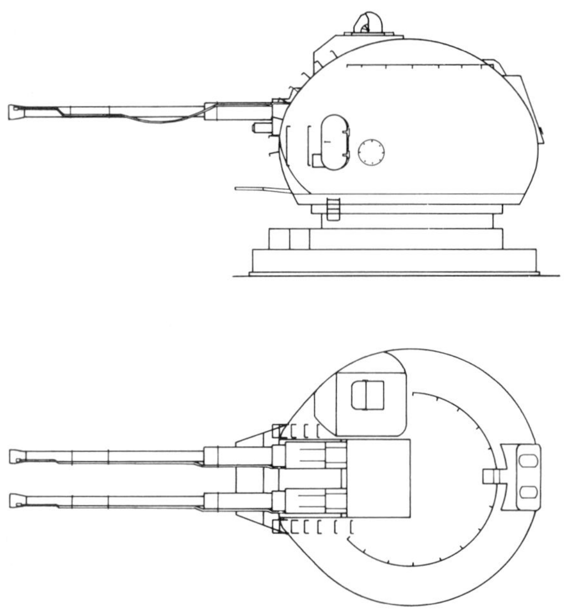

In service: 1981
Angle of elevation: +85/−15 degrees
Rate of fire: 30rpm per barrel
Projectile weight: 35kg?(77lb)
Muzzle velocity: 1,000m/sec(3,280ft/sec)
Range: 28km(15nm) max.
Fire control: Kite Screech

amphibious operations in the wake of the Vietnam War were to result in the abandonment of the programme, and attention turned instead to developing improved munitions for the existing standard lightweight 5-inch gun.

The Soviet Navy found itself in a similar situation to the US Navy, but the crisis point would occur some ten years later because of the large conventional post-war construction programme. The last of the 6-inch gun cruisers of the *Sverdlov* class had been completed in 1956 and the last of the 'Kotlins' in 1957. Given the Soviet propensity for squeezing the maximum service life out of their ships, these vessels would not be due for retirement until the late 1980s.

The twin 130mm gun mounting developed for *Sovremenny* is in effect a linear successor to the twin mounting of the same calibre installed in 'Kotlin'. Unlike its US Navy counterpart, the single 5-inch Mk 45 lightweight gun, it is a true dual-purpose weapon, with a high angle of elevation, water-cooled barrels and a high rate of fire. There is provision for local control using an optical sight similar to that of the single 100mm mounting installed in the later 'Krivaks' and other major surface vessels.

Fire control against both surface and aerial targets is provided by a single Kite Screech radar located on the bridge. Previous Soviet surface ships with 'double-ended' installations have been provided with two such radars, but there was clearly a problem in fitting in a second fire control radar on the centre-line, given the adoption of a telescopic hangar amidships and the absence of any clearly defined after superstructure. The single installation is adequate for engaging targets on the beam, but is at a distinct disadvantage when aerial targets are engaged, particularly if their approach is from the stern quarter.

Perhaps the most interesting feature of the gun fire control system associated with the 130mm mountings is an electro-optical device mounted on the deckhouse forward of the bridge, which has received the NATO designation Squeeze Box. The device can be fully rotated in azimuth, and is believed to incorporate a low-light television system, a laser range-finder and infra-red sensors. Squeeze Box has also been fitted in some Soviet amphibious units where it is associated with the 140mm barrage rocket-launcher.

The laser range-finder would be particularly useful for gunfire support, and the development of projectiles with semi-active guidance using laser illumination would improve accuracies still further. If the projectiles were rocket-assisted (RAP), maximum range could be extended beyond the normal 15nm to perhaps 20–25nm. The new Soviet 130mm gun can therefore be assumed to have a potent fire support capability.

Anti-Submarine Warfare

The anti-submarine capabilities of *Sovremenny* are of a secondary nature. They provide only for close-range self-defence and would add little to the overall capabilities of a task group. Conspicuous by their absence are stand-off missiles and the now-customary variable depth sonar; the single helicopter carried is probably for over-the-horizon (OTH) missile targeting, although one published photograph shows what is clearly the Ka-27 Helix A anti-submarine variant on the helicopter deck.

The configuration of the stem suggests a bow-mounted sonar dome, but the wave pattern of the ship indicates that it is comparatively small.[1] It has been suggested that it may be a multifunctional model operating at medium frequency, performing both the search and tracking functions as well as providing fire control inputs for the ship's anti-submarine weapons. The latter comprise two RBU-1000 rocket-launchers, mounted below the after end of the helicopter deck,

▲Close-up of the twin 130mm gun mounting, with the after SA-N-7 missile-launcher at the top right of the picture. Note the water-cooled barrels which can be elevated to 85 degrees, and the local control station. The photograph also provides an excellent view of the mine rails and the twin-barrelled chaff-launchers, mounted on the outer corners of the transom stern. (MoD)

▶On-deck view of *Bezuprechny*, taken during a visit to Portsmouth, UK, in July 1990. Note the characteristic numbering of the quadruple launch tubes for the SS-N-22 anti-ship missile. (Author)

▼ *Osmotritelny*, the first of the *Sovremenny* class to deploy to the Pacific, photographed with a Ka-27 Helix A helicopter on deck. Substituting the Helix A for the Ka-25 Hormone B missile targeting helicopter would provide a dramatic boost to anti-submarine capabilities, which are otherwise somewhat limited. (RAAF)

and homing torpedoes fired from twin trainable 533mm tubes located beneath the shelter deck forward of the funnel.

Udaloy

In contrast to *Sovremenny*, *Udaloy* is the most capable anti-submarine warfare vessel of its size in the Soviet Navy. The design is essentially conservative; the hull-form is derived from the 'Krivak', the gas-turbine propulsion system from the 'Kashin' and the anti-submarine weaponry is little different from that of the 'Kresta' and 'Kara'. However, *Udaloy* reveals a growing stability and maturity in Soviet design and construction. The ship is more spacious, less cramped than earlier BPKs; topweight has been reduced to a minimum by the adoption of comparatively low superstructures and lattice masts, and sea-keeping is reported to be outstanding.

The ship is powered by four large gas turbines in a COGAG (COmbined Gas And Gas) configuration, driving two large-diameter fixed-pitch propellers via twin shafts. The turbine is almost certainly the same model which propels the 'Kara' and *Slava*. However, whereas the latter vessels have two turbine rooms but only a single set of exhaust uptakes, *Udaloy* has reverted to the 'Kashin' arrangement of two sets of paired uptakes with the lattice mainmast between them. This consumes more centre-line space above-decks – less crucial in a ship without an area defence system – but reduces the amount of exhaust trunking required between decks, thereby freeing hull and superstructure volume for additional accommodation and electronics.

The configuration of the after section of the hull, which has an exceptionally broad waterplane, suggests that the twin shafts are mounted parallel to each other. This would facilitate a mechanical cross-connection to enable a single turbine to drive both shafts at cruising speed, with consequent fuel economies. Endurance is reported to be superior to that of earlier types.

Anti-Submarine Warfare

The primary anti-submarine weapons are the SS-N-14 stand-off missile and the Ka-27 Helix A helicopter. The missiles are accommodated in identical fashion with those of the 'Kresta' and 'Kara', in quadruple box launchers beneath cantilevered bridge wings. Since there are no Head Lights radars available to provide command guidance, dimimutive Eye Bowl radars are mounted on the navigation bridge to port and to starboard.

The Ka-27 Helix helicopter ran its first sea trials from *Udaloy*. Derived from the Ka-25 Hormone, with which it shares the classic Kamov coaxial rotor configuration, the Helix shows a number of incremental improvements over its predecessor. It has a higher cruise speed and greater patrol endurance. Weapons capacity is similar, but the Helix carries a more sophisticated electronics and avionics suite. The most advanced feature of the helicopter is its automatic hovering capability which enables it to use its dipping sonar at night or in adverse weather.

Udaloy breaks with previous Soviet Navy practice in carrying two helicopters which are accommodated and maintained in twin side-by-side hangars with a flight control station between them. This change was undoubtedly influenced by developments in Western navies, which from the mid-1970s began to deploy anti-submarine frigates and destroyers equipped with two helicopters in an effort to improve availability and to sustain near-continuous flight operations. The square transom stern of *Udaloy* is particularly well suited to accommodating both the twin hangars and the broad flight deck required.

As in earlier BPKs the hangar floors are at quarterdeck level in order to reduce the overall height of the hangar, while the helicopter deck is one level above. Each of the twin hangars has a roof comprising telescoping panels which slide back to enable the Helix to be wheeled in. The traversing system uses twin tracks and chains. Once inside the hangar, the helicopter is secured to the hangar floor which then descends to quarterdeck level. The hangar roof panels move back again and the hinged doors are closed.

In *Udaloy* and the second ship, *Vitse-Admiral Kulakov*, the outboard edge of the helicopter deck runs in line with the twin hangars, but later units have had

▼A bow view of *Vitse-Admiral Kulakov*, the second ship of the *Udaloy* class. *Udaloy*, with her broad hull-form and comparatively low profile, is reputed to have exceptionally good sea-keeping qualities which enable it to conduct round-the-clock anti-submarine operations in the most difficult weather. (US Navy, 1988)

Udaloy Class: Profile and Plan

The profile drawing is of *Admiral Zakharov*, with the prototype SA-N-9 installation in place. The plan view is of *Udaloy* as completed, with empty circular platforms forward and aft for Cross Sword. The inset shows the revised flight deck arrangements for later units.

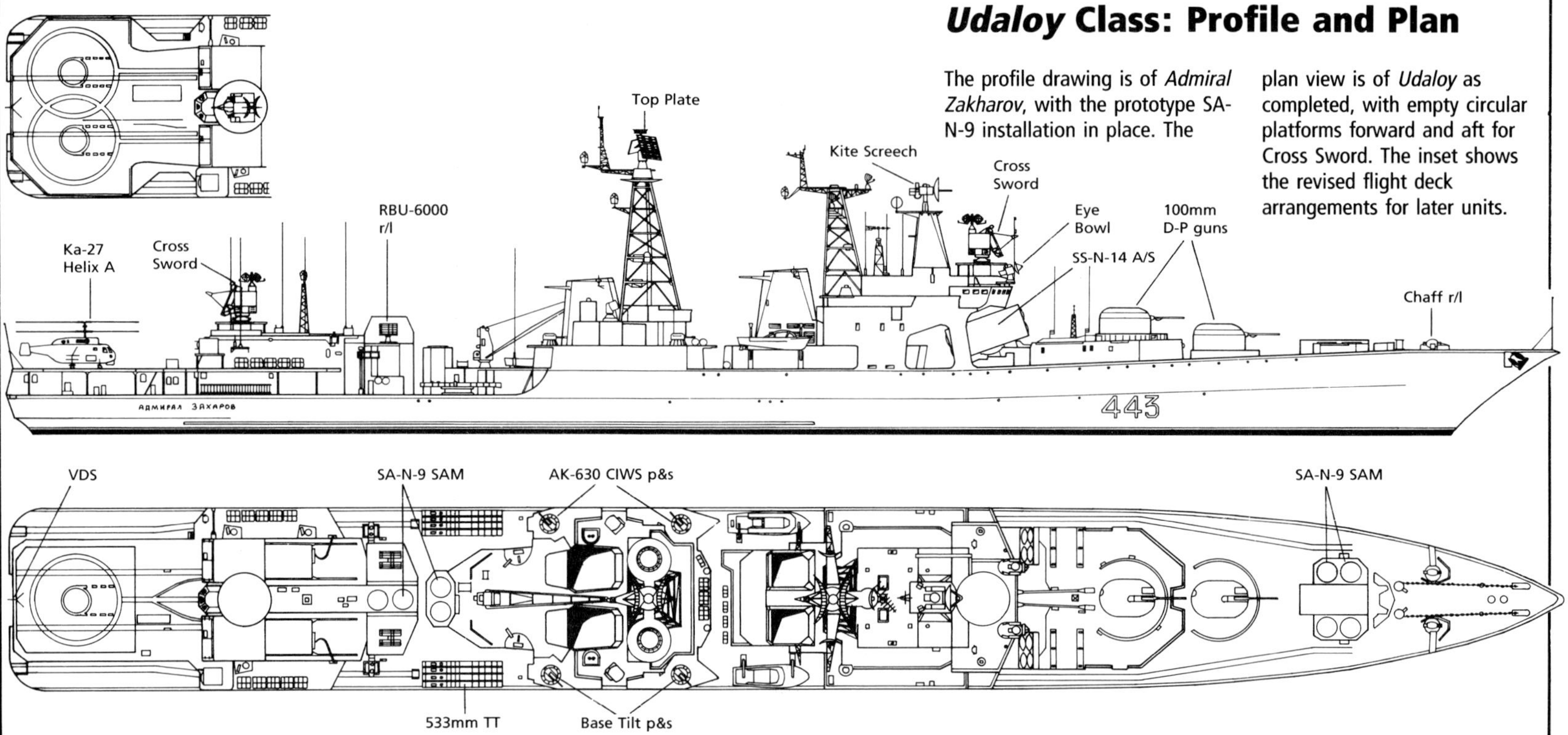

The Ka-27 Helix A ASW Helicopter

In service: 1981
Length: 11.3m(37ft) fuselage
Width: 4m(13ft)
Height: 5.4m(17ft 8in)
Weight: 11,000kg(24,250lb) normal take-off
Engines: two Glushenkov TV3-117V turboshafts, each 1,660kW(2,225shp)
Speed: 230km/h(124kts) cruising; 250km/h(135kts) max.
Ceiling: 6,000m(19,500ft)
Combat radius: 375km(200nm); endurance 4.5hr
Armament: two 400mm A/S torpedoes; nuclear/conventional depth-bombs in internal weapons bay
Sensors: chin surveillance radar; dunking sonar; towed MAD boom; sonobuoys from internal weapons bay

Udaloy Class

Soviet Designation: Bol'shoy Protivolodochny Korabl' (BPK) – Large Anti-Submarine Ship

Construction

Name	Builder	Laid down	Launched	In service
Udaloy	Kaliningrad Shipyard	1978	Feb 1980	Nov 1980
Vitse-Admiral Kulakov	Zhdanov Shipyard	1978	Apr 1980	Sept 1981
Marshal Vasilevsky	Zhdanov Shipyard	1979	Jan 1982	June 1983
Admiral Zakharov	Kaliningrad Shipyard	1979	Apr 1982	Oct 1983
Admiral Spiridonov	Kaliningrad Shipyard	1981	Nov 1983	Sept 1984
Admiral Tributs	Zhdanov Shipyard	1980	Apr 1983	Aug 1985
Admiral Shaposhnikov	Kaliningrad Shipyard	1983	Jan 1985	Oct 1986
Simferopol	Kaliningrad Shipyard	1983	Feb 1985	Dec 1986
Admiral Levchenko	Zhdanov Shipyard	1983	Mar 1985	Jan 1988
Admiral Vinogradov	Kaliningrad Shipyard	1985	June 1987	Oct 1988
Admiral Kharlamov	Kaliningrad Shipyard	1985	June 1988	Sept 1989
...	Kaliningrad Shipyard	1987	Feb 1990	
...	Kaliningrad Shipyard	1988		

Displacement (tons)
Standard: 6,700
Full Load: 8,100

Dimensions
Length: 162m(531ft) overall
Beam: 19.3m(63ft)
Draught: 6.2m(20ft)

Propulsion
2-shaft COGAG; 4 gas turbines each 31,250hp; 125,000hp=32 knots max.

Complement
250

Weapons
ASW: 2 quadruple SS-N-14 launchers (8 Silex missiles); 2 Ka-27 Helix A helicopters; 2 RBU-6000 rocket-launchers; 2 quadruple banks of 533mm(21in) torpedo tubes
AAW: 2 groups each of four SA-N-9 vertical launch magazines (64 missiles); 4 AK-630 gatlings; 4 twin 12.7mm AA in *Admiral Vinogradov* only
ASuW: 2 single 100mm(3.9in) D-P guns; rails for 30 mines

Sensors
Surveillance radars: 2 Strut Pair 3-D in *Udaloy, V-Ad Kulakov*; Top Plate 3-D in others, with Strut Pair in *Simferopol* onwards; 3 Palm Frond
Fire control: 2 Cross Sword (SA-N-9); 2 Eye Bowl (SS-N-14); 1 Kite Screech (100mm); 2 Bass Tilt (AK-630)
Sonars: bow-mounted Horse Jaw LF; Horse Tail LF VDS
Electronic warfare: 4 Foot Ball (not fitted in early units); 2 Bell Shroud, 2 Bell Squat; 2 twin-barrelled chaff-launchers

◀ The Ka-27 Helix A anti-submarine helicopter. The Helix has greater endurance than the Hormone, together with an automatic hovering capability which enables it to use its dipping sonar at night or in adverse weather. (US Navy)

the helicopter deck broadened to extend the full width of the quarterdeck. The single large helo-circle of the first two ships has been superseded by twin overlapping circles broadly in line with each of the twin hangars. This presumably serves to minimize lateral movement of the helicopters when being spotted for take-off or stowed.

In addition to her three Palm Frond navigation radars which can be used to track the helicopters, *Udaloy* has the Fly Screen B microwave landing radar first observed on *Kirov*.

The minor anti-submarine weaponry is identical with that of the 'Krivak'. Quadruple banks of 533mm torpedo tubes are mounted just abaft the break in the forecastle, and directly above them are twin RBU-6000 rocket-launchers. The latter are mounted on pedestals at the forward end of the hangar with prominent blast deflectors behind them. The RBU-6000 rocket-launcher has never before been mounted so far aft on a major surface vessel, but given the layout of the weapons forward, the only available position would have been the forecastle where the launchers would have been exposed to the worst effects of weather and heavy seas. Undoubtedly the forward-mounted launchers of earlier BPKs have proved increasingly difficult to maintain because of corrosion.

Udaloy has the Horse Jaw low-frequency bow sonar and the matching Horse Tail LF variable depth sonar which are fitted in the *Kirov* class. The large 'fish' for the variable depth sonar is deployed, as on earlier BPKs, via a large hinged door set into the stern.

Air Defence

Unlike the earlier BPKs of the 'Kresta II' and 'Kara' classes, the *Udaloy*s do not have an area defence missile system. In its place is the new SA-N-9 missile system which is capable of defending the ship against aircraft – and presumably missile – attack at short and medium ranges and which employs vertical launch technology.

It would have been difficult to fit the standard Soviet double-ended area defence missile system into a design so clearly optimized for anti-submarine warfare, given the priority accorded to helicopter operation. Moreover, the bulky long-range air surveillance and tracking radars associated with such systems would have required at least one tower mast, and the increase in topweight would almost certainly have resulted in stability problems in a ship driven by relatively lightweight gas turbines. The SA-N-6 Grumble missile system installed in the *Kirov*s would have been out of the question for a ship of this size, while the SA-N-7 Gadfly system fitted in the contemporary destroyers of the *Sovremenny* class would have displaced one of the

The SA-N-9 Missile System

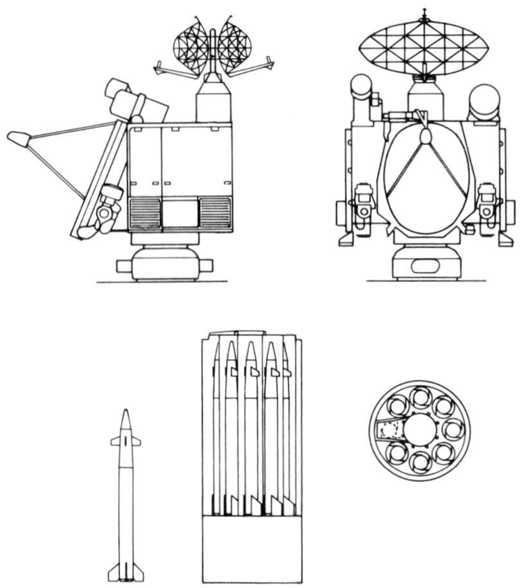

Soviet designation: unknown
In service: 1986?
Length: 3.5m(11ft 6in) approx.
Diameter: 0.22m(1ft 2in)
Span: 0.5m
Weight: 170kg(375lb) approx.
Warhead: 15kg(33lb)
Propulsion: dual-thrust, solid-fuel rocket

Speed: Mach 2.5
Range: 12km(6.5nm)
Ceiling: 6,000m(20,000ft)
Min. altitude: 10m(35ft)
Guidance: radar/command
Fire control: Cross Sword

100mm mountings forward, and would have been difficult to accommodate in the after part of the ship where both deck-space and hull volume are constrained by the layout of the propulsion machinery and the helicopter facilities.

Again the weapons choices made in *Udaloy* mark out the design as an outgrowth of the 'Krivak' which was similarly equipped with only short- and close-range air defence systems. The SA-N-9 is clearly regarded by the Soviets as the linear successor to the SA-N-4 Gecko missile system which it has replaced on the last two units of the *Kiev* class.

Problems appear to have been experienced with the development of the SA-N-9 system; the first two units of the *Udaloy* class deploying with empty radar platforms, their magazine silos being sealed by circular blanking plates. Moreover, these two ships were fitted from the outset with what was clearly an interim outfit of air surveillance antennae comprising twin Strut Pair back-to-back radars, derived from the Strut Curve S-Band radar fitted in smaller Soviet surface vessels since the mid-1960s but reportedly employing pulse-compression technology.

In 1983 the third ship of the class, *Marshal Vasilevsky*, entered service with a new 3-D target designation and tracking radar which subsequently received the NATO designation Top Plate. The antennae was installed at the top of the mainmast, leaving the foremast free, apparently for the installation of a new type of air surveillance radar.

Top Plate comprises two planar antennae mounted back-to-back and inclined at an angle of about 30 degrees relative to each other. The larger of the two antennae is broadly square and of open mesh construction; the other is rectangular and of solid construction. Atop the central mounting pillar is a small lattice aerial for IFF. Top Plate is the first planar radar to have been developed for the Soviet Navy, and employs the electronic frequency scanning techniques favoured by the US Navy since the early 1960s, but the relative angle of the Top Mesh and Top Plate antennae, reminiscent of the Head Net C and Strut Pair radars, suggests continued reliance on V-beam techniques to obtain accurate height data.

The next unit of the class, *Admiral Zakharov*, was fitted not only with Top Plate but with the prototype SA-N-9 missile installation. Four cylindrical magazines were installed on the forecastle, with a further four between the after funnels and the helicopter hangar, and Cross Sword fire control radars were installed forward and aft. *Zakharov* ran extensive trials in the Baltic before being transferred to the Far East in 1987–8.

The vertical launch magazine associated with the SA-N-9 missile system is similar in conception to the SA-N-6 installation aboard the Rocket Cruisers of the *Slava* class. Eight missile rounds are stowed in a rotating magazine ring and are aligned with a single missile port for firing. Each of the missiles is housed within its own individual launch cylinder, which confers a high degree of readiness. The magazine cylinder is designed as a self-contained module for ease of installation at the shipyard, and is approximately 2m in diameter and about five metres deep.

The SA-N-9 missile, which like the SA-N-4/SA-8 Gecko is a joint development between the Soviet Navy and the PVO/SV (the variant employed by the Ground Forces is designated SA-15), is cold-launched using a gas ejection system to a height of 18–20m. Thruster jets then turn the missile over towards the target and the solid-fuel motor is ignited. The adoption of vertical

▶ A stern quarter view of *Vitse-Admiral Kulakov* of the *Udaloy* class. (US Navy, 1988)

launch enables the missile to engage at relatively close range, given positive control at launch and the high speed attained by the time the missile enters the guidance beam of the tracking radar.

The missile is essentially similar in configuration to the earlier Gecko, with a relatively slim body, prominent cruciform tail surfaces and smaller canard fins indexed in line close to the nose. The version offered for export carries a 15kg HE fragmentation warhead activated by a proximity fuze, and overall weight is estimated at 170kg. Maximum range is 12,000m, with a ceiling of 6,000m – half the initial Western estimate. However, the missile can engage targets flying as low as 10m above the surface and this would make it particularly effective against carrier-based attack aircraft and strike fighters.

The Cross Sword fire control radar is significantly larger than the Pop Group radar associated with the SA-N-4 system. Two large boxes housing the radar electronics are mounted on each side of a central pillar. Between them and above them at the rear end of the installation are two small elliptical antennae of lattice construction mounted back-to-back to provide a high data rate against fast-moving targets such as low-flying aircraft and sea-skimming missiles. The antennae, which probably operate in the C-Band, provide range, azimuth, elevation and automatic threat evaluation for multiple targets out to a range of 25km, and can also initiate tracking of up to ten targets.

The large monopulse tracking dish is angled at about 22.5 degrees to the vertical and has a large feed to the front. It operates in the K-Band and, if performance is comparable to that of its land-mobile counterpart, can track and engage two targets simultaneously.

To the right of the solid reflector, mounted atop one of the two boxes housing the electronics, is a drum-shaped antenna for missile tracking. On the other box, on the left-hand side, is a missile command link antenna similar to that associated with the Top Dome guidance radar for the SA-N-6 missile. At the base of the main tracking antenna, on each side, are electro-optical TV trackers which can be used independently for missile guidance in a difficult ECM environment.

Development and production problems have continued to plague the SA-N-9, and shortages have been exacerbated by the adoption of large-scale installations

◀ **Overhead of *Marshal Vasilevsky*, showing the enlarged helicopter landing pad with its twin helo-circles. Note the paired uptakes for the gas turbines. (MoD Bonn, 1983)**

▶ ***Marshal Vasilevsky*, with the prototype Top Plate 3-D antenna at the top of the mainmast and an empty foremast platform. The vertical launch magazines for the SA-N-9 air defence system have been installed, but not the associated Cross Sword fire control radars for which there are empty platforms forward and aft. (US Navy, 1986)**

▶ ***Admiral Zakharov*, the fourth ship, running trials in the Baltic with the prototype SA-N-9 installation. The next three ships would be completed with only the after Cross Sword radar in place. (MoD, Bonn)**

for major new surface units such as *Baku* and *Tbilisi*, which are fitted with no fewer than 24 missile magazines (the equivalent of three complete *Udaloy* installations) and four Cross Sword fire control radars (*see* Chapter 8). The solution adopted has been to complete ships with both sets of vertical launch systems but only the after Cross Sword radar in place. The first ship to deploy with a complete operational SA-N-9 installation was *Simferopol* (Unit no. 8). The first three ships of the class have yet to be retro-fitted with their missile systems, and *Udaloy* and *Vitse-Admiral Kulakov* retain their original Strut Pair air surveillance radars up to the time of writing.

Paired AK-630 gatlings are located on their own magazine deckhouse abeam the second pair of funnels, with their Bass Tilt fire control radars and optical sights for back-up above and between them. Twin-barrelled chaff-launchers of the standard pattern are located close to the bow, but the electronic countermeasures outfit remains otherwise incomplete. As with the *Sovremenny* class, the early ships have been completed with only the diminutive Bell Shroud/Bell Squat combination of the 'Krivaks'. There are, however, empty platforms at the base of the mainmast intended to accommodate four Foot Ball ECM antennae as and when these become available. The first unit to be fitted with Foot Ball was again *Simferopol*, which entered service in late 1986.

Anti-Surface Warfare

In common with earlier Soviet BPKs, the *Udaloys* have no long-range anti-surface weaponry, but they have

◀ **Stern view of *Marshal Vasilevsky*, showing the large hinged door which conceals the hoist mechanism for the Horse Tail variable depth sonar. Note the indentations for mine rails in the corners of the stern. (MoD, Bonn)**

645

443

▲▼ *Admiral Spriridonov* and *Admiral Tributs*, two of the mid-series *Udaloy*s which deployed to the Pacific Fleet in 1988. They have the Top Plate 3-D radar at the top of the mainmast and both groups of SA-N-9 vertical launch magazines, but only the after Cross Sword guidance radar is fitted. They have their full ESM outfit, comprising Bell Shroud and Bell Squat antennae at the base of the foremast, and a pair of unidentified Bell-series antennae on the mainmast, but the larger mainmast platforms intended for the Foot Ball combined receiver/jammers remain empty. (MoD, Bonn)

▲**The first unit of the class with the complete electronics outfit is *Simferopol*, which ran trials from late 1986 before transferring to the Soviet Northern Fleet. She has both Top Plate and Strut Pair, and the four Foot Ball ECM radomes have been installed on the mainmast platforms. The plinths of the latter are painted black, the radome itself being picked out in white. (MoD, Bonn)**

the two single 100mm dual-purpose gun mountings of the later 'Krivaks'. These are mounted forward and are controlled by a single Kite Screech radar located on a tall pylon on the bridge.

A 'destroyer' feature which they share with the 'Krivak' and the *Sovremenny* is the ability to lay mines. The mine-rails run parallel to one another on each side of the quarterdeck. On the first two ships they ran outboard of the helicopter deck, but from *Marshal Vasilevsky* onwards the flight deck has been extended over the mine-rails. It is difficult to estimate mine capacity because the rails also serve as part of the deck transport system for solid stores and munitions. They terminate immediately abaft the quadruple banks of torpedo tubes where they run beneath constant-tension transfer stations. There are similar sets of rails serving the forecastle area; these terminate close to the forward SA-N-9 vertical launch magazines. A crane, which is normally stowed between the after pair of funnels, serves the after rails and the ship's boats, and is presumably also used to reload the after SA-N-9 magazines when the ship is alongside.

Service Life

The destroyers of the *Sovremenny* and *Udaloy* classes have deployed exclusively to the Northern and Pacific Fleets where increasingly they constitute the backbone of the bastion defence forces. The general-purpose armament of the *Sovremenny* class makes these ships well suited both to the defence of Soviet anti-submarine task groups against air or surface attack, and to the support of amphibious assault forces within their respective operational areas. They can also combine effectively with the larger Rocket Cruisers to create surface action groups to counter offensive operations by Western carrier battle groups.

The *Udaloys*, on the other hand, would be deployed with the older BPKs in anti-submarine task groups some of which would be centred around a PKR of the *Kiev* class. The suitability of these ships for carrier escort duties has also been noted and they will undoubtedly be employed for this purpose when the new Soviet carriers of the *Kuznetsov* class (*see* Chapter 8) enter service in the early 1990s.

Deployments have followed a similar pattern. The first three ships of both classes were transferred to the Northern Fleet following sea trials and work-up in the Baltic. The fourth unit of the *Udaloy* class, *Admiral Zakharov*, remained in the Baltic to conduct trials with the SA-N-9 missile system but the fifth ship, *Admiral Spiridonov*, joined company with the fourth ship of the *Sovremenny* class, *Osmotritelny*, to escort the Rocket Cruiser *Frunze* to the Far East in the autumn of 1985.

The next batch of major transfers took place in 1987–8 when the Iran–Iraq War provided the stimulus for a number of deployments to the Arabian Gulf to ensure the safe passage of Soviet merchant ships and to provide a political counterweight to the presence of the

493

ВИЦЕ-АДМИРАЛ КУЛАКОВ
401

◀▲ *Admiral Vinogradov*, the tenth ship of the class, photographed in the Baltic shortly after completion. She carries a complete outfit of weapons and sensors. She deployed to the Pacific Fleet in the autumn of 1989. (MoD, Bonn)

◀The first three units of the *Udaloy* class deployed to the Soviet Northern Fleet. Subsequent units have deployed to both the Northern and the Pacific Fleets, the latter receiving substantial reinforcements during the late 1980s. (RDN)

▲*Boyevoy* during her deployment to the Gulf, which preceded her transfer to the Pacific Fleet. She has the Top Plate radar antenna of later units of the class, together with four Foot Ball receiver/jammers, which can be seen between the plated foremast and the funnel. (Courtesy Gary Davies)

Western navies. The first to deploy were *Boyevoy* and *Stoyky* of the *Sovremenny* class which were operating close to the Straits of Hormuz during the latter part of 1987, before continuing their transit to the Far East. They were relieved by *Admiral Zakharov*, *Marshal Shaposhnikov* and *Admiral Tributs*, all of the *Udaloy* class, which in turn transferred to the Soviet Pacific Fleet.

In the meantime *Bezuprechny*, the fifth ship of the *Sovremenny* class, had deployed to the Northern Fleet area where she was joined in late 1988 by her sister *Okrylenny* and the eighth ship of the *Udaloy* class, *Simferopol*. The year 1989 saw the deployment of *Admiral Levchenko* (*Udaloy* class) to the Northern Fleet and her sister *Admiral Vinogradov* to the Pacific, while the *Sovremenny*-class destroyers *Gremyashchy* and *Burny* joined the Northern and Pacific Fleets respectively. Their sister *Bystry* transferred to the Pacific Fleet in company with the *Slava*-class cruiser *Chervona Ukraina* in the autumn of 1990, a year which also saw the deployment of *Rastoropny* (*Sovremenny* class) and *Admiral Levchenko* (*Udaloy* class) to the Northern Fleet.

Series production of *Sovremenny* is continuing at the rate of two hulls per year at the Zhdanov Shipyard, Leningrad. Thirteen ships had been completed by early 1991 and it is reported that the class will eventually number 22 units.

Eleven units of the *Udaloy* class had been completed by early 1991 with a twelfth due to complete later in the year. Hull no. 13 is to be of a modified design with updated weapons systems. The single 100mm mountings will be superseded by the twin 130mm fitted in *Sovremenny*, and CADS-N-1 combined missile/gun systems will replace the AK-630 CIWS. The first of the new type is due to enter service in 1993.

The 'Krivak III'

Udaloy proved to be too large and too complex to be built at the Kamysh-Burun Shipyard, Kerch, which had participated alongside Kaliningrad in the 'Krivak' construction programme. With the completion of *Poryvisty* in 1982, production at the smaller of the two yards was switched to a new variant of the 'Krivak' intended for the Maritime Border Guard branch of the KGB.

Designated *Pogranichny Storozhevoy Korabl'* or PSKR (Border Patrol Ship), the new vessels are fitted with the full range of military equipment, and would be a valuable adjunct to the regular naval forces in the north-west Pacific area to which they have been deployed. In design terms, they would constitute an intermediate stage between the original 'Krivak' and a new type of BPK or SKR intended to complement the larger *Udaloy*, and which would enter production in the late 1980s.

The recasting of the original design is much more dramatic than was the case with the 'Krivak II'. The principal change is the substitution of a hangar and flight deck for a Ka-27 Helix helicopter for the after weapons systems, and the replacement of the quadruple bank of SS-N-14 launchers forward by a single 100mm gun. There are a number of other important modifications, however, including the installation of a close-in weapons system for anti-missile defence. This alone serves to provide an indication of the emphasis on military qualities in the design.

In contrast to the BPKs, the hangar floor is on the same level as the helicopter deck, which is essentially a continuation of the forecastle deck and sits on top of the

'Krivak III' Class: Profile and Plan

The drawing is of the first unit, *Dzerzhinsky*.

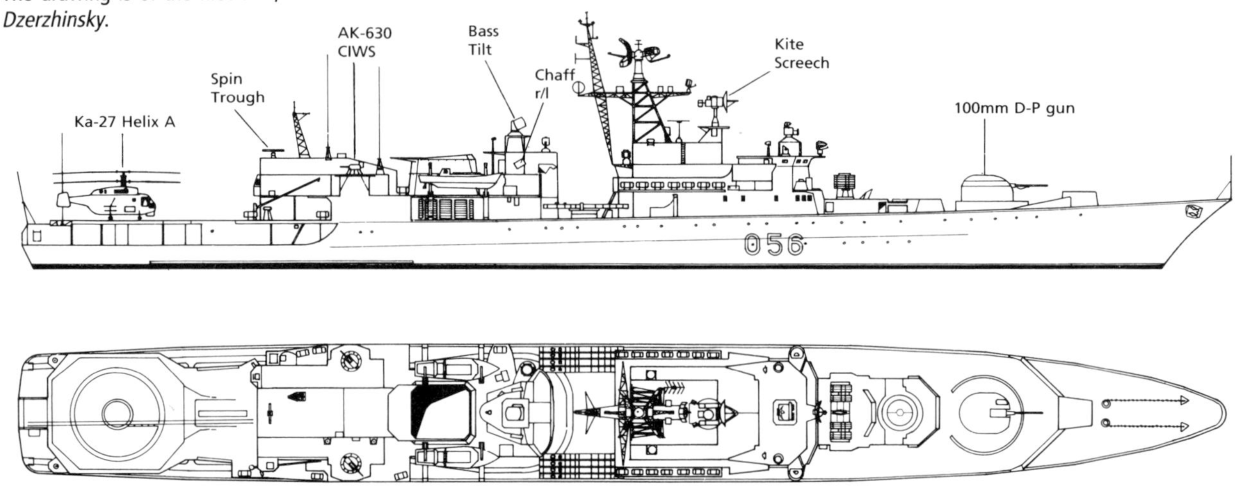

'Krivak III' Class

Soviet Designation: Storozhevoy Korabl' (SKR) – Patrol Ship

Construction

Name	Builder	In service
Menzhinsky	Kamysh-Burun Shipyard	Aug 1984
Dzerzhinsky	Kamysh-Burun Shipyard	Aug 1985
Imeni XXVII Syezda KPSS	Kamysh-Burun Shipyard	Feb 1987
Imeni LXX Letiya Vuk KGB	Kamysh-Burun Shipyard	April 1988
Imeni LXX Letiya Pogranvoysk	Kamysh-Burun Shipyard	Apr 1989
Kedrov	Kamysh-Burun Shipyard	1989
...		
...		

Displacement (tons)

Standard: 3,300
Full Load: 3,900

Dimensions

Length: 125m(410ft) overall
Beam: 14.1m(46ft)
Draught: 4.9m(16ft)

Propulsion

2-shaft COGAG; 2 gas turbines each 24,300hp; 48,600hp=30 knots max.

Complement

200

Weapons

ASW: 1 Ka-27 Helix A helicopter;* 2 RBU-6000 rocket-launchers; 2 quadruple banks of 533mm(21in) torpedo tubes
AAW: 1 twin-arm SA-N-4 launcher (20 Gecko missiles); 2 AK-630 gatlings
ASuW: 1 single 100mm(3.9in) D-P gun; rails for 20 mines

Sensors

Surveillance radars: Head Net C 3-D in *Menzhinsky*, *Dzerzhinsky*; Top Plate in later units; 1 Don-Kay, 1 Palm Frond, 1 Spin Trough
Fire control: 1 Pop Group (SA-N-4); 1 Kite Screech (76.2mm); 1 Bass Tilt (AK-630)
Sonars: bow-mounted Bull Nose MF; Mare Tail MF VDS
Electronic warfare: 2 Bell Shroud, 2 Bell Squat; 4 16-tube chaff-launchers

*Ka-25 Helix D SAR variant in peacetime.

▲*Menzhinsky*, the second of the KBG-manned 'Krivak IIIs'. A hangar and flight deck for a Ka-27 Helix helicopter have replaced the after gun mountings, one of which has been re-sited on the forecastle in place of the quadruple SS-N-14 launcher of the earlier 'Krivaks'.

housing for the variable depth sonar. This simplifies helicopter stowage arrangements, but brings the hangar roof level with the top of the funnel, which is unmodified. The helicopter deck is fitted with a chain-operated traverse system similar to that of *Udaloy*. There is a flying control station to port of the hangar, but these ships do not have the Fly Screen microwave landing radar of *Kirov* and *Udaloy*. Instead there is a small Spin Trough radar on top of the after end of the hangar to assist with helicopter control.

The helicopter carried by these ships for their current duties of border surveillance, search and rescue is the Helix D variant, a general-purpose model equipped with a surface surveillance radar. However, in the event of hostilities the Helix D could easily be exchanged for the standard Helix A model, and the 'Krivak III' would then realize its considerable potential for anti-submarine warfare.

On each side of the hangar an AK-630 gatling is mounted on a pedestal which houses its magazine. The single Bass Tilt fire control radar is mounted on the tall deckhouse forward of the funnel, where it replaces the Pop Group and Owl/Kite Screech fire control radars of earlier 'Krivaks'. A searchlight is mounted immediately below this position and this is presumably related to the ships' current coastal security duties.

The electronic countermeasures outfit is identical with that of the earlier series, although the position of some equipment has had to be adjusted to compensate the revised layout of the after part of the ship. The 16-tube chaff-launchers, which in the original series were mounted at a fixed angle on the sides of the deckhouse housing the VDS towing mechanism, have had to be relocated amidships. They are now mounted around the tall deckhouse forward of the funnel, the deckhouse being remodelled to provide the appropriate angles so that chaff clouds can be thrown up on either beam and on the bow quarters.

There are fewer modifications at the forward end of the ship. The quadruple SS-N-14 missile-launcher has been replaced by a single 100mm gun mounting, but the forward SA-N-4 air defence missile-launcher and the two RBU-6000 anti-submarine rocket-launchers remain in place. The navigation bridge has been raised slightly in order to improve the view forward over the 100mm gun housing, but the bridge structure is otherwise unmodified. The Kite Screech fire control radar for the 100mm gun, mounted on a short pylon as in *Udaloy*, has replaced the two Eye Bowl radars of the earlier sub-groups.

▲This close-up of the midships section of the third unit, *Imeni XXVII Syezda KPSS*, shows clearly the repositioned 16-tube chaff-launchers grouped around the deckhouse which houses the gas-turbine intakes. From this ship onwards Top Plate is fitted in place of Head Net C. Note the height of the hangar roof, which is level with the top of the funnel. (US Navy, 1987)

The first two units of the 'Krivak III' class, *Menzhinsky* and *Dzerzhinsky*, retained the original Head Net C 'V-beam' air surveillance radar, backed up by Don-Kay and Palm Frond radars for surface surveillance and navigation. From the third ship onwards, however, Head Net C was superseded by the Top Plate planar radar of the later ships of the *Udaloy* and *Sovremenny* classes.

Service Life

The first ship, *Menzhinsky*, was completed in 1984 and following trials and work-up in the Black Sea deployed to the Far East via the Indian Ocean. She was followed a year later by *Dzerzhinsky*. The third ship, *Imeni XXVII Syezda K.P.S.S.*, arrived in the Far East in 1987, and three further ships have followed at fourteen-month intervals. Two units are still under construction at Kamysh-Burun.

All the ships of the class are believed to operate out of Nakhodka, whence they conduct regular patrols in the Sea of Okhotsk and the northern part of the Sea of Japan. The strategic importance of these waters to the Soviet Union is such that KGB maritime operations in the region are inextricably bound up with military security rather than with support of the civil authorities. Not only is the Sea of Okhotsk one of the primary operational areas for the Soviet Navy's ballistic missile submarines, but testing of the missiles themselves is generally conducted in these waters.

Neustrashimy

The smaller, 'second-rate' counterpart to *Udaloy* began construction in 1986 at Yantar, Kaliningrad. A second unit was laid down in May 1988 on the same slipway as the first, and a third hull followed in 1989. The class is in competition with *Udaloy* both for building space and for shipyard labour. This suggests either a further expansion in the capacity of the Yantar Shipyard or a tailing off of the *Udaloy* construction programme. With at least thirteen units of the latter class laid down between 1978 and 1988, this would not be altogether surprising.

Neustrashimy, the name-ship of the new class, began her sea trials in the Baltic in December 1990. She is a large, capable vessel, comparable in size to the latest NATO frigates built in western Europe, with a length overall of 136m and a full load displacement estimated at 3,800 tons. In terms of capabilities she is a scaled-down *Udaloy*, with approximately half the weapon-load of the latter, but with an upgraded close-in weapon system.

General Configuration

The hull-form of *Neustrashimy* features a long forecastle, breaking only beneath the helicopter deck some

20m from the stern. It is a hull designed for good sea-keeping with considerable sheer from just abaft amidships and substantial flare at the bow. The straight stem, which is angled at almost 55 degrees to the vertical, suggests the presence of a large bow sonar dome. The bow is also noteworthy for its 'reverse sheer', a feature of contemporary French and Dutch warships which enables the forecastle gun to fire at low angles of elevation.

The superstructures are arranged as two separate blocks linked by a single-storey superstructure deck amidships. The forward block incorporates the navigation bridge and has a short lattice mast at its after end carrying the single Cross Sword fire control radar for the SA-N-9 missile system. It is linked by a deckhouse housing the intakes for the forward set of gas turbines to a broad funnel of rectangular configuration.

The after superstructure block incorporates the second funnel, which is similar in size to the first, and the helicopter hangar. At its forward end there is a tall lattice mainmast on which is mounted a Top Plate back-to-back planar array for air surveillance and tracking.

The superstructures of *Neustrashimy* are noteworthy for their 'stealth' characteristics which are clearly intended to reduce the radar signature of the ship at all angles of approach. Soviet surface ships have long featured superstructure sides and faces angled inboard from the vertical. The disadvantages of such design are that superstructure volume declines with height, and that internal fittings have to be matched to the angle. The ingenious expedient adopted for the forward superstructure block of *Neustrashimy* is that the sides of each deck are angled both ways, resulting in a distinctive 'concertina' effect (*see* photographs).

The broad, square, cutaway stern is reminiscent of earlier Soviet corvette-size craft. At its centre is a raised housing for the variable depth sonar which deploys through a hinged door, some five metres across. The housing is more substantial than that aboard *Krivak* or *Kara*, and may conceal the Horse Tail model installed in *Udaloy* and the *Kirovs*.

▼ *Neustrashimy* photographed on sea trials in the Baltic in early 1991. Like her larger BPK counterpart, *Udaloy*, she is designed primarily for anti-submarine warfare. Note the large VDS housing in the centre of the stern. (MoD, Bonn)

Neustrashimy Class: Profile and Plan

Note: The plan view must be regarded as provisional pending the availability of overhead photographs.

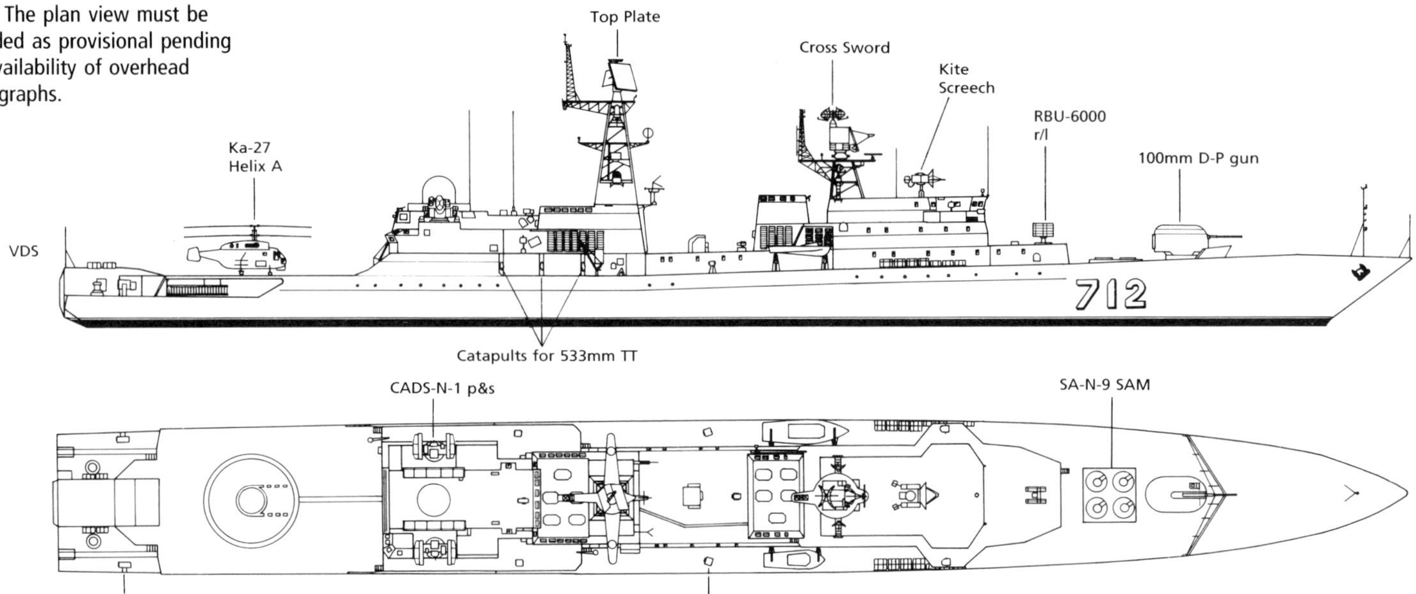

Neustrashimy Class

Soviet Designation: Bol'shoy Protivolodochny Korabl' (BPK) – Large Anti-Submarine Ship

Construction

Name	Builder	Laid down	Launched	In service
Neustrashimy	Kaliningrad Shipyard	Apr 1986	May 1988	1991
...	Kaliningrad Shipyard	May 1988	1990	
...	Kaliningrad Shipyard	1989		

Displacement (tons)
Standard: 3,100
Full Load: 3,800

Dimensions
Length: 136m(446ft)
Beam: 15.5m(50ft)
Draught: 5.6m(18ft)

Propulsion
2-shaft COGAG; 4 gas turbines each 15,000hp; 60,000hp=30kts

Complement
200

Weapons
ASW: 1 Ka-27 Helix A helicopter; 1 RBU-6000 rocket-launcher; 6 catapults for 533mm torpedoes
AAW: 1 group of four SA-N-9 vertical launch magazines (32 missiles); 2 CADS-N-1 combined missile/gun CIWS
ASuW: 1 single 100mm(3.9in) D-P gun; rails for mines

Sensors
Surveillance radars: Top Plate 3-D; 2 Palm Frond, 1 Spin Trough
Fire control radars: 1 Cross Sword (SA-N-9); 1 Kite Screech (100mm); 1 unidentified+2 Hot Flash (CADS-N-1)
Sonars: bow-mounted LF-MF(?); LF/MF(?) VDS
Electronic warfare: 2 Mod. Foot Ball; 2 Bell ?; 2 unidentified ESM; 4 16-tube chaff-launchers, 2 10-tube chaff-launchers

◀The hull-form of *Neustrashimy* is noteworthy for its marked sheer. The vertical launch magazines for the SA-N-9 missile system are countersunk into the forecastle immediately abaft the single 100mm D-P gun. (MoD, Bonn)

To each side of the VDS housing, mine-rails run forward for approximately 18m beneath the broad helicopter deck.

Propulsion

Neustrashimy is reported to be powered exclusively by gas turbines. This appears to be confirmed by the configuration of the funnels and by the multiplicity of air intakes and cooling vents characteristic of earlier Soviet gas-turbine powered ships.

Rather more difficult to establish, in the absence of good-quality aerial photographs, is the precise layout and configuration of the propulsion machinery. The after funnel appears to incorporate five (possibly six) uptakes, of which four are clearly offset to starboard. If this arrangement is reversed in the first funnel, as one might reasonably expect, it is probable that there are four main turbines and that these are paired *en echelon* as in the US *Spruance* class, with the forward pair driving the port shaft in a COGAG configuration and the after pair driving the starboard shaft. The remaining uptakes presumably serve auxiliary diesel (or gas-turbine) generators distributed between the two engine rooms.

It is as yet unclear which gas turbine forms the basis for the propulsion machinery. The 31,250hp model employed in *Udaloy* and *Slava* is a candidate only if two turbines are installed, in which case a single turbine would be uneconomical for prolonged anti-submarine patrols at cruise speed. On the other hand, four NK-12 turbines of the type employed for the missile corvettes of the 'Tarantul' class would provide a total horsepower of only 48,400hp, barely sufficient to drive a ship of this size at 30 knots. It may be that these vessels incorporate a newly developed turbine comparable in power to the British Rolls-Royce Spey, delivering 15,000–18,000hp. This would give a maximum speed in excess of 30 knots, and would provide the required flexibility for fuel economy at lower speeds.

Weapons and Sensors

The primary mission of *Neustrashimy* is anti-submarine warfare. The sonar outfit is at least as capable as that of the 'Kara' and, judging from the bow overhang and the comparatively large size of the VDS housing, may be comparable to that of *Udaloy*.

The primary anti-submarine weapon system is the single Ka-27 Helix A helicopter, operated from the broad flight deck immediately forward of the VDS housing. The hangar is fully integrated into the after superstructure and the hangar floor is on the same level as the flight deck, thereby facilitating stowage and deployment. In contrast to earlier Soviet BPKs the hangar door features a standard up-and-over roller mechanism similar to that generally employed on Western surface warships. There is a helicopter control cabin to starboard, but the Fly Screen B microwave landing radar is not fitted.

Anti-submarine weaponry is completed by a single RBU-6000 rocket-launcher and 533mm homing torpedoes. The RBU-6000 rocket-launcher occupies the conventional position forward of the bridge. The torpedoes, however, are housed not in the standard quadruple or quintuple trainable mountings, but in six fitted catapults angled out from the sides of the after superstructure. Although pioneered by the French Navy during the 1960s and subsequently adopted for the Dutch/German 'Standard' frigate, torpedo catapults have never before been employed by the Soviet Navy.

The theoretical advantages of torpedo catapults are that they can be reloaded internally and that associated firing and control mechanisms are better protected from the elements. Their adoption by the Soviet Navy for this class may indicate the development of a wire-guided torpedo, the pay-out mechanism being concealed within the superstructure. There may also be a reload magazine which would be conveniently located beneath the hangar, and would also house the smaller 400mm weapons carried by the Ka-27 Helix helicopter.

An alternative theory put forward for the six angled tubes is that they house not torpedo catapults, but a new SSM similar in configuration to the Western Harpoon. A small-diameter Soviet SSM designated SS-NX-25 has been trialled aboard a variant of the 'Nanuchka' Small Rocket Ship, 'Nanuchka IV', since 1987, and this has presumably influenced Western Intelligence thinking in favour of a missile solution for the tubes aboard *Neustrashimy*. However, the 'Nanuchka IV' installation, which comprises two angled launchers each housing six missile tubes, is far less volume-intensive. Moreover, the tubes aboard *Neustrashimy* are angled in azimuth but not in elevation, and there appear to be no arrangements for the dispersal of the rocket exhausts. Should the Soviets have required an SSM capability, the mounting of missile canisters on open ramps between the funnels would seem to be a superior arrangement.

Neustrashimy is well protected against attack by hostile aircraft or missiles. Four vertical launch bins for the SA-N-9 missile are countersunk into the forecastle deck immediately abaft the single 100mm gun mounting, which provides back-up in the anti-aircraft mode as well as a useful anti-surface capability against small craft such as missile-armed FPBs. Fire control for the SA-N-9 missile system is provided by the single Cross Sword radar located at the top of the lattice foremast. The height at which the latter is mounted ensures rapid reaction to low-flying aircraft or missiles. Top Plate provides tracking and target designation. The Kite Screech fire control radar for the 100mm dual-purpose gun is mounted atop the bridge structure.

In place of the AK-630 CIWS of earlier BPKs, *Neustrashimy* is equipped with two CADS-N-1 combined missile/gun systems which are mounted on each side of the hangar. Between them is a 3.25m radome previously seen only on the Rocket Cruiser *Kalinin*, which also has the CADS-N-1 system. The position of the radome on *Neustrashimy* appears to confirm that it is indeed an integral part of the CIWS. Presumably it provides initial tracking and target designation to the on-mount fire controls system of the CADS-N-1 in the absence of a 360-degree phased array radar system such as Sky Watch, which is installed in the two Soviet carriers fitted with the combined gun/missile system.

The ECM outfit of *Neustrashimy* comprises both passive and active countermeasures. ESM radomes of the Bell series are carried on yard-arms projecting from the upper mainmast. Beneath them are box-shaped ESM antennae of new design which are complemented by a pair of ECM combined receiver/jammers located at the foot of the mainmast. These closely resemble the Foot Ball radomes installed in the *Udaloy* and *Sovremenny* classes, but without the plinth or back-plate. *Neustrashimy* appears to have four of the 16-tube chaff-launchers previously seen on 'Krivak'. The launchers are located on each side of the superstructure abeam the after funnel, and on the short quarterdeck. There are also what appear to be circular multi-tube launchers of new design mounted on each side of the forecastle deck amidships, and these may also be for launching chaff rounds.

The sensor outfit is completed by two Palm Frond surface surveillance/navigation radars mounted on platforms projecting from the sides of the lattice foremast, and a single Spin Trough navigation radar.

Missions and Possible Deployment

Neustrashimy does not appear to have the elaborate liquid/solid stores transfer stations which are a feature of recent Soviet large surface combatants. This suggests that the ship is designed primarily for missions within or on the perimeters of the Soviet maritime zones, and that oceanic deployments, whether in peacetime or in conflict, will be left to the *Udaloy* and *Sovremenny* classes. It will be interesting to see whether the Soviet Navy will classify these vessels as Large Anti-Submarine Ships (BPK), like *Udaloy*, or as Patrol Ships (SKR), like the 'Krivak'.

Deployments can be expected to both the major Soviet Fleets, in the Arctic and the Pacific, where these ships will complement their larger sisters of the *Udaloy* class, but it may be that some units will be retained in the Baltic where they would provide a useful anti-submarine capability.

In contrast to the 'Krivak' programme, it appears that construction of the *Neustrashimy* class may be concentrated on the Kaliningrad Shipyard. Kamysh-Burun, currently occupied with the 'Krivak III', is probably incapable of building a hull of this size. If there is a requirement for additional capacity to build the *Neustrashimy* class, it appears more likely that any outstanding orders for the *Udaloy* class will be shifted across to the Zhdanov Shipyard which has built five of the first thirteen hulls.

▶ **The broad funnels house the exhaust uptakes for the main gas turbines and the auxiliary generators. The second funnel is integrated into the after superstructure. Directly beneath it are the fixed torpedo catapults angled out from the ship's axis. Note the angled sides and faces of the funnels and superstructures, designed to reduce the radar signature of the ship. (MoD, Bonn)**

▶ **The single helicopter hangar is flanked by two CADS-N-1 combined gun/missile systems. The starboard installation has the missile canisters and twin 30mm multi-barrel guns elevated to 90 degrees. The port installation does not have the missile canisters mounted. Note the square cutaway stern and the indentation for mine rails. The stanchion-mounted device to starboard appears to be a 16-tube chaff-launcher similar to those fitted in 'Krivak'. (MoD, Bonn)**

◀The last unit of the *Kiev* class, *Baku*, was extensively modified to enable the Soviet Navy to conduct shipboard trials of the new radar and command/control systems intended for the new generation of through-deck carriers. Note the revised layout of the forecastle deck in this early bow view. The anti-submarine rocket-launcher is a new model, the RBU-12000, which fires a larger rocket with significantly greater range than its predecessors. (MoD, 1988)

8. The New Carriers

FROM 1972 UNTIL 1974 *Morskoy Sbornik*, the Soviet Navy's official magazine, published a series of articles by Admiral Gorshkov under the title 'Navies in Peace and War'. The views expressed in these articles appear to have angered certain figures in the non-navy military establishment, there being evidence that publication of some of the later articles in the series was delayed by the censors.[1]

Gorshkov's key thesis was that the Soviet Union lacked a coherent policy for the deployment of the Navy in peace and war, and that the development of the fleet was being hampered by tailoring it to a single, restrictive defensive mission. This led to an 'unbalanced' fleet composed of mission-specific types which lacked the flexibility for a proper application of seapower in support of Soviet foreign and defence policies.

The implication of this argument was that naval operations should constitute an independent area of study within the field of 'military science', and that the Soviet Navy should no longer be a mere adjunct to the land forces, charged with guarding the seaward flank, but should have an expanded role which might on occasions involve the primacy of naval operations in maritime theatres. The heretical nature of this argument within the context of traditional Marxist–Leninist thinking is self-evident.

It would appear that a compromise of sorts was reached in 1974 when Marshal Grechko acknowledged that the 'external' function of the armed forces had been expanded, and Gorshkov in turn acknowledged that operations against land (rather than enemy fleets) now constituted the primary mission of the Soviet Navy in wartime. The argument was to rumble on throughout the 1970s and into the early 1980s, however, both within the broader military establishment and within the Soviet Navy itself.

Gorshkov was to continue to develop the Navy's case in his book *The Seapower of the State*, published in early 1976. In a new section entitled 'Fleet against Fleet and Fleet against Shore', he attempted to divide 'fleet against fleet' operations into two distinct forms: the traditional 'pure' form favoured by Western naval strategists, the purpose of which was to establish and maintain command of the sea for its own sake; and operations that are 'tied to the simultaneous accomplishment of other missions'. The latter type of operation could in fact be used to support operations against the shore.

The outspoken nature of Gorshkov's early criticism of traditional Soviet military thinking was almost certainly provoked by the move in the late 1960s towards forward deployment, for which the Soviet Navy was prepared neither in its tactical thinking nor in its training, and for which many of its ships were fundamentally unsuited. However, it was also an undoubted piece of political opportunism in that Gorshkov was quick to see the extent to which increased demands on the Navy to support Soviet foreign policy in the Mediterranean and the Indian Ocean could be converted into new surface ship construction.

Ultimately, Gorshkov's advocacy of these views would lead to his political demise. An unexpected (and unpleasant) side-effect of a larger, more powerful and more conspicuous Soviet surface fleet was the leverage it would provide in the United States for the '600-ship Navy', and the aggressive Maritime Strategy that would provide the rationale for its deployment. This provoked alarm in the upper echelons of the Soviet military and political establishments.

All references to 'naval science' would be deleted from the second edition of *The Seapower of the State*, and a separate doctorate in 'naval science' was discontinued in April 1979. In a further series of articles which appeared between April 1981 and July 1983 the role of the Soviet Navy was again the subject of discussion, and many of the ideas previously advanced by Gorshkov came under attack from senior serving naval officers including the then C in C of the Northern Fleet, Admiral Chernavin.[2]

The idea that 'naval science' should constitute an independent branch of military science was singled out for particular attention, with Chernavin in particular pleading for 'unified views' and an adherence to Marxist–Leninist philosophy which recognizes 'immutable' laws of society and war. Although some Western observers have seen in this an element of political manœuvring in anticipation of the struggle for Gorshkov's succession at a time when his days in office were clearly numbered, it is equally clear that opposition to Gorshkov's advocacy of a wider role for the Soviet Navy was not confined to the other armed services, and that there were a number of officers in the higher echelons of the Soviet Navy itself who viewed these developments with unease.

An Expanded Naval Programme

The impact of the debate on Soviet naval programmes of the 1970s and 1980s is difficult to quantify. What is clear is that the construction programme approved under the 1976 five-year plan was the most ambitious yet in terms of the number and the individual displacement of the surface ships projected.

The first of the large nuclear-powered Rocket Cruisers, *Kirov*, was already under construction as were two further units of the *Kiev* class. Under the 1976 programme the naval construction programme was boosted by three new types of surface ship, all with a full load displacement in excess of 7,500 tons; the Rocket Cruisers of the *Slava* class, and the large destroyers of the *Sovremenny* and *Udaloy* classes (laid down in 1976 and 1978 respectively). The second unit of the *Kirov* class and the fourth (and last) of the *Kievs* would be laid down under the same programme.

All of these types were designed for operations outside the traditional maritime defence zones. The large destroyers of the *Sovremenny* and *Udaloy* classes, in line with Gorshkov's 'balanced fleet' proposals, were to be given complementary missions. The primary mission of the *Kirovs* and the *Slavas*, and of their submarine counterparts of the 'Oscar' class, was to engage enemy carrier battle groups in what would essentially constitute a 'fleet to fleet' engagement, although the purpose of this engagement would be to protect Soviet territory, not to establish command of the sea.

It has also become clear from recent revelations in the Soviet press that preliminary designs for a 'defensive' aircraft carrier were begun in 1974, official approval for the concept being obtained two years later.

The 'Defensive' Aircraft Carrier

The concept of a 'defensive' aircraft carrier was very different from that of *Kiev*, which is essentially an air-capable anti-submarine cruiser with extensive command facilities, relying primarily on surface-to-air missiles for protection against air attack.

It was also very different from that of Western attack carriers whose primary mission revolves around their ability to deploy squadrons of attack aircraft against both naval and land targets. The fighters aboard these carriers are intended both to escort the attack aircraft on their offensive missions and to protect the carrier task force against attack. Dual-capable fighter aircraft such as the US F/A-18 Hornet can also perform offensive missions in their own right.

The rationale for the Western attack carriers is that they carry their striking and defensive power with them wherever they go. They are therefore the ideal weapons system for the projection of air power against distant or isolated targets beyond the reach of conventional land-based air power.

The concept of the 'defensive' carrier espoused by the Soviets could not be more different. Far from envisaging distant deployments, the Soviets regard the 'defensive' carrier as an extension of the land-based air umbrella which has traditionally covered naval operations in the maritime defence zones. The mission envisaged for this ship is to provide fighter cover for Soviet Rocket Cruisers and anti-submarine vessels operating beyond the range of shore-based combat air patrols.

This task could not have been adequately performed by the only existing ship-based naval fixed-wing aircraft, the VTOL Yak-38 Forger, nor by any projected successor. Soviet VTOL technology was as yet immature and was unlikely to deliver dramatic performance improvements in the foreseeable future. The Forger itself was proving to be barely adequate as a strike fighter because of its severe range and payload limitations. Given that the threat to its surface forces would come primarily from high-performance carrier-borne strike aircraft, the Soviet Navy would require a large fleet air defence fighter comparable in size to the US Navy's F-14 Tomcat. Only a Soviet equivalent of the F-14 would have the necessary combat radius and endurance to fly prolonged combat air patrols. The aircraft would also need a high-performance look-down, shoot-down radar plus a powerful battery of radar-guided air-to-air missiles to protect Soviet surface forces from the US Navy's low-flying attack squadrons.

Such an aircraft was already under development for the strategic air defence forces by the mid-1970s. The prototype 'A'-model of the Sukhoi Su-27 Flanker was to embark on its first test flight in May 1977. One of the primary missions for which the Su-27 was designed appears to have been strategic air defence of the 'oceanic' maritime defence zones, because production aircraft have subsequently been deployed both to the Kola Peninsula and to the Far East, where they replaced older types such as the Yak-28P, Su-15 and Tu-28P 128. Operating in conjunction with the Mainstay AEW command aircraft, they have been particularly active in simulated interceptions of NATO aircraft over the Barents Sea. In conception the Flanker is therefore particularly well suited to adaptation to the fleet air defence role, and it seems reasonable to assume that the aircraft entered consideration as the primary Soviet carrier-based fighter at a relatively early stage in the carrier design process.

The new carrier (initially christened *Tbilisi*) would be laid down in January 1983, so the design must have been finalized in about 1980, in time for the Party Congress which would consider the new five-year plan. In configuration it would resemble the big US Navy carriers much more closely than its anti-submarine predecessors of the *Kiev* class. An aircraft the size of the Su-27 Flanker would require a full-size flight deck some 300m in length to enable it to take off and land. It would also require larger lifts and a significant increase in hangar and deck parking space.

The hybrid carrier/cruiser configuration of *Kiev* was out of the question for the new vessel and this would inevitably result in changes in Soviet operational thinking. The new carrier would not be 'self-escorting' because long-range surface-to-air missile systems could not be satisfactorily accommodated in a ship with a full-length flight deck. This in turn implied that the ship would have to be accompanied by 'carrier escorts' on the US Navy pattern, which would provide an area defence missile capability and some degree of anti-submarine protection. The impact of the new carrier on traditional Soviet operating concepts would therefore be considerable. The carrier would provide the cement which would bind the building blocks of complementary surface vessels in Gorshkov's 'balanced fleet' into an integrated whole, a force capable of mutually self-supporting operations in sea areas outside the reach of land-based air.

Conventional fixed-wing carrier operations were new to the Soviet Navy although detailed observations of Western carrier operations had been made since the late 1960s. By the mid-1980s a simulated carrier flight deck, some 320m in length, had been built at the Saki naval airfield in the Crimea where both the Su-27 Flanker and its smaller land-based counterpart, the Mig-29 Fulcrum, were to undertake an intensive programme of short take-offs and arrested landings. The

The Mig-29 Fulcrum D Interceptor

In service: 1990
Length: 17.3m(56ft 9in) incl. nose probe
Span: 11.4m(37ft 3in); 8m(26ft 3in) wings folded
Height: 4.7m(15ft 6in)
Weight: 15,000kg(33,050lb) normal 18,000kg(39,700lb)
Engines: two Isotov RD-33 turbofans, each 5,100kg(11,240lb) thrust, 8,300kg(18,300lb) with afterburning
Speed: 2,440km/h(1,320kts=Mach 2.3) at high level; 1,300km/h(700kts=Mach 1.06) at sea-level
Ceiling: 17,000m(56,000ft)
Combat radius: 1,250km(675nm)
Armament: one 30mm cannon port wingroot extension; six pylons for AA-10A Alamo medium-range AAMs or AA-11 Archer/AA-8 Aphid short-range AAMs; in strike role some AAMs replaced by bombs or rocket pods
Sensors: search-while-scan pulse Doppler radar; forward-looking IRST sensor; Sirena-3 passive warning

▶An Su-27 Flanker D on the flight deck of *Tbilisi* (later renamed *Admiral Kuznetsov*). Note the folding wings to facilitate stowage. (TASS, 1989)

The Su-27 Flanker D Fleet Air Defence Fighter

In service: 1990
Length: 21m(69ft) approx.
Span: 14.7m(48ft 3in); 10m(33ft) wings folded
Height: 6m(19ft 6in)
Weight: 22,000kg(44,500lb) normal; 30,000kg(66,135lb) max.
Engines: two Lyulka AL-31F turbofans, each 12,500kg(27,500lb) thrust
Speed: 2,500km/h(1,345kts=Mach 2.35) at high level; 1,345km/h(725kts=Mach 1.1) at sea-level
Ceiling: 18,000m(59,000ft)
Combat radius: 1,500km(810nm)
Armament: one 30mm cannon in starboard wingroot extension; six inner pylons for AA-10A/B/C Alamo medium-range AAMs plus four outer pylons for AA-11 Archer or AA-8 Aphid close-range AAMs
Sensors: track-while-scan pulse Doppler radar with search range of 240km(130nm) and tracking range of 185km(100nm); forward-looking IRST sensor; Sirena-3 passive warning

'flight deck' incorporated two ski-jumps on the British pattern and carrier-style arrestor wires. The programme served not only to train pilots (the two-seat Su-25UB Frogfoot was employed as the initial trainer), but also to develop the naval variants of the Su-27 and Mig-29 aircraft which would require a number of special modifications to equip them for carrier operation. It is also reported that efforts were made to develop steam catapults, although these appear to have encountered serious technical problems.

The size of *Tbilisi* was such that the hull had to be built as two separate sections at the Black Sea Shipyard, the bow and stern sections being joined together after launch prior to fitting out. The second hull could be laid down only after the first had been launched.

These measures give an indication of the scale of the undertaking and of the immense cost of the carrier programme. The new combat system developed for *Tbilisi* would be tested separately aboard the last ship of the *Kiev* class, *Baku*. This would result in an island superstructure that was radically different from that of earlier units of the class, and a completely revised outfit of weapons and sensors. Further changes were made to the flight deck layout. These modifications delayed construction and must have added considerably to the overall cost of the ship; *Baku* took a full ten years to build, six years of which were spent fitting out.

As completed *Baku* constitutes an intermediate type with the hull, air group and surface-to-surface missiles of the *Kiev* class, but with the combat system of *Tbilisi*. The changes to the original specifications are so extensive as to merit a separate description of the ship.

▲The after bank of SS-N-12 missile launch tubes, with the single 100mm gun mountings to starboard. The missile reloading tray has been retained, together with the athwartships rails on which it moves, but the centre-line elevator has been suppressed. Note the AK-630 gatlings grouped together with their Bass Tilt fire control radars at the forward end of the island superstructure. (US Navy, 1988)

Baku

Perhaps the most surprising aspect of *Baku*'s weapon/sensor outfit is that so few of the new systems can be regarded as linear successors of those which constituted the 'main armament' of her earlier sisters. *Kiev* carried three long-range weapon systems: the SA-N-3 area defence missile system, the FRAS-1 nuclear-tipped anti-submarine missile system and the SS-N-12 anti-surface missile system. Of these only the SS-N-12 system has survived aboard *Baku*.

There is no direct replacement for the SA-N-3 area defence system. *Baku* is fitted instead with the short-range SA-N-9 which is essentially a self-defence

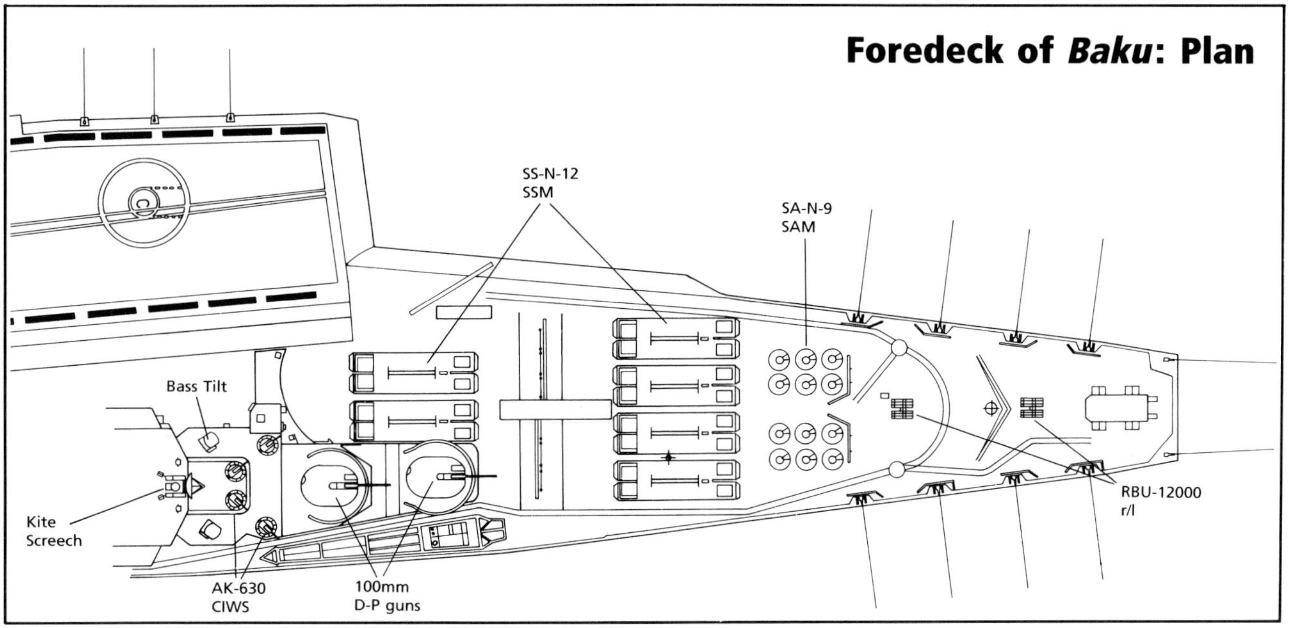

system incapable of engaging aircraft armed with stand-off guided weapons. The ship would therefore need to be accompanied by air defence escorts when operating outside the Soviet maritime defence zones.

The other weapon system which has disappeared without trace is the FRAS-1 anti-submarine missile. This weapon was carried only by the three earlier *Kiev*s and their two predecessors of the *Moskva* class, and has been superseded on other Soviet anti-submarine units by the SS-N-14 missile which carries a homing torpedo with a conventional warhead.

At a time when Soviet military strategy was moving rapidly away from the projected use of tactical nuclear weapons, it appears logical that the inertially guided FRAS-1 should be abandoned, particularly as *Baku* carries an estimated eighteen Ka-27 Helix A anti-submarine helicopters which carry a more flexible payload.

The Foredeck

Elimination of the above weapons systems has permitted a major revision of the layout of the foredeck (*see* drawing). The forward two groups of SA-N-9 vertical launch bins, each comprising six cylindrical magazines holding eight missiles, are located in the space previously occupied by the SUW-N-1 launcher and its accompanying magazine for the FRAS-1 missile.

Immediately abaft the launch bins is a continuous bank of four double elevating launchers for the SS-N-12 anti-ship missile. The elimination of the two centre-line deckhouses which accommodated the 76.2mm twin D-P gun mounting and the forward SA-N-3 launcher in the first three units of the class has made it possible to insert two additional twin launchers in the forward bank, and to compress the after pair of double launchers into a single bank. This has enabled the Soviet constructors to create a prominent 20-metre extension to the island superstructure to starboard, on which two single 100mm D-P mountings are located.

Between the two banks of SS-N-12 launchers the reloading tray, which runs on a system of athwartships rails, has been retained, but there is no evidence of the centre-line munitions lift which was a feature of *Kiev* and *Minsk*. It therefore appears that, as in *Novorossiysk*, the between-decks magazine has been suppressed in order to create more internal volume, and that the increase in the number of above-decks launchers is in part to compensate a reduction in the overall number of missiles carried. The large crane to starboard, which handles the missiles when the ship is alongside, has been moved farther aft.

The arrangements at the forward end of the forecastle remain unchanged, although the RBU-6000 has been superseded by a new model, the RBU-12000, which has ten barrels of larger diameter and launches a rocket with substantially greater range. It therefore matches more closely the improved detection capabilities of the latest Soviet sonars and helps to compensate the abandonment of the FRAS-1 anti-submarine missile.

Finally, the port-side sponson, which caused persistent problems in the earlier ships because of their relatively low freeboard, has been suppressed altogether. The two AK-630 gatlings have been relocated, together with their Bass Tilt fire control radar, immediately forward of the bridge where they are now grouped together with the original starboard-side gatlings.

Baku: The Island Superstructure

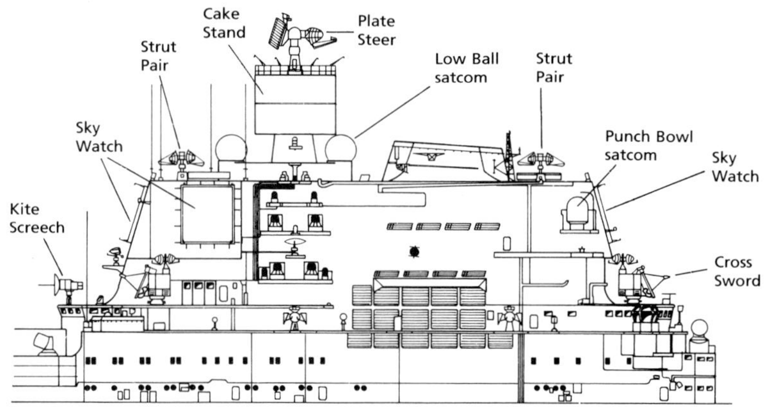

Baku

Soviet Designation: Avionosny Kreyser (AKR) – Aircraft-Carrying Cruiser

Construction

Name	Builder	Laid down	Launched	In service
Baku	Black Sea Shipyard	Dec 1978	Apr 1983	June 1988

Displacement (tons)
Standard: 38,000 standard
Full Load: 45,000 full load

Dimensions
Length: 273m(895ft) overall
249.5m(818ft) waterline
Beam: 53m(174ft) flight deck
32.7m(107ft) waterline
Draught: 10m(33ft)

Propulsion
4-shaft geared steam turbines; 8 turbo-pressurized boilers; 200,000shp=32 knots max.

Complement
1,600 (incl. air group)

Aircraft
12 Yak-38 Forger A VTOL aircraft; 18 Ka-27 Helix A anti-submarine helicopters; 2–3 Ka-25 Hormone B missile targeting helicopters; 2 Ka-32 Helix D SAR/utility helicopters

Weapons
ASW: 2 RBU-12000 rocket-launchers
AAW: 4 groups each of six SA-N-9 vertical launch magazines (196 missiles); 2 single 100mm(3.9in) D-P guns; 8 AK-630 gatlings
ASuW: 6 paired SS-N-12 launchers (12 Sandbox missiles)

Sensors
Surveillance radars; Sky Watch 3-D, Plate Steer 3-D; 2 Strut Pair; 3 Palm Frond
Fire control radars: 1 Trap Door C (SS-N-12); 4 Cross Sword (SA-N-9); 1 Kite Screech (100mm); 4 Bass Tilt (AK-630)
Sonars: bow-mounted Horse Jaw LF; Horse Tail LF VDS
Electronic warfare: 8 Foot Ball; 4 Wine Flask ESM; 4 Bell Bash, 4 Bell Thump, 2 Bell Nip; 2 twin-barrelled chaff-launchers

The Island Superstructure

The island has been completely remodelled to accommodate the electronics suite intended for the new through-deck carriers. Width and height are virtually unchanged, as is the position of the distinctive machinery-room vents. However, the island is markedly shorter, 'cleaner' and more compact, without the prominent 'steps' of the original model.

The bulky rotating air surveillance and tracking radars and the Head Lights missile guidance radars of the earlier ships have been superseded by a new phased-array system which has received the NATO designation Sky Watch. Sky Watch comprises four electronically scanned planar arrays of rectangular configuration. The arrays measure approximately 6m by 5.25m and are disposed in similar fashion to the SPY-1 Aegis radars of contemporary US missile cruisers to provide full 360-degree coverage. The forward and after edges of the island have been angled at approximately 15 degrees in order to accommodate two of the arrays, while those installed *en echelon* on the sides of the island are mounted on large steel plates fixed at a similar angle.

Atop the island is a short tower mast carrying a distinctive 'dustbin' some 8m in diameter (NATO designation: Cake Stand), which houses the carrier-controlled approach radar and is the linear successor to Top Knot. Mounted on Cake Stand is a conventional rotating Plate Steer air surveillance and tracking radar, and there are Strut Pair air/surface surveillance radars at either end of the island.

Directly beneath Cake Stand is a housing for two Low Ball satellite communications radomes, and Punch Bowl satcom radomes associated with targeting for the SS-N-12 missile are located on platforms opposite the two side-mounted Sky Watch antennae.

The ECM antennae occupy platforms projecting from the superstructure sides immediately forward of the engine-room vents, as in the earlier ships. The

▶**The island superstructure of *Baku* was completely rebuilt. It is dominated by the four electronically scanned planar arrays of the Sky Watch radar system and the distinctive Cake Stand radome which houses the carrier-controlled approach radar. (US Navy, 1988)**

arrays are dominated by eight Foot Ball plinth-mounted ECM radomes (four per side) which have superseded Side Globe on the later *Kirovs* and on the destroyers of the *Sovremenny* and *Udaloy* classes. They are completed, as on *Frunze* and *Kalinin*, by Wine Flask ESM antennae and smaller Bell-type radomes. Between the platforms for Foot Ball are mounted Palm Frond surface surveillance radar antennae.

At the lower level of the island there are paired Cross Sword radars at each end, each of which provides guidance for a group of six SA-N-9 surface-to-air missile-launchers. The forward pair of Cross Swords control the two groups located on the forecastle, while the after Cross Swords control the SA-N-9 launcher groups installed on each side of the flight deck. The Kite Screech fire control radar for the 100mm guns is located directly above the navigation bridge, and there are Tin Man electro-optical devices to port and to starboard.

At the after end of the island, beneath the flying control station, the Bob Tail radio sextant has been retained and there are the customary microwave landing aids.

Flight Deck

In addition to the changes in *Baku*'s weapon/sensor outfit, a number of important modifications have been made to the flight deck arrangements.

The decision to dispense with the forward port-side sponson has made it possible to extend the angled portion of the deck by some five metres. This has provided more protection for the foremost (plane guard) helicopter spot in rough weather, and a longer rolling take-off run for the latest version of the Yak-38 Forger. It has also enabled the Soviet designers to incorporate a smoother, more rounded leading edge to improve airflow over the flight deck. The air-flow pattern has been further improved by building out the port deck-edge of the forecastle and rounding it off in similar fashion.

Whereas the third unit of the class, *Novorossiysk*, incorporated three fixed-angle wind deflection plates abeam the second rank of SS-N-12 launchers on the port side of the forecastle (these were subsequently retro-fitted to the first two ships), *Baku* has a single large deflector plate, 9.5m long, which is hinged at its outboard edge, enabling the angle to be varied to

Flight Deck of *Baku*: Plan

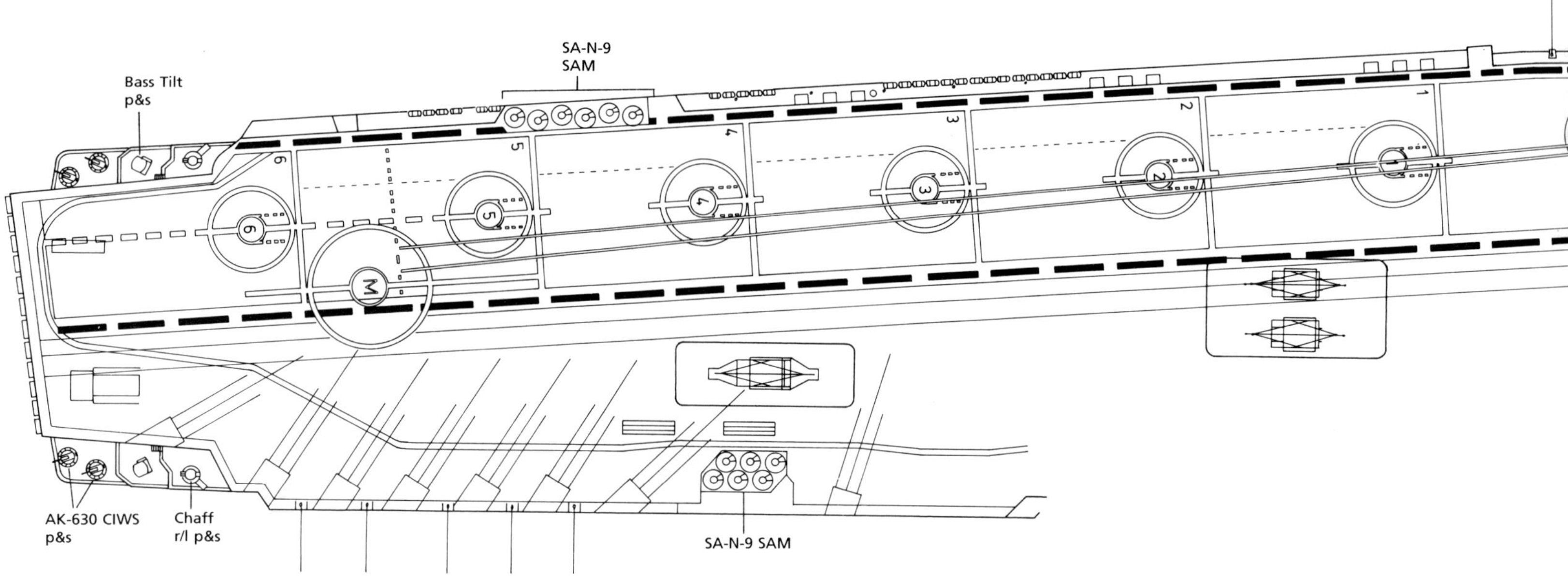

create the most favourable air-flow pattern over the forward end of the flight deck.

The layout of the helicopter spots on the flight deck is identical with that of the earlier ships, with six numbered spots plus a seventh spot forward offset to port and marked with the Cyrillic letter 'C' (roman 'S'), which is used by the helicopter on plane guard duty (generally a Ka-27 Helix D). *Baku* also retains the larger (13m radius) circle marked with an 'M' offset to the starboard side of the angled deck between helocircles 5 and 6, which is generally used for Forger landings.

There are two major innovations as regards deck markings. The broken line which on the first three ships marked the centre-line of the angled deck is now restricted to the after end of the flight deck, covering only the last two segments, and there are convergent lines running from the large circle marked 'M' to the forward edge of the flight deck.

Kiev featured a similar inverted 'V' marking following her major refit in the Black Sea during the mid-1980s, and it was during her transit from the Mediterranean to the Northern Fleet area in May 1985 that the first rolling take-offs of the Forger were observed (*see* Chapter 5).

The inverted 'V' marked out on *Baku*'s flight deck is significantly narrower than that on *Kiev* and begins slightly farther forward (although this is compensated by *Baku*'s longer flight deck). On *Baku* it is approximately 145m in length and is angled at 8.5 degrees to the ship's centre-line, while on *Kiev* it is just under 150m (the angle is identical).

The broken line which previously marked the centre-line of the angled deck may have confused pilots as it crosses the inverted 'V' at an angle of approximately five degrees and appears to serve little useful purpose. It has presumably been retained at the after end of the deck (where it is clear of the 'V' marking) to assist pilots approaching the carrier from astern in lining up with the angled deck prior to landing.

Deck Park

One of the principal defects of the original *Kiev* design was the lack of attention given to providing an adequate deck park. The primary constraints on the latter were the length of the island superstructures and the layout of the aircraft and munitions lifts. In *Novorossiysk* the deck park abaft the superstructure was extended over the starboard side of the ship, increasing flight deck width by 2.5m at this point. The third munitions lift was also suppressed to provide an additional aircraft parking space, and the railway track was extended only to the second elevator.

The arrangements for *Baku* are similar to those introduced in *Novorossiysk*, except that by eliminating all major weapons systems from the after end of the

▶The port-side sponson, which was responsible for sea-keeping problems in the earlier ships, has been eliminated altogether in *Baku*. This has enabled the angled deck to be extended by some five metres. (MoD, 1988)

island a further substantial deck park area approximately 17.5m long has been created forward of the second aircraft lift. Parking spaces for seven aircraft are marked out abaft of the lift, and there is a single space marked out between the lift and the after end of the island superstructure, together with an unmarked square capable of accommodating a further two aircraft.

To starboard, directly above the stern, *Baku* has the position for Forger warm-up introduced by *Novorossiysk* and subsequently retro-fitted to *Kiev*. However, the 'railway track' has not only been revived but has been extended to encompass the after end of the flight deck, terminating close to the port-side group of SA-N-9 launchers.

Both aircraft lifts have been fitted with an elaborate system of tie-down points and the configuration of these serves to confirm that the forward lift was

designed from the outset to accommodate two Forger VTOL aircraft side by side. It is of interest to note that the after lift, which on the earlier units was only half as wide as the main lift, has had its width increased from about 5m to 6.75m. This provides safer clearance for the aircraft currently embarked, but it may also be an indication that the Yak-38 VTOL successor reported to be under development, the Yak-41, is somewhat larger than its predecessor.

Service Life

Baku conducted intensive trials of the new Sky Watch phased-array system in the Black Sea before she entered service in June 1988. Some problems, reported by the Soviet press, appear to have been experienced during these trials and it remains unclear whether these have been resolved.

Following commissioning, the ship exited from the Black Sea and made the lengthy transit to the Soviet Northern Fleet where she joined her half-sister *Kiev*. She has since taken part in defensive exercises in the Barents Sea, but the low profile adopted by major Soviet surface units in these waters is now quite marked. In 1990, amid major civil unrest in the Soviet republics, the ship was renamed *Admiral Gorshkov*.

▲The after end of the island superstructure, with two Cross Sword FC radars above the flying control station, which is located at a lower level than in earlier units of the class. Note the spherical Bob Tail radio sextant, visible in the raised position. Other sensors prominent in this view are the port-side Punch Bowl satcom radome, and one of the three Tin Man electro-optical devices. The two Forger strike fighters, cocooned in canvas covers, occupy the new 17.5m deck park area made available by shortening the island. (US Navy, 1988)

▶Stern quarter view of *Baku*, showing the revised flight deck markings and the significant increase in the size of the deck park area to starboard. Note the projection housing the port-side v/l magazines for the SA-N-9 missile system, which are arranged in a single line on the outer edge of the angled deck. (MoD, 1988)

Tbilisi

Construction of the new through-deck carrier was begun under the 1981 Five-Year-Plan, the first keel plates being assembled in January 1983. The two sections of the hull were launched in December 1985. The keel of a second ship, to be christened *Riga*, was laid down the following day.

Admiral Kuznetsov: Profile and Plan

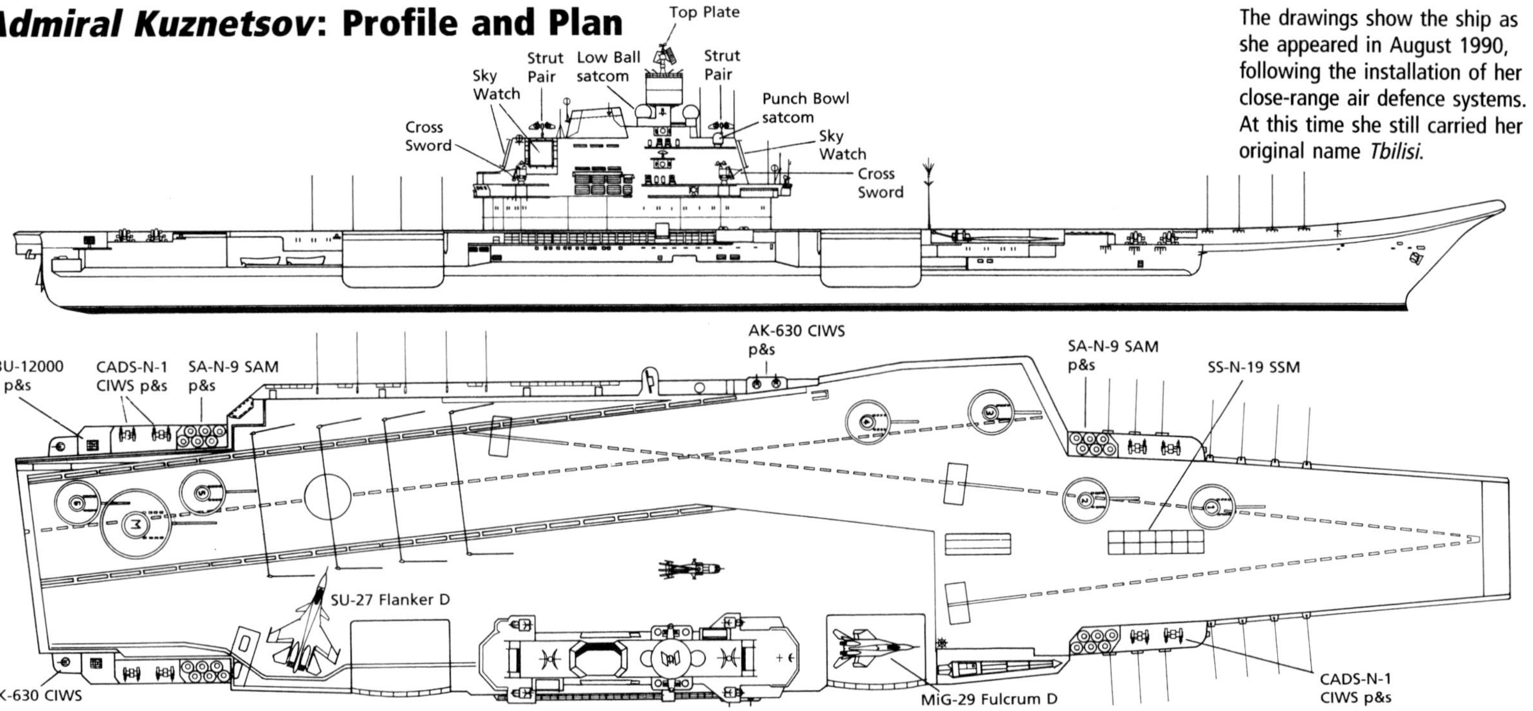

The drawings show the ship as she appeared in August 1990, following the installation of her close-range air defence systems. At this time she still carried her original name *Tbilisi*.

Following completion, *Tbilisi* ran sea trials from 19 to 24 October 1989. The first arrested landings and take-offs were reported in the Soviet Press on 22 November, less than a month later. Western Intelligence appears to have been taken by surprise by the speed of these developments, and there are strong indications that the Soviet Navy was anxious to demonstrate its new capabilities to the Soviet people and, more importantly, to state officials increasingly concerned at the high cost of the programme.

Trials of new Soviet ship-types, including those of *Kiev* and *Kirov*, have traditionally been conducted in the greatest secrecy, and yet in this particular instance the Soviet Navy appears to have gone out of its way to stimulate media interest. Press releases on the first carrier landings and ski-jump-assisted take-offs were accompanied by liberal hand-outs of photographs. One month previously, on the first day *Tbilisi* took to sea, the ground was already being prepared with an extended interview with Admiral Chernavin himself, who took great pains to explain the essentially 'defensive' mission of the new carrier, to contrast the role of the Soviet ship with that of her American counterparts and to justify development as a logical step from the *Moskva* and *Kiev* classes. Chernavin took particular exception to *Pravda*'s description of *Tbilisi* as an 'aircraft-carrier', and a hasty correction was published three days after the original interview appeared in which the ship was redefined as a 'heavy aircraft-carrying cruiser' (*tyazholy avionosny kreyser*).

There are further indications of the Soviet Navy's anxiety to secure favourable early publicity for the ship. Early photographs show that *Tbilisi* was barely ready when she first went to sea. In particular, the SA-N-9 missile system lacked its fourth Cross Sword radar and not one of the eight projected CADS-N-1 combined missile/gun systems had been installed. These would be fitted when the ship next entered dockyard hands. The next, and more extensive, series of flight trials would take place some eight months later in the summer of 1990.

Available evidence therefore points to an attempt to convince wavering Soviet politicians of the military potential of the new carrier and to secure funding for the full projected programme, estimated by Western Intelligence sources to have been at least four ships.

These efforts appear to have been only partially successful. Following the launch of *Riga* in December 1988, the first of an even larger type, *Ulyanovsk*, was laid down. Construction of this ship, which has an estimated displacement of 75,000 tons and will be powered by nuclear reactors (possibly in conjunction with conventional steam propulsion machinery), is continuing. However, it appears that a projected sister-

Admiral Kuznetsov (ex-*Tbilisi*) Class

Soviet Designation: Tyazholy Avionosny Kresyer (TAKR) – Heavy Aircraft-Carrying Cruiser

Construction

Name	Builder	Laid down	Launched	In service
Admiral Kuznetsov (ex-*Tbilisi*)	Black Sea Shipyard	Jan 1983	Dec 1985	Jan 1991
Varyag (ex-*Riga*)	Black Sea Shipyard	Dec 1985	Nov 1988	

Displacement (tons)
Standard:
Full Load: 65,000 full load

Dimensions
Length: 300m(985ft) overall
Beam: 71m(233ft) flight deck
40m(130ft) waterline
Draught: 11m(36ft)

Propulsion
4-shaft geared steam turbines; 8 turbo-pressurized boilers; 200,000shp=30 knots max.

Complement
2,100 (incl. air group)

Aircraft
(projected) 9 Su-27 (Su-27 Flanker K) fleet air defence fighters, 9 Mig-29 Fulcrum D interceptors; 18 Ka-27 Helix A anti-submarine helicopters, 2–3 Ka-27 Helix D SAR/utility helicopters

Weapons
ASW: 2 RBU-12000 rocket-launchers
AAW: 4 groups each of six SA-N-9 vertical launch magazines (196 missiles); 8 CADS-N-1 combined missile/gun CIWS; 6 AK-630 gatlings

Sensors
Surveillance radars: Sky Watch 3-D, Top Plate 3-D; 2 Strut Pair; 3 Palm Frond
Fire control radars: 4 Cross Sword (SA-N-9); 8 Hot Flash (CADS-N-1/AK-630)
Sonars: bow-mounted Horse Jaw LF
Electronic warfare: 8 Foot Ball; 8 unidentified ESM radomes; 4 unidentified ESM fixed arrays; 2 Bell Push; 2 twin-barrelled chaff-launchers

ship has been cancelled as the Soviets have now stated that *Ulyanovsk* will be the last ship of this type to be built. The Soviet Navy will therefore be left with only three fixed-wing carriers, and this will have an inevitable impact on the development of suitable aircraft because it may prove prohibitively expensive to provide air groups comprising specialized carrier-borne types beyond those already in service for such a small force.

New Names

Growing political and economic problems in the USSR were responsible for further, more immediate changes. In July 1990 it was announced by the Soviet Navy that both ships of the *Tbilisi* class were to be renamed. The official reason given was that seamen of the Soviet Navy did not want its major surface ships to carry the names of cities in which there was open opposition to the Soviet Union.

Tbilisi, which as late as July 1990 had carried this name emblazoned on each side of the bow, has now been renamed *Admiral Sovietskugo Soyuza Flota N. G. Kuznetsov* (Admiral of the Soviet Fleet Kuznetsov), in memory of the architect of the Soviet post-war fleet.

This constituted something of a volte-face because Kuznetsov's name had been under a cloud ever since he took early retirement in 1956. The large conventional fleet which he had championed during the early 1950s had been immediately dismantled by Khrushchev on his accession to power, and it took a lengthy campaign by the wife and supporters of Kuznetsov to secure his rehabilitation and to restore him to his former rank of Admiral of the Fleet.

The second ship, *Riga*, was reported in autumn 1990 to have been renamed *Varyag*, after a recently stricken rocket cruiser of the 'Kynda' class. This suggests that the new carrier will deploy to the Pacific on completion.

Overall Configuration

Admiral Kuznetsov is the first Soviet carrier designed to operate conventional fixed-wing aircraft. Her overall configuration therefore closely resembles that of contemporary Western carriers, and her design incorporates a number of features which are new to the Soviet Navy but which are standard on the latest French and US Navy carriers. These include: a fully angled deck with substantial overhang, equipped with arrestor wires and a mirror landing sight; deck-edge lifts to maximize hangar volume; a comparatively small island superstructure, located outboard of the hull on sponsons to provide a counterweight to the angled deck; and weaponry sufficient only for close-range defence against aircraft and missiles.

The major innovation made by *Admiral Kuznetsov* as compared with her Western fixed-wing counterparts is the incorporation of a British-style ski jump which, allied to STO modifications to the fixed-wing aircraft types embarked, has effectively replaced the steam catapults of conventional large-deck carriers. This has been possible only because the ship is intended to operate lightly armed, high-performance fighters, not attack types with heavy payloads of bombs and rockets.

Admiral Kuznetsov has an overall length of approximately 300m. The beam at the waterline is about 40m, while the flight deck has a maximum width of just over 70m. There is therefore substantial overhang on both sides of the ship, the island superstructure to starboard and the forward section of the angled deck to port being supported on prominent sponsons.

The angled portion of the flight deck is set at 5.5 degrees to the ship's axis. By comparison, *Kiev*'s angled deck is at 3.5 degrees to the centre-line, but the angled decks of the French *Charles de Gaulle* and the US supercarriers are fully 8 degrees, the after end of the angled deck being located on the starboard deck edge in order to create a triangular parking area to port (served by the fourth aircraft lift in the US ships). This area is virtually unusable in *Admiral Kuznetsov* because of its proximity to the port edge of the angled deck, although this does mean that there is a corresponding increase in parking space to starboard of the angled deck.

The angled deck on *Admiral Kuznetsov* is about 205m long and 23m wide (the angled deck on the USS *Nimitz*, which operates aircraft of comparable size, is 230m by 25m). It is crossed by four arrestor wires spaced at the comparatively wide interval of 14m, and there is a Western-style mirror landing aid to port.

At the four 'corners' of the ship, where the main body of the flight deck narrows, deep sponsons support the defensive weapons, comprising both short-range missiles and point-defence combined missile/gun systems.

The major surveillance and tracking radars and electronic countermeasures systems are mounted on a comparatively small island superstructure some 58m long at its base, with prominent overhang both for the navigation bridge and for the flying control station located at its after end.

The two large deck-edge lifts, approximately 20m by 15m, are disposed forward and aft of the island superstructure to starboard. They are comparable in size to those of the US supercarriers, although capacity is probably somewhat lower at about 35–40 tonnes. They can therefore handle the Mig-29 Fulcrum (maximum take-off weight 18,000kg; 17.3m by 8m with wings folded) quite comfortably. The Su-27 Flanker (maximum take-off weight 30,000kg; 22.5m approx. by 10m with wings folded) is a different proposition, the size of this aircraft being such that it can be accommodated on the lift only if the nose or tail-plane overhangs the outer edge.

Some sources have stated that there is a third (inboard) lift located to port of the island superstructure, but the photographs so far released show no evidence of this, nor is there any compelling reason why a third lift should be needed given the com-

▲Port bow view of *Admiral Kuznetsov*, carrying her original name *Tbilisi*. Note the 12-degree ski jump and the considerable overhang of the angled deck. The sharply curved forefoot of the stem suggests the presence of a large bow bulb for a low-frequency hull sonar transducer. (TASS, 1990)

▼The view forward from the navigation bridge. The forward deck-edge lift is at hangar deck level, and the starboard blast deflection panel is in the raised position. The third lift can be seen to port of the blast deflection panel; it may be capable of handling the Helix helicopter in addition to deck tractors, stores and munitions. Forward of the lift the flush-deck hatches for the SS-N-19 missile can be just made out. The helicopter to the left of the picture is a new, as yet unidentified, variant of the Ka-29TB assault helicopter, which may be for ESM support or for AEW. (TASS, 1990)

▲ ***Admiral Kuznetsov* as she appeared in mid-1990, with her full complement of air defence weapons. The original flight deck markings have been partially painted out during her spell in dockyard hands. Note the semi-circular indentation of the stern flanked by vertical guide rails. (TASS, 1990)**

paratively small size of the air group. Moreover, a lift in this position would hinder the movement of aircraft fore and aft along the deck park to starboard of the angled deck, and would reduce available hangar volume – an important consideration if Soviet operating practice in the Northern Fleet area continues to require the entire air group to be accommodated in the hangar, with no permanent deck park.

Photographs recently released by the Soviets do indeed show a third lift, but this is located on the centre-line forward between the hinged blast-deflection flaps which mark the beginning of short take-off runs using the ski jump. Parallax makes it difficult to be precise about the dimensions of this lift, but it appears to be about 13m long and 4m wide, and is therefore more substantial than one would have expected had its function simply been the transfer of munitions between a below-decks magazine, the hangar deck and the flight deck.

The position of the lift suggests that it is above the forward end of the hangar. One of its functions may therefore be the movement of deck tractors between the hangar and the flight deck, but the layout of guide-rails and tie-down points on the upper surface of the lift suggests that it could accommodate a Ka-27 Helix helicopter, thereby freeing the larger deck-edge lifts for the movement of the Mig-29 Fulcrum interceptors and the bulky Su-27 Flanker fleet air defence fighters.

The 12-degree ski jump begins some 60m from the ship's bow, at the point where the forward sponsons end. It begins from a relatively broad base and narrows to some 23/24m at the ship's bow, which is rounded off in order to minimize turbulence. The fore-foot of the stem curves sharply forward at the waterline, suggesting a large bow bulb for a powerful low-frequency sonar.

At the after end of the ski jump, on the centre-line, are two rows of hatches set flush with the flight deck. Each of these measures approximately 3m by 2m and conceals an angled launch tube for the SS-N-19 anti-ship missile. The hatches are only just visible in published photographs, which makes it difficult to be precise about their number. US Intelligence sources credit *Kuznetsov* with sixteen missile tubes, British sources with twelve. Whichever figure is correct the missiles give the ship an anti-surface capability on a par with the *Kiev*s, thereby confirming that the air group of *Kuznetsov* will be essentially defensive, and that offensive strike operations against hostile surface warships will continue to be performed by long-range missiles.

Immediately forward of the first deck-edge lift is a heavy, 20m crane similar in configuration to that installed on the forecastle of the *Kiev*s. The ship's boats are set into recesses in the hull abaft the second lift, and cocooned inflatable life-rafts are disposed port and starboard along the outer edge of the angled deck and directly beneath the island superstructure. Now that the full complement of life-rafts has been installed there appear to be 77 to starboard in three continuous rows superimposed one above another, and 50 to port in double rows, broken into smaller groups, for a total of 127. The ship's complement including air group, has been estimated at 2,100.

The square transom stern is reminiscent of that of *Kiev*, but there are a number of hitherto unexplained features. Most noteworthy are the prominent semi-circular indentation in the centre, and the vertical guide-rails to port and starboard, the purpose of which has yet to be determined. In contrast to other Soviet large surface combatants there appears to be no variable depth sonar.

The Hangar

The configuration of the hull and flight deck suggests a hangar 32–35m wide and about 170m in length. A hangar of this size could accommodate two 9-aircraft squadrons of Su-27 Flanker and Mig-29 Fulcrum fixed-wing fighters together with a substantial complement of Ka-27 Helix helicopters (perhaps 15–21, including SAR variants). If the Yak-141 VTOL Forger successor were also to be embarked, however, there would have to be some reduction in the numbers of the other types carried.

A key difference between *Kiev* and *Admiral Kuznetsov* is the greater freeboard of the latter – some 16.5m as compared with only 13m for the VTOL carriers. This increase was necessary in order to incorporate deck-edge lifts into the design because the latter must be well clear of the ship's waterline if they are to be operable in all weather. It was also influenced by the need to provide greater hangar clearance so as to accommodate the Su-27 Flanker, which has an estimated height of almost 6m. The author estimates hangar clearance to be about 6.5m. This is still well short of the 7.8m of the latest US Navy carriers, but is a significant improvement on *Kiev*.

When speculating on the possible future embarkation of airborne early warning (AEW) types by *Admiral Kuznetsov* it is important not to ignore the constraints imposed by maximum hangar clearance on the size and configuration of such aircraft. Hangar height on the latest US super-carriers is largely dictated by the configuration of the E-2C Hawkeye, which has its distinctive 'saucer' radome mounted above the fuselage. On older carriers of the *Midway* class the Hawkeye can be struck down in the hangar only because of a modification which enables the radome to be retracted. Latest US Intelligence reports state that trials have been carried out with an AEW variant of the Ka-29 Helix,[3] which would be less capable than a fixed-wing aircraft but would be comparatively easy to accommodate.

Flight Deck Operations

During her first aircraft trials *Admiral Kuznetsov* had flight deck markings sufficient only to enable those trials to take place. The 23m-wide angled deck was painted in the customary dark green, with its centre-line marked by a broken white line and its outer edges also highlighted in white. A noteworthy innovation was

▲ **In this view, dating from mid-August 1990, *Kuznetsov* is conducting further flight trials off Sevastopol. She displays her new flight deck markings which include six helo-circles. Immediately abaft the ski jump are an Su-27 Flanker D fleet air defence fighter and a Mig-29 Fulcrum D interceptor, both with wings folded. A second Mig-29 is making a landing approach. (TASS, 1990)**

a solid white circle 9m in diameter centred on the second arrestor wire to serve as a 'target' for aircraft landing on.

Both the Mig-29 and the Su-27 use the ski jump for take-off. The standard take-off run begins immediately forward of the first aircraft lift from one of two positions some 24m apart. Each is marked by a large hinged blast-deflection panel measuring 8m by 4m. Quick-action fold-down launch restrainers built into the flight deck immediately forward of the blast panels permit the engines to run up to their maximum power prior to release for take-off over the ski jump. The take-off runs are marked by broken white lines and converge at the forward end of the ski jump. The Mig-29 Fulcrum is known to have been launched successfully using these 105m take-off runs, but the larger Su-27 Flanker generally uses a longer, 195m, take-off run which extends aft from the port-side run on to the angled deck, and is marked in similar fashion.

The only other markings visible during the initial aircraft trials were a single 9m circle at the forward end of the angled deck for the Ka-27 Helix D SAR version employed for 'plane guard' duties, and a larger 14m circle aft identical to the one used for VTOL Yak-38 Forger landings on the *Kievs*. It has now been confirmed that the Forger has conducted familiarization trials aboard the ship,[4] although the development of its Yak-141 successor appears to have been seriously delayed by technical problems.

Following extensive dockyard work in the spring of 1990, which included the installation of the CADS-N-1 point-defence missile/gun systems, *Admiral Kuznetsov* conducted further aircraft trials in the Black Sea with a new set of flight deck markings.

The markings associated with fixed-wing operations are unchanged, but there are now six numbered helo-circles in place of the single 'plane guard' circle of the early trials – an indication that the ship is now close to operational status. In contrast to the *Kievs*, which had their helo-circles marked at regular intervals along the centre-line of the angled deck, they are grouped forward and aft, leaving the central section of the angled deck completely free for fixed-wing landing operations. There are two circles immediately abaft the ski jump, to port, and a further two at the forward end of the angled deck. The spacing between their respective centres is about 25m, as on the *Kievs*. The fifth and sixth circles are in the after end of the angled deck, to port.

Positioning the island superstructure close to the edge of the flight deck has created substantial deck parking areas to starboard of the angled deck. The after deck park is served by the customary 'railway track' which also crosses the second deck-edge lift and is used by trolleys for aircraft ordnance. Immediately forward of the starboard after SA-N-9 launchers, in the centre of the after deck park, are twin studded doors which presumably cover vertical hoists leading from the ordnance magazines. Each of the doors is about 3m by 1.75m.

Outside the port after edge of the angled deck, in close proximity to the after helo-circle, there appears to be a third munitions lift similar to those on the *Kievs*, measuring approximately 5.5m. by 1.25m. There may be further munitions lifts on the inboard side of the island superstructure.

Weapons and Electronics

Admiral Kuznetsov ran her first sea trials without her full weapons outfit. Four groups each of six SA-N-9 vertical launch magazines were fitted from the outset, but it is doubtful whether the system was fully operational at this stage because initial sea trials were run with only three of the four Cross Sword guidance radars in place.

There were also two of the new ten-barrelled RBU-12000 anti-submarine rocket-launchers first observed on *Baku* and the Rocket Cruiser *Kalinin*. These are located on sponsons at the after end of the ship, and not in the customary forward firing position. This is partly due to considerations of space. However, the BPKs of the *Udaloy* class have their RBU-6000 rocket-launchers similarly mounted, thereby reinforcing the theory put forward in some quarters that they are not simply for use against submarines, but can also be employed for anti-torpedo protection.

The only other weapons aboard during initial sea trials were six AK-630 30mm anti-missile gatlings. These are mounted in pairs on sponsons projecting from the port and starboard edges of the flight deck level with the forward edge of the island superstructure, and singly on sponsons on either side of the stern. There was initially some doubts as to the identity of these installations because the gun mountings were protected with canvas covers and there was no sign of the tell-tale Bass Tilt FC radars generally associated with the AK-630.

The full weapons outfit was embarked only in mid-1990, and the latest photographs show not only the fourth Cross Sword radar in place, but eight of the new CADS-N-1 combined gun/missile systems first observed on *Kalinin* (*see* Chapter 6), disposed symmetrically in pairs forward and aft.

Interestingly, there are still no Bass Tilt radars for the AK-630 anti-missile guns, nor is there any space provision for future installation. It therefore seems probable that these guns are tied to the CADS-N-1 mountings for fire control purposes.

The electronics outfit of *Admiral Kuznetsov* is virtually identical with that trialled in *Baku*. There are four Sky Watch planar arrays for 3-D multiple-target tracking, twin Strut Pair F-band radars for air/surface surveillance and the customary three Palm Frond navigation radars. However, the back-up 'rotating' radar for Sky Watch is Top Plate, not Plate Steer.

Satellite communications arrays are identical with those of *Baku*, with 3m Low Ball satellite navigation radomes beneath the distinctive Cake Stand TACAN 'dustbin', and Punch Bowl satcom radomes disposed *en echelon* opposite the side-mounted Sky Watch arrays. These are associated with over-the-horizon targeting for the SS-N-19 anti-ship missiles.

The satellite communications arrays are complemented by a variety of conventional loop arrays and an exceptionally complete outfit of whip, wire and cage communications antennae covering the LF, MF and VHF/UHF radio frequencies.

The electronic countermeasures outfit, although bearing a superficial resemblance to that of *Baku*, incorporates a number of important new elements. The eight Foot Ball jammers have been retained, but whereas on *Baku* these were paired with Bell Thump ESM antennae to provide accurate direction-finding, on *Admiral Kuznetsov* Bell Thump has been superseded by a larger, unidentified antenna located outside the upper Foot Ball radomes and between the lower ones.

Above each of the two main platforms on which these antennae are located is a new fixed ESM array which appears to be the replacement for Rum Tub. Each array is three-sided, the outer faces angled at 45 degrees to the front face to give all-round coverage. Each of the three faces has a fixed circular plate at its centre. The configuration of this array, as for its predecessor, Rum Tub, suggests the adoption of electronic scanning techniques similar to those employed in the US SLQ-32.

Fighter Aircraft

The Su-27 Flanker and Mig-29 Fulcrum models which have conducted preliminary trials aboard *Admiral Kuznetsov* have been specially adapted for maritime operations. Modifications have centred on: maximizing engine thrust to enable the aircraft to use the shortest-possible take-off run, assisted by a ski jump; strengthening the landing gear to resist the impact if landing at high speed on a pitching flight deck; fitting arrestor hooks; and providing folding wings for economical use of deck and hangar space.

▲Beyond the deck crew members in fire-fighting gear is the compact island superstructure, dominated by the Sky Watch planar array and the Cake Stand TACAN 'dustbin'. The forward starboard Cross Sword FC radar has yet to be installed. The helicopter to the left of the picture is the Helix D SAR variant of the Ka-27. (TASS, 1989)

The Su-27 Flanker, which is an exceptionally large aircraft comparable in size to the US Navy's F-14 Tomcat, has been given movable canard foreplanes to give enhanced STOL performance. The nose landing gear has been strengthened with twin wheels, and the prominent tailcone of the land-based model has been eliminated to prevent tail-scrapes on landing. The thrust-to-weight ratio of the powerful Lyulka AL-31F turbofans is such that in favourable weather conditions the Flanker can get into the air using the standard short 105m take-off run. With maximum fuel load and a full complement of air-to-air missiles, however, the aircraft would use the longer, 195m run.

Modifications to the naval Mig-29 Fulcrum include the removal of the distinctive hinged doors which prevent the ingress of foreign objects into the underwing engine intakes on the land-based models, and of the louvres and ducts through which the air is supplied during take-off and landing, when the doors are closed. This modification serves to maximize engine thrust, and the elimination of internal ducting in the centre-section of the fuselage has enabled an additional 2,550 litres of fuel to be carried. As with the Su-27, the landing gear has been strengthened and an arrestor hook fitted. The Fulcrum naval variant has an inflight refuelling capability and carries a modified avionics outfit including a different infra-red search/track (IRST) sensor, which is presumably more closely tailored to maritime operations.

The air defence capabilities of these two aircraft are of a completely different order from those of the

ECM Arrangement

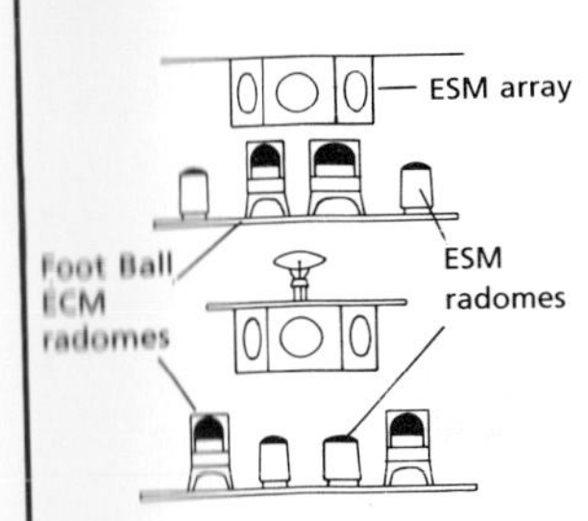

Yak-38 Forger. The Flanker has sufficient internal fuel for a range of 4,000km, enabling it to conduct prolonged combat air patrols at considerable distances from the carrier. It has been developed specifically for all-weather operation and has a sophisticated look-down shoot-down radar with a reported search range of 130nm (240km) and a tracking range of 100nm (185km). Weapons include a 30mm cannon in the starboard wing root extension and up to ten air-to-air missiles, carried on tandem pylons under the fuselage between the engine ducts. Missiles include short-range AAMs with infra-red homing and AA-10A/C Alamo semi-active types, the long-burn AA-10C having a beyond-visual-range capability. Latest reports suggest the development of a reconnaissance pod and a 'buddy pack' for inflight refuelling.

The Mig-29 is comparable in size to the US F/A-18 Hornet. It is agile and has good combat manœuvrability. Like the Su-27, it has a large pulse Doppler look-down shoot-down radar although the latter is limited to search-while-scan rather than track-while-scan. Its primary role in a maritime context would be that of close-range interceptor. However, the land-based Fulcrum is a dual-role aircraft with a substantial light attack capability, and the inflight refuelling capability of the naval variant would enable it to carry out long-range strikes against either surface ships or land targets carrying a light underwing stores load.

It is not yet clear whether *Admiral Kuznetsov* will operate the Yak-38 Forger VTOL successor, the Yak-141, which is reported to have been delayed by technical problems. Some commentators have expressed the view that the Yak-141 is not intended to operate from the new through-deck carriers, but will simply replace the Forger aboard *Kiev* and her sisters.

While there would be nothing to prevent the Yak-141 from operating from *Admiral Kuznetsov*, and deck markings appear to have taken the requirements of this aircraft into account, the embarkation of Yak-141s would necessarily displace another component of the current air group (possibly a 9-aircraft squadron of Ka-27 Helix A ASW helicopters). It therefore seems more likely that the aircraft would be embarked for specific missions unrelated to the ship's primary mission of fleet air defence, such as the support of the Naval Infantry in an amphibious assault.

Aircraft trials aboard *Kuznetsov* appear to confirm that the amphibious assault mission has figured prominently in Soviet thinking. In addition to the Ka-27 Helix A and D variants, employed for ASW and for plane guard/SAR duties respectively, recent on-deck photographs have featured the Ka-29TB Helix B assault helicopter which was first displayed at the Frunze air show in Moscow in August 1989, together with an unidentified model of the Helix derived from the Ka-29 airframe.

The Ka-29TB has a wider flight deck than the standard Ka-27 Helix and can easily be distinguished by its three-panel windscreen. Beneath the windscreen a snub nose contains operational equipment related to the assault mission, including a four-barrel gatling-type machine-gun. Four pylons under the outriggers can be fitted with 57mm/80mm rocket pods or AT-6 Spiral air-to-surface missiles, guidance for the latter being provided by an undernose sensor pod. Sensors include an ESM radome mounted above the rear of the engine bay fairing and an infra-red jamming pod.

The cabin has folding seats along the sidewalls for twelve assault troops who disembark via a two-part upward/downward rear-facing cabin door to facilitate rapid exit. The flight deck and engine bay are heavily armoured.

The new variant of the Ka-29 has neither armour nor the downward articulated gun door of the TB model. A large fairing beneath the nose appears to extend the full length of the fuselage pod and there are further large box-like fairings on the sides of the fuselage, forward and aft of the main landing gear. It has been suggested that this new aircraft may be an electronic warfare (EW) variant of the Helix, intended to provide ECM support and communications capabilities for an assault force of Ka-29TBs.[6] Alternatively it may be the AEW variant reported by US Intelligence, relying on passive ESM sensors rather than on active search radar to provide early warning of hostile surface forces.

Further speculation has centred around a possible naval mission for the new Kamov Hokum combat helicopter which is due to enter service in 1991. Armed with packs of unguided rockets, a podded undernose multi-barrel gun and wing-tip air-to-air missiles, the Hokum would be well suited to an escort role in support of the Ka-29TB assault helicopter. Its survivability is enhanced by extensive armouring, infra-red suppressors and infra-red decoy dispensers.

Clearly amphibious assault is very much a subsidiary mission for the new carriers which in the event of a major outbreak of hostilities would be fully occupied with their primary mission of fleet air defence. However, *Admiral Kuznetsov* is the first Soviet air-capable ship to have this degree of operational flexibility, the range of possible missions being dictated by the composition of the air group rather than by the configuration of the ship. She will be a major asset to the Soviet Northern Fleet when she joins it as flagship, probably late in 1991.

References

Introduction
1. *Red Star*, 26 March 1991.
2. Report in *Red Star*, January 1991.
3. Admiral Chernavin, quoted in *Red Star*, 2 March 1991.

1. The Post-war Era
1. Siegfried Breyer, *Guide to the Soviet Navy*, 2nd ed., p. 196, Naval Institute Press, Annapolis, 1977.
2. Norman Friedman, *All the World's Fighting Ships, 1947–82*, vol. 2, p. 465, Conway Maritime Press, 1983.
3. Ibid., p. 471.

2. The Anti-Carrier Strategy
1. Norman Friedman, *All the World's Fighting Ships, 1947–82*, vol. 2, p. 482.
2. Norman Polmar, *Guide to the Soviet Navy*, 4th ed., p. 428, Naval Institute Press, Annapolis, 1986.
3. Milan Vego, 'Their Torpedoes', p. 139, in USNI *Proceedings*, July 1984.
4. Ibid.
5. Kehoe, Brower and Meier, 'US and Soviet Design Practices, 1950–1980', p. 125, in USNI *Proceedings*, May 1982.
6. Norman Friedman, *World Naval Weapons Systems*, p. 170, Naval Institute Press, Annapolis, 1990.
7. Kehoe, East and Brower, 'Their *Slava*-Class Strike Cruiser', p. 151, in USNI *Proceedings*, August 1984.
8. Ibid.
9. Norman Friedman, *Naval Radar*, p. 187, Conway Maritime Press, 1981.

3. Anti-Surface and Anti-Submarine
1. Michael McCGwire, 'The Rationale for the Development of Soviet Seapower', p. 164, in USNI *Proceedings*, May 1980.
2. Steven J. Zaloga, *Soviet Air Defence Missiles*, p. 251, Jane's Information Group, 1989.
3. Ibid., p. 251.
4. Ibid., p. 252.

4. The Large Anti-Submarine Ships
1. Admiral Gorshkov, *Red Star Rising at Sea*, p. 103, Naval Institute Press, Annapolis, 1974.
2. Michael McCGwire, 'The Rationale for the Development of Soviet Seapower', p. 167, in USNI *Proceedings*, May 1980.
3. Michael McCGwire, 'Gorshkov's Navy', Part 1, p. 47, in USNI *Proceedings*, August 1989.
4. Roy Corlett, 'Advances in Soviet underwater capability and weapons', p. 203, in *Maritime Defence*, June 1987.
5. Michael McCGwire, 'Gorshkov's Navy', Part 1, p. 47, in USNI *Proceedings*, August 1989.
6. Steven J. Zaloga, op. cit., pp. 256–7.

5. The *Kiev* Class
1. Michael McCGwire, 'Gorshkov's Navy', Part 1, p. 47, in USNI *Proceedings*, August 1989.
2. John D. Gresham, 'Their Sea-Based Fighter', p. 122, in USNI *Proceedings*, February 1985.
3. Ibid., p. 122.

6. The New Rocket Cruisers
1. Kehoe, Brower and Meier, 'US and Soviet Ship Design Practices, 1950–1980', p. 127, in USNI *Proceedings*, May 1982.
2. Steven J. Zaloga, op. cit., p. 112.
3. Norman Polmar, 'A Friendly Assault on Norfolk', p. 137, in USNI *Proceedings*, November 1989.
4. Kehoe, East and Brower, 'Their *Slava*-Class Strike Cruiser', p. 148, in USNI *Proceedings*, August 1984.
5. Ibid., p. 151. Also *Combat Fleets of the World, 1988–89*, ed. A. D. Baker III, Naval Institute Press, Annapolis, 1988.

7. The New Destroyers
1. Kehoe and Brower, 'One of Their New Destroyers: *Sovremennyy*', p. 123, in USNI *Proceedings*, June 1981.

8. The New Carriers
1. Michael McCGwire, 'The Rationale for the Development of Soviet Seapower', p. 174, in USNI *Proceedings*, May 1980.
2. Robert C. Suggs, 'The Soviet Navy: Changing of the Guard', in USNI *Proceedings*, April 1982.
3. Rear-Admiral Thomas Brooks, Annual intelligence assessment presented to the House Armed Services Committee, reported in *Jane's Defence Weekly*, 16 March 1991.
4. Ibid.
5. Ibid.
6. John W. R. Taylor, 'Deck Trials of Three "Helix" Variants', p. 80, in *Jane's Soviet Intelligence Review*, February 1991.

ECM Arrangement

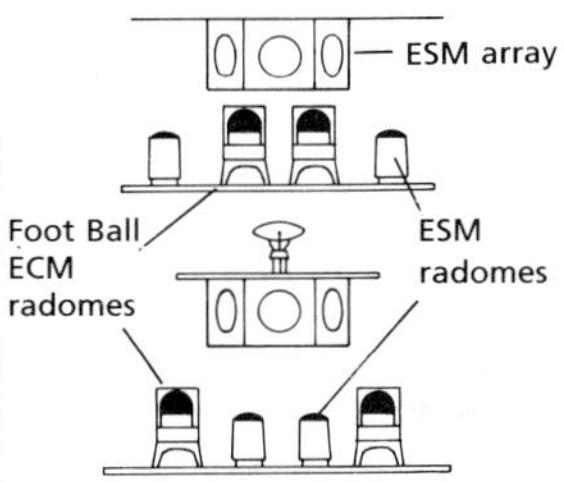

Yak-38 Forger. The Flanker has sufficient internal fuel for a range of 4,000km, enabling it to conduct prolonged combat air patrols at considerable distances from the carrier. It has been developed specifically for all-weather operation and has a sophisticated look-down shoot-down radar with a reported search range of 130nm (240km) and a tracking range of 100nm (185km). Weapons include a 30mm cannon in the starboard wing root extension and up to ten air-to-air missiles, carried on tandem pylons under the fuselage between the engine ducts. Missiles include short-range AAMs with infra-red homing and AA-10A/C Alamo semi-active types, the long-burn AA-10C having a beyond-visual-range capability. Latest reports suggest the development of a reconnaissance pod and a 'buddy pack' for inflight refuelling.

The Mig-29 is comparable in size to the US F/A-18 Hornet. It is agile and has good combat manœuvrability. Like the Su-27, it has a large pulse Doppler look-down shoot-down radar although the latter is limited to search-while-scan rather than track-while-scan. Its primary role in a maritime context would be that of close-range interceptor. However, the land-based Fulcrum is a dual-role aircraft with a substantial light attack capability, and the inflight refuelling capability of the naval variant would enable it to carry out long-range strikes against either surface ships or land targets carrying a light underwing stores load.

It is not yet clear whether *Admiral Kuznetsov* will operate the Yak-38 Forger VTOL successor, the Yak-141, which is reported to have been delayed by technical problems. Some commentators have expressed the view that the Yak-141 is not intended to operate from the new through-deck carriers, but will simply replace the Forger aboard *Kiev* and her sisters.

While there would be nothing to prevent the Yak-141 from operating from *Admiral Kuznetsov*, and deck markings appear to have taken the requirements of this aircraft into account, the embarkation of Yak-141s would necessarily displace another component of the current air group (possibly a 9-aircraft squadron of Ka-27 Helix A ASW helicopters). It therefore seems more likely that the aircraft would be embarked for specific missions unrelated to the ship's primary mission of fleet air defence, such as the support of the Naval Infantry in an amphibious assault.

Aircraft trials aboard *Kuznetsov* appear to confirm that the amphibious assault mission has figured prominently in Soviet thinking. In addition to the Ka-27 Helix A and D variants, employed for ASW and for plane guard/SAR duties respectively, recent on-deck photographs have featured the Ka-29TB Helix B assault helicopter which was first displayed at the Frunze air show in Moscow in August 1989, together with an unidentified model of the Helix derived from the Ka-29 airframe.

The Ka-29TB has a wider flight deck than the standard Ka-27 Helix and can easily be distinguished by its three-panel windscreen. Beneath the windscreen a snub nose contains operational equipment related to the assault mission, including a four-barrel gatling-type machine-gun. Four pylons under the outriggers can be fitted with 57mm/80mm rocket pods or AT-6 Spiral air-to-surface missiles, guidance for the latter being provided by an undernose sensor pod. Sensors include an ESM radome mounted above the rear of the engine bay fairing and an infra-red jamming pod.

The cabin has folding seats along the sidewalls for twelve assault troops who disembark via a two-part upward/downward rear-facing cabin door to facilitate rapid exit. The flight deck and engine bay are heavily armoured.

The new variant of the Ka-29 has neither armour nor the downward articulated gun door of the TB model. A large fairing beneath the nose appears to extend the full length of the fuselage pod and there are further large box-like fairings on the sides of the fuselage, forward and aft of the main landing gear. It has been suggested that this new aircraft may be an electronic warfare (EW) variant of the Helix, intended to provide ECM support and communications capabilities for an assault force of Ka-29TBs.[6] Alternatively it may be the AEW variant reported by US Intelligence, relying on passive ESM sensors rather than on active search radar to provide early warning of hostile surface forces.

Further speculation has centred around a possible naval mission for the new Kamov Hokum combat helicopter which is due to enter service in 1991. Armed with packs of unguided rockets, a podded undernose multi-barrel gun and wing-tip air-to-air missiles, the Hokum would be well suited to an escort role in support of the Ka-29TB assault helicopter. Its survivability is enhanced by extensive armouring, infra-red suppressors and infra-red decoy dispensers.

Clearly amphibious assault is very much a subsidiary mission for the new carriers which in the event of a major outbreak of hostilities would be fully occupied with their primary mission of fleet air defence. However, *Admiral Kuznetsov* is the first Soviet air-capable ship to have this degree of operational flexibility, the range of possible missions being dictated by the composition of the air group rather than by the configuration of the ship. She will be a major asset to the Soviet Northern Fleet when she joins it as flagship, probably late in 1991.

References

Introduction
1. *Red Star*, 26 March 1991.
2. Report in *Red Star*, January 1991.
3. Admiral Chernavin, quoted in *Red Star*, 2 March 1991.

1. The Post-war Era
1. Siegfried Breyer, *Guide to the Soviet Navy*, 2nd ed., p. 196, Naval Institute Press, Annapolis, 1977.
2. Norman Friedman, *All the World's Fighting Ships, 1947–82*, vol. 2, p. 465, Conway Maritime Press, 1983.
3. Ibid., p. 471.

2. The Anti-Carrier Strategy
1. Norman Friedman, *All the World's Fighting Ships, 1947–82*, vol. 2, p. 482.
2. Norman Polmar, *Guide to the Soviet Navy*, 4th ed., p. 428, Naval Institute Press, Annapolis, 1986.
3. Milan Vego, 'Their Torpedoes', p. 139, in USNI *Proceedings*, July 1984.
4. Ibid.
5. Kehoe, Brower and Meier, 'US and Soviet Design Practices, 1950–1980', p. 125, in USNI *Proceedings*, May 1982.
6. Norman Friedman, *World Naval Weapons Systems*, p. 170, Naval Institute Press, Annapolis, 1990.
7. Kehoe, East and Brower, 'Their *Slava*-Class Strike Cruiser', p. 151, in USNI *Proceedings*, August 1984.
8. Ibid.
9. Norman Friedman, *Naval Radar*, p. 187, Conway Maritime Press, 1981.

3. Anti-Surface and Anti-Submarine
1. Michael MccGwire, 'The Rationale for the Development of Soviet Seapower', p. 164, in USNI *Proceedings*, May 1980.
2. Steven J. Zaloga, *Soviet Air Defence Missiles*, p. 251, Jane's Information Group, 1989.
3. Ibid., p. 251.
4. Ibid., p. 252.

4. The Large Anti-Submarine Ships
1. Admiral Gorshkov, *Red Star Rising at Sea*, p. 103, Naval Institute Press, Annapolis, 1974.
2. Michael MccGwire, 'The Rationale for the Development of Soviet Seapower', p. 167, in USNI *Proceedings*, May 1980.
3. Michael MccGwire, 'Gorshkov's Navy', Part 1, p. 47, in USNI *Proceedings*, August 1989.
4. Roy Corlett, 'Advances in Soviet underwater capability and weapons', p. 203, in *Maritime Defence*, June 1987.
5. Michael MccGwire, 'Gorshkov's Navy', Part 1, p. 47, in USNI *Proceedings*, August 1989.
6. Steven J. Zaloga, op. cit., pp. 256–7.

5. The *Kiev* Class
1. Michael MccGwire, 'Gorshkov's Navy', Part 1, p. 47, in USNI *Proceedings*, August 1989.
2. John D. Gresham, 'Their Sea-Based Fighter', p. 122, in USNI *Proceedings*, February 1985.
3. Ibid., p. 122.

6. The New Rocket Cruisers
1. Kehoe, Brower and Meier, 'US and Soviet Ship Design Practices, 1950–1980', p. 127, in USNI *Proceedings*, May 1982.
2. Steven J. Zaloga, op. cit., p. 112.
3. Norman Polmar, 'A Friendly Assault on Norfolk', p. 137, in USNI *Proceedings*, November 1989.
4. Kehoe, East and Brower, 'Their *Slava*-Class Strike Cruiser', p. 148, in USNI *Proceedings*, August 1984.
5. Ibid., p. 151. Also *Combat Fleets of the World, 1988–89*, ed. A. D. Baker III, Naval Institute Press, Annapolis, 1988.

7. The New Destroyers
1. Kehoe and Brower, 'One of Their New Destroyers: *Sovremennyy*', p. 123, in USNI *Proceedings*, June 1981.

8. The New Carriers
1. Michael MccGwire, 'The Rationale for the Development of Soviet Seapower', p. 174, in USNI *Proceedings*, May 1980.
2. Robert C. Suggs, 'The Soviet Navy: Changing of the Guard', in USNI *Proceedings*, April 1982.
3. Rear-Admiral Thomas Brooks, Annual intelligence assessment presented to the House Armed Services Committee, reported in *Jane's Defence Weekly*, 16 March 1991.
4. Ibid.
5. Ibid.
6. John W. R. Taylor, 'Deck Trials of Three "Helix" Variants', p. 80, in *Jane's Soviet Intelligence Review*, February 1991.